PHYSICAL GEOGRAPHY OF CHINA

Zhao Songqiao

Institute of Geography, Chinese Academy of Sciences

Christopher L. Salter

Consulting Editor For Wiley-Science Press Co-publications

SCIENCE PRESS
Beijing, China, 1986

JOHN WILEY & SONS
New York · Chichester · Brisbane · Toronto · Singapore

Science Press
137 Chaoyangmennei Street
Beijing, China

John Wiley & Sons
605 Third Avenue
New York, New York 10158

Responsible Editor Yao Suihan

Science Press Book No. 4928-63
Library of Congress Cataloging-in-Publication Data
Zhao Songqiao
Physical geography of China.

Includes index.
1. Physical geography–China. I. Title.
GB316. C465 1986 915.102 85-15574
ISBN 0-471-09597-4

PHYSICAL GEOGRAPHY OF CHINA

Acknowledgements

I wish to express my deep gratitude to the Science Press and John Wiley & Sons Inc., whose joint sponsorship makes this publication possible. I am especially grateful to Mr. Yao Suihan, Geography Editor of the Science Press, and Miss Katie Vignery, Geography Editor of John Wiley & Sons Inc., for their excellent and laborious editorship.At the same time, I must pay homage to Professor Huang Bingwei, Director of Institute of Geography, Chinese Academy of Sciences, and President of the Geographical Society of China, for his support and encouragement in writing this textbook.

I also wish to thank my 45 colleagues in the writing of the Chinese edition of *Physical Geography of China: General Survey,* from which the basic scientific data for this English edition have been freely drawn. The list is too lengthy to include all their names and contributions; yet, Professor Lin Chao (Peking University) must be particularly mentioned here for his good advice in devising the scheme for China's comprehensive physical regionalization as well as in preparing the chapter on North China. Contributions of Professors Xing Jiaming on geomorphology, Wang Dehui on climatology, Xiong Yi and Tang Qicheng on water resources, Huang Rongjin on soil geography, Wang Hesheng and Zhang Rongzu on biogeography, Jing Guihe on Northeast China, Sun Jinzhu on Nei Mongol, and Zheng Du on the Qinghai-Xizang Plateau, are also affectionately remembered. I should again thank Dr. Christopher L. Salter (University of California at Los Angeles), Dr. Clifton Pannell (University of Georgia), Dr. Edward Derbyshire (University of Keele) and Mr. Richard W. Cooper (Beijing Language Institute) for constructively reviewing and revising the manuscript.

I am also grateful to all the people who have helped to draw maps, to type the manuscript and finally to all who have put this textbook into print. Special thanks are due to Mr. Duanmu Jie for his able designing of all figures and to Qian Jinkai and Shi Zuhui for their topography map of China, and to Huang Xuan, Li Huiguo and Zhang Shengkai for their composite falsecolor Landsat imageries.

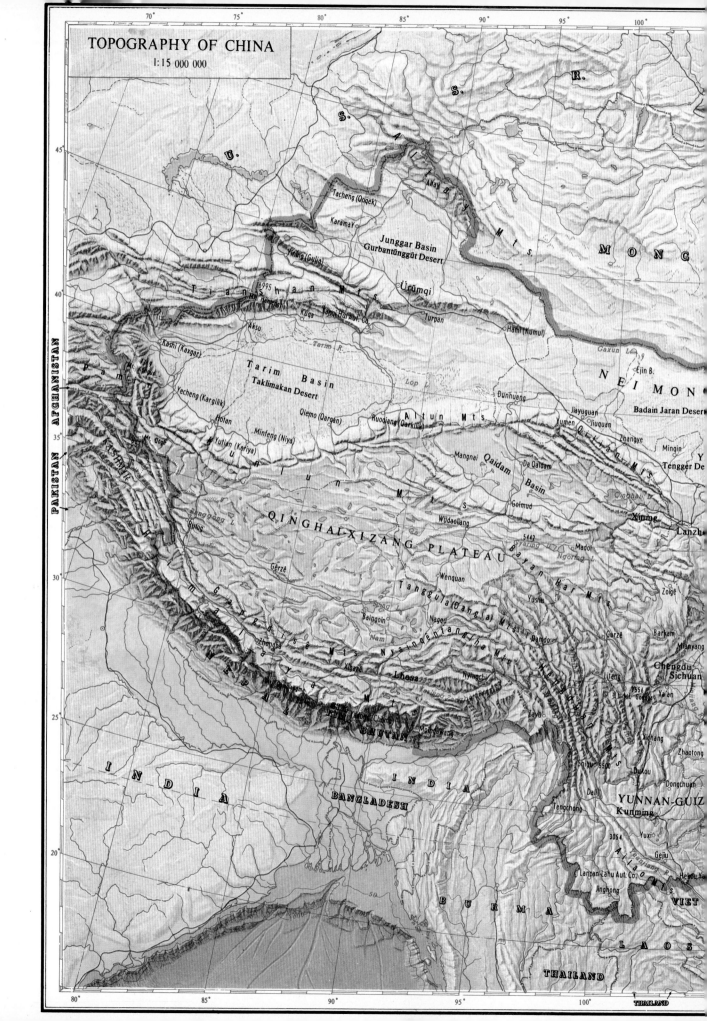

TOPOGRAPHY OF CHINA
1:15 000 000

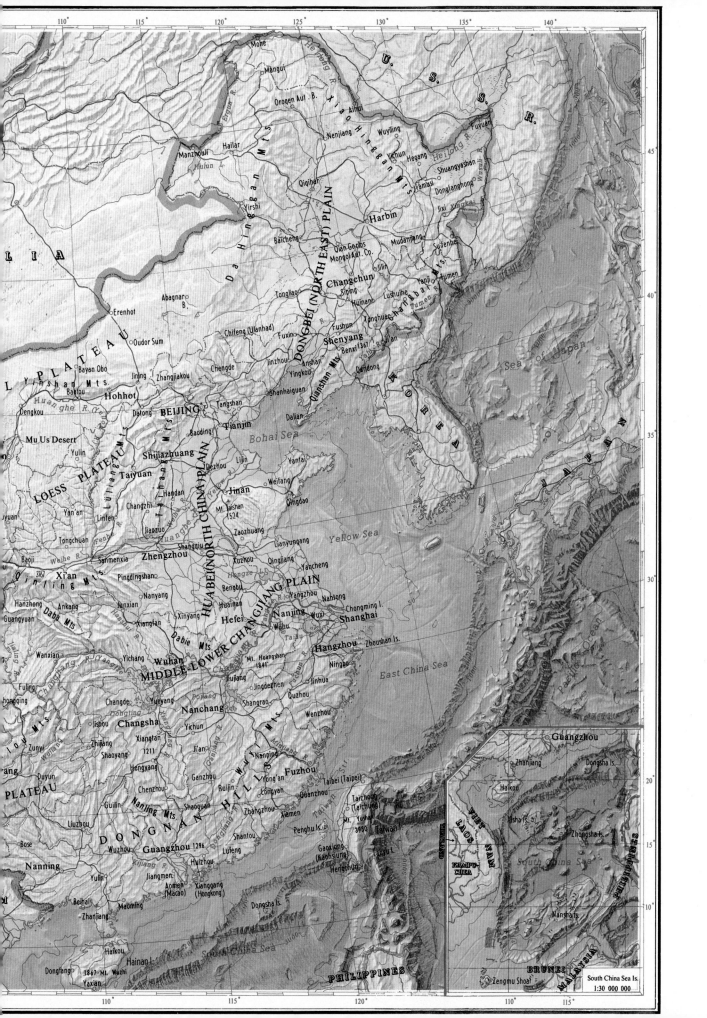

CONTENTS

Photographs and Landsat Imageries

INTRODUCTION

China is an immense country. It has an area of about 9.6 million sq km which comprises about 6.5 per cent of the world's total land area. Its population of more than 1 billion accounts for approximately 23 per cent of the total human population. In addition, China is the world's oldest continuous civilization; its agricultural tradition has survived uninterrupted for perhaps 7000 to 8000 years. Literally billions of Chinese have helped to shape the country's physical environment, and nearly every aspect of that environment is deeply connected with the cultural heritage of China. An understanding of the land and its people is of great significance to all students of civilization. The long-standing lack of current information available to English-speaking readers makes such an understanding all the more urgent.

The physical geographical environment is the foundation of Chinese civilization. This environment is an integration of all the physical elements, including climate, landforms, surface and ground water, neighboring seas, soils, flora and fauna, and the impact of all past and present human activities. It is clear, therefore, that any understanding of China requires a clear grasp of its physical geographical environment.

Geography is an old as well as a new science in China. As early as the fifth century B.C., the classic *The Tribute of Yu (Yugong, 禹贡)* subdivided China into nine regions and contained, for each region, an inventory of mountains, rivers, lakes, swamps, soils, and chief economic products. At about the same time *Record of the Rites of Zhou (Zhouli, 周礼)* classified China's land into five major types, whereas the *Book of Master Kwan (Kwanzi, 管子)* offered a systematic hierarchical classification of the land with three first-level types (plains, hills and mountains) and 25 second-level types (determined chiefly on the basis of ground surface materials and soils). These schemes are certainly among the earliest systems of physical regionaliza-

tion and land classification in the world. Later, a series of well-known studies emerged, including *the Geographic Book in the Han Chronology* (汉书·地理志), *Yuan Hu Reign Period Regional Geography* (元和郡县志) *Diary of the Travels of Xu Xia-ke* (徐霞客游记), *Comprehensive Geography of the Chinese Empire under the Qing Dynasty* (大清一统志), and *Essentials of Historical Geography* (读史方舆记要). In practically every ancient dynastic chronology, there is a chapter on the geographical environment; for every province, prefecture, and district, there is at least one regional geographical study. In all, more than 9000 regional studies are known to have been produced by Chinese chroniclers.

From the middle of the 19th century, modern physical geography began to be gradually introduced in China. Many topical studies of different physical elements as well as modern regional studies have since been conducted.

The founding of the People's Republic of China in 1949 added a tremendous impetus to the development of science, including geography. Not only did social revolution promote expanded interest in science and technology, but also socialist construction and production required an evaluation of natural conditions and resources as well as large-scale planning for different regions. Hence, a series of integrated investigations into resource inventories and agricultural potential as well as comprehensive regional plans were launched for many parts of China. This is especially significant in the frontier areas where scientific information had been hitherto unavailable. Many basic and theoretical studies, including *Physical Regionalization of China* (8 volumes, compiled by a special committee of Chinese Academy of Sciences, 1953—1959), have also been produced. During the past 35 years, the present author has had the opportunity to perform fieldwork throughout China, spending more than 15 field seasons in Northwest China and more than 7 in the Northeast. In addition, the present author has served as a coauthor of the monograph *Comprehensive Physical Regionalization of China.* International progress in geographical science and technology has also played an important role in helping Chinese geographers update and revise their data.

At present, it is estimated that there are more than 40000 geographers with a university education in China, of whom about 70% are working in the area of physical geography. Most of these physical geographers are concentrating their research in topical studies, particularly in geomorphology and climatology. However, integrated physical geography and hydrology have recently been gaining ground, and a group of energetic geographers is now working hard in the areas of paleogeography and historical physical geography.

In 1973, the *Physical Geography of China* Compilation Committee of Chinese Academy of Sciences was established with a view to summing up the chief features of China's physical geographical environment. The 12 volumes of the study (in Chinese) are (1) *General Survey,* (2) *Geomorphology,* (3) *Climatology,* (4) *Surface Water,* (5) *Ground Water,* (6) *Soil Geography,* (7) *Vegetation,* (8) *Animal Geography,* (9) *Paleogeography,* (10) *Historical Physical Geography,*(11) *Physical Conditions and Agricultural Development,* and (12) *Geography of Neighboring Seas.* The present author, who wrote and edited the first volume, is optimistic that the study will prove useful both for scientific research and agricultural production.

This book is essentially an English version of the *Physical*

Table I-1 Three Natural Realms and 30 First-level Administrative Units in China

Natural realms	Traditional divisions	First-level administrative units		
		Provinces	Autonomous regions	Municipalities
Eastern Monsoon China	Northeast China （东北）	Liaoning （辽宁）		
		Jilin （吉林）		
		Heilongjiang （黑龙江）		
	North China （华北）	Hebei （河北）		Beijing (Peking) （北京）
		Henan （河南）		
		Shandong （山东）		Tianjin (Tientsin) （天津）
		Shanxi （山西）		
		Shaanxi （陕西）		
	Central China （华中）	Jiangsu （江苏）		Shanghai （上海）
		Zhejiang （浙江）		
		Anhui （安徽）		
		Jiangxi （江西）		
		Hubei （湖北）		
		Hunan （湖南）		
	South China （华南）	Fujian （福建）	Guangxi （广西）	
		Guangdong （广东）		
		Taiwan （台湾）		
	Southwest China （西南）	Sichuan （四川）		
		Yunnan （云南）		
		Guizhou（贵州）		
Northwest Arid China	Northwest China （西北）	Guansu （甘肃）	Nei Mongol (Inner Mongolia) （内蒙古）	
			Ningxia （宁夏）	
			Xinjiang （新疆）	
Qinghai-Xizang Frigid Plateau	Qinghai-Xizang (Tibetan) Plateau （青藏）	Qinghai （青海）	Xizang (Tibet) （西藏）	

Table I-2 Commonly Used Chinese Words With Their Romanization and English Meaning

Chinese characters	Romanization		English meaning
	Pinyin	Wade-Giles	
一	yi	e	one
二	er	erh	two
三	zan	san	three
四	si	sze	four
五	wu	wu	five
六	liu	liu	six
七	qi	tsi	seven
八	ba	pa	eight
九	jiu	kiu	nine
十	shi	shi	ten
百	bai	pai	hundred
千	qian	tsian	thousand
万	wan	wan	ten thousand
东	dong	tung	east
南	nan	nan	south
西	xi	si	west
北	bei	pei	north
上	shang	shang	up
下	xia	sia	down
大	da	ta	large
小	xiao	siao	small
天	tian	tien	heaven
地	di	ti	earth
中	zhong	chung	center
山	shan	shan	mountain
岭	ling	ling	range
峰	feng	feng	peak
砂	sha	sha	sand
石	shi	shih	stone
河	he	ho	river
川	chuan	chwan	river
江	jiang	kiang	river
水	shui	shui	water
湖	hu	hu	lake
海	hai	hai	sea
洋	yang	yang	ocean
桥	qiao	chiao	bridge
口	kou	kou	mouth
省	sheng	sheng	province
县	xian	hsien	county, district
市	shi	shi	municipality
京	jing	king	capital
红	hong	hung	red
黄	huang	huang	yellow
蓝	lan	lan	blue
白	bai	pai	white
黑	hei	hei	black

Geography of China: General Survey. The presentation has been modified in some places and some new material has been added. The book consists of two parts. The first, containing seven chapters, is an integrated study of the whole country. It opens with a chapter that discusses the major factors which have shaped China's physical geographical environment. Then, successive chapters explore climatology, geomorphology, surface and ground water, soil geography, biogeography, and comprehensive physical regionalization and land classification. The differentiation of physical features and areas as well as the economic evaluation of these features and regions are analyzed. Because climate and landforms are two basic elements in the geographical environment, and because vegetation and soil reflect the total physical geographical characteristics, considerable emphasis has been placed on these elements. In addition, the question of surface water receives detailed attention because it comprises one of the most active physical elements and is of special importance in agricultural development. The second part of the book consists of regional studies. It devotes a chapter to each of seven natural divisions of China. Thirty-three natural regions have been subdivided from these seven divisions. Each region, together with its natural subregions, is identified in terms of its physical features and areal differentiation as well as its potential for agricultural development.

In addition, it might be helpful briefly to introduce English readers to China's present administrative units and some commonly used terms. Broadly speaking, China can be divided into three natural realms: Eastern Monsoon China, Northwest Arid China, and the Tibetan Frigid Plateau. The first realm can again be subdivided into five traditional divisions: Northeast China, North China, Central China, South China, and Southwest China. There are now 30 first-level administrative units (provinces, autonomous regions, and national municipalities) in China. They may be roughly grouped within these natural realms and divisions as shown in Table I-1.

The romanization of Chinese words can sometimes be confusing. For English-language readers the official Pinyin system predominates although some places are introduced with their Wade-Giles spellings. Chinese names can be

Table I-3 Conversion of Some Commonly Used Chinese Units into English Systems

1 mow (亩) = 1/6 acre = 1/15 hectare
1500 mows = 640 acres = 1 square kilometer
1 li (里) = 1/3 mile = 1/2 kilometer
1 chi (尺) = 1.09 foot = 1/3 meter
1 tael (两) = 0.13 pound = 50 grams
1 catty (斤) = 10 taels = 1.33 pound = 1/2 kilogram
1 picul (担) = 100 cattes = 133 pounds = 50 kilograms

rendered both phonetically and literally. There are many traditional place names such as the Yangtze River (phonetically Chang Jiang or literally Long River), the Yellow River (phonetically Huang He), the Tibetan Plateau (Qinghai-Xizang Plateau in Pinyin), Inner Mongolia (Nei Mongol in pinyin), the Great Hinggan Mountains (Da Hinggan Ling in pinyin), and so on. Furthermore, a decision has to be made about Chinese nomenclature, particularly in the terms for (1) mountains (in Chinese: *shan, ling, feng,* etc.; in Mongolian: *ula;* in Tibetan: *daban;* in Uygur: *tag*); (2) seas (in Chinese: *hai, yang*); (3) rivers (in Chinese: *he, jiang, shui, xi,* etc.; in Mongolian: *moron*); and (4) lakes (in Chinese: *hu, hai;* in Mongolian: *nor;* in Tibetan: *co;* in Uigur:*kol*). For the convenience of English-language readers, they are essentially converted to English terms, although, for the phonetic consideration, any place name with only one Chinese character has its Chinese term repeated in English, for example, Qinling Mountains, Tianshan Mountains, Huaihe River, Hanshui River, Bohai Sea, and so on (See the locational glossary in Appendix I). As not all of the place names mentioned in the text could be included in the small-scale maps in the book, the reader is advised to use an up-to-date Chinese atlas (in pinyin) for additional reference.

Table I-2 lists some commonly used Chinese words with their romanization and English meaning. Table I-3 contains some conversions of measurement.

References

[1]   *Physical Geography of China* Compilation Committee, Chinese Academy of Sciences, 1979 – 1985, *Physical Geography of China,* 12 vols, Science Press, Beijing. (In Chinese)

[2]   *Physical Regionalization of China* Working Committee, Chinese Academy of Sciences, 1958 – 1959, *Physical Regionalization of China,* 8 vols, Science Press, Beijing. (In Chinese)

[3]   Ren Mei'e (Jen, Mei-O) et al., 1979, *An Outline of the Physical Geography of China,* Commercial Press, Beijing. (In Chinese)

[4]   Cressey, G.B., 1934, *China's Geographic Foundations: A Survey of the Land and Its People,* McGraw-Hill Book Co., New York.

[5]   Cressey, G.B., 1955, *Land of the 500 Million,* McGraw-Hill Book Co., New York.

[6]   Buchanan, K., 1970, *The Transformation of the Chinese Earth,* Bell, London.

[7]   Tregear, T.R., 1980, *China: A Geographical Survey,* John Wiley and Sons, New York.

[8]   Sion J., 1928, *Asie des moussons.* Géographie universelle, Tome IX. Libraire Armand Colin, Paris. (In French)

[9]   Bouterwek, K. et al., 1937, *Nordasien, zentralund Ostasien, Handbuch der Géographischen Wissenschaft,* Akademische Verlagsgessellschaft Athenaien, Potsdam. (In German)

[10]   Institute of Geography, USSR, 1964, *Physical Geography of China,* Moscòw. (In Russian)

[11]   New Light Society, 1939, *China,* Japanese World Geography Series, vols. 2, 3, Tokyo. (In Japanese)

[12]   Commission of Agricultural Natural Resources and Agricultural Regionalization, 1984, *An Outline Comprehensive Regionalization of China,* Science Press, Beijing. (In Chinese)

[13]   Institute of Geography, Chinese Academy of Sciences, 1980, *Agricultural Geography of China: A General Survey,* Science Press, Beijing. (In Chinese)

[14]   Zhao Songqiao, et al., 1979, "Thirty Years in Comprehehensive Physical Geography in China," *Acta Geographica Sinica,* Vol. 34, No. 3.(In Chinese, with English abstract)

[15]   Zhao Songqiao, 1981, "Thirty Years in Physical Geography in the People's Republic of China", *Chili,* Vol. 26, No. 3, Tokyo. (In Japanese)

[16]   Zhao Songqiao, 1983, "A New Scheme for Comprehensive Physical regionalization in China." *Acta Geographica Sinica,* Vol. 38, No. 1. (In Chinese, with English abstract)

[17]   Zhao Songqiao, 1983, "Land Classification and Mapping in China", Land Resources of the People's Republic of China, The United Nations University, Tokyo.

Chapter 1

Major Factors Shaping China's Physical Geographical Environment

China is one of the largest countries in the world, with a rich and varied physical geographical environment. This, in turn, has been shaped by the interaction and integration of the following major factors: vast area, midlatitudinal and east coastal location, mountainous topography, complex geological history, and the significant impact of human activity. The influence of these major factors has resulted in the dominant features of Chinese physical geographical environment.

VAST AREA

China has a land area of about 9.6 million sq km, occupying 6.5 per cent of the total land area of the world. Only the USSR and Canada have larger land areas; yet, both are located adjacent to the frigid Arctic environment. Only the United States has an approximately similar land area as well as similar favorable natural conditions. Table 1-1 shows some selected countries' land areas and their percentage of the world's total land area.

From the confluence of the Heilong River and its tributary, the Wusuli River, westward to the Pamir Plateau, the distance is more than 5200 km; when the noon sun shines brilliantly over the Wusuli, it is still early morning in the Pamir. From midstream of the Heilong River north of Mohe, southward to the Nansha Islands near the equator, the distance in more than 5500 km; when blizzards still threaten the Heilong River in the winter, spring sowing is under way on Hainan Island, whereas the Nan-

Table 1-1 Land Areas of China and Some Selected Countries in the World

Country	Land area (in 10000 km²)	Percentage of total world land area
USSR	2240	15.0
Canada	995	6.7
China	960	6.5
United States	936	6.3
Brazil	851	5.7
Australia	770	5.2
India	295	2.0
Saudi Arabia	240	1.6
Indonesia	190	1.3
France	55	0.37
Japan	37	0.25
United Kingdom	24	0.16

sha Archipelago remains hot and humid all year yound.

Besides a vast land area, there are also extensive neighboring seas and numerous islands. One territorial sea – the Bohai Sea, and three neighboring seas – the Yellow Sea (Huanghai Sea), the East China Sea (Donghai Sea) and the South China Sea (Nanhai Sea), have altogether an area of about 4.73 million sq km.,

Both the vast land and sea area provide a large spatial

Table 1-2 Distribution of Cropland and Undeveloped Arable Land in Seven Major Agricultural Reclamation Regions in China*

Regions	Cropland (in million ha)	Undeveloped arable land (in million ha)
North China	31.0	2.8
Northeast China	18.4	10.0
Northwest China	10.7	17.5
Central China	21.5	3.6
South China	7.0	3.7
Southwest China	11.5	6.5
Qinghai-Xizang Plateau	0.8	3.1

*Agricultural reclamation regions are grouped by administrative provinces and autonomous regions; croplands and undeveloped arable lands are compiled from statistics and estimates by the author.

basis for different geographical elements and different geographical processes. Given similar climatic and geomorphological conditions, the larger the area the better the opportunity for possessing a more varied physical geographical environment and richer natural resources. This is the case in China. It has not only a varied physical geographical environment but also it is rich in natural resources — land, water, mineral, energy, and biological resources.

Consider for example arable land resources. China has now about 100 million hectares (ha) of cropland, which have been developed ever since 7000 B.P. During historical times, both cropland and population increased when the country was united and, consequently, relatively prosperous; they decreased sharply when the country was divided and suffered from civil wars and poverty. On the whole, however, cropland and population have been increasing steadily. In A.D. 2 (West Han dynasty), when China had her first census, there were about 38 million ha of cropland and a population of an estimated 60 million. The cropland has now increased 2.5 times and the population by more than 18 times.

Most of the arable lands in China have now been cultivated. Cropland is distributed extensively in Eastern Monsoon China, especially in some highly developed regions such as the North China Plain, the middle and lower Changjiang (Yangtze) Valley, the Sichuan Basin, and the Zhujiang (Pearl River) Delta, with croplands occupying 50 to 60 per cent of the total land area in those regions. Yet, owing to certain physical or social conditions unfavorable to the agricultural use of land, there are still about 47 million ha of potentially arable land throughout China

that is uncultivated. This land is chiefly distributed in Northeast and Northwest China, the Sanjiang (Three Rivers) plain in Northeast China (Landsat image 2) is now the most important reclamation area in China. From the viewpoint of agricultural reclamation, seven major regions may be identified, as shown in Table 1-2.

MIDLATITUDINAL AND EAST COASTAL LOCATION

China also has an excellent geographic location. Most parts of her vast area are situated at midlatitudes, hence, climates are mostly temperate or subtropical. China is sandwiched between the largest continent (Eurasia) and the largest ocean (Pacific), consequently, allowing the monsoons to become well-developed. During the summer, high temperatures together with plentiful rainfall make agriculture quite productive.

Latitudinal Location

The northern edge of Mohe in the far Northeast has a latitude of about 53°31' N, while Zengmu Shoal of the Nansha Islands is located at about 3°50' N. Between these northernmost and southernmost points, angles of solar incidence are quite different, hence, there is a great diversity in climate, especially in temperature conditions during winter. Climate together with vegetation, soil, and other physical factors is, thus, arranged in latitudinal zones. This is the so-called latitudinal zonation, which is particularly conspicuous in Eastern Monsoon China. In 1958, the Physical Regionalization Working Committee of Chinese Academy of Sciences took the accumulated temperature during the ≥ 10°C period as the chief criterion for dividing China (not including the Qinghai-Xizang Plateau) from north to south into six temperature zones as shown in Table 1-3 and Figure 1-2.

China has about 98 percent of its land area located between 20° to 50°N, hence, the temperate zones (including warm-temperate and cool-temperate) and the subtropical zone are most extensively distributed, accounting for 45.6 per cent and 26.1 per cent of the total land area respectively. In planetary wind systems, latitudes between 20° to 35° are generally referred to as the subtropical high-pressure belt or horse latitudes, characterized by variable winds and calms and, consequently, by little precipitation. But in China, monsoons dominate both the subtropical zone and the eastern part of the temperate zone, resulting in plentiful rainfall during the summer. Thus, the extensive subtropical zone in China instead of being a horse latitude desert, turns out to be an area with favorable

Table 1-3 Temperature Zones in China (according to the Physical Regionalization Working Committee of Chinese Academy of Sciences, 1958)

Temperture zones	Accumulated temperature during ⩾ 10°C period (°C)	Physical features
Equatorial	Around 9500	Hot and humid all year round; rainforest; laterite
Tropical	8000–9000	Coldest month > 16°C; monsoon forest; lateritic soil
Subtropical	4500–8000	Coldest month 0–16°C; evergreen broadleaved forest
Warm-temperate	3200–4500	Coldest month −8 to 0°C; deciduous broadleaved forest
Temperate	1700–3200	Coldest month −24 to −8°C; Mixed broad- and needle-leaved forest
Cool-temperate	<1700	Coldest month < −24°C; Taiga forest

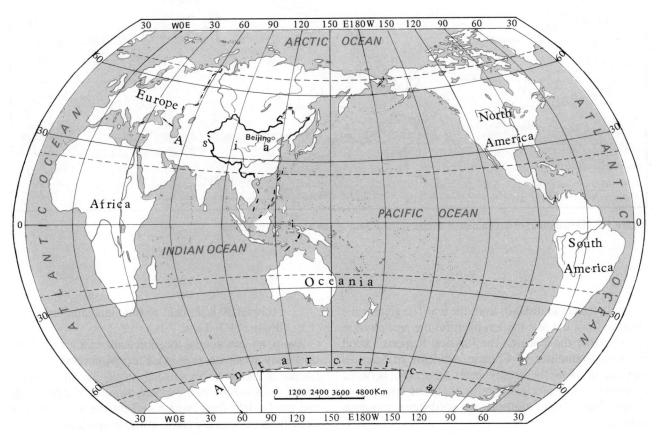

Figure 1-1 Location of China in the World

temperature and moisture conditions for agricultural production.

Distribution of Land and Sea

As mentioned above, China is located at the east coast of the largest continent (Eurasia) as well as the western margin of the largest ocean (Pacific). It has also the largest and highest plateau — the Qinghai-Xizang Plateau —

uplifted and located in its southwestern territory. Owing to the interaction of these three major components, the all-important monsoon climate has been well developed ever since late Pleistocene times, with a dry, cold, northwestern continental monsoon dominating during the winter and a moist, warm, southeastern maritime monsoon prevailing during the summer.

As moisture in the atmosphere over China comes mainly from a warm and moist maritime monsoon, precipitation

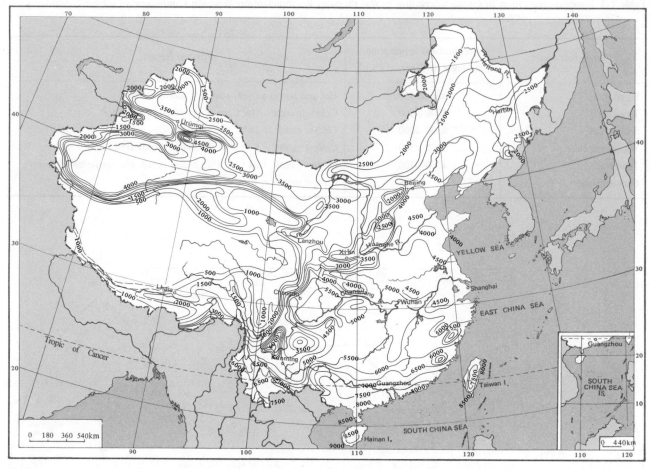

Figure 1-2   Distribution of accumulated temperature during ≥ 10°C period in China

is closely related to distance from the sea. The greater the distance from the sea, the less plentiful the precipitation and the drier the climate. The Physical Regionalization Working Committee of Chinese Academy of Sciences adopted in 1958 the aridity index (the ratio between evaporation and precipitation) as the chief criterion to divide China, from southeast to northwest, into four moisture zones.

1) Humid: aridity index $(k) < 1.0$, dominant natural vegetation being forest, occupying 32.2 per cent of the total land area;

2) Subhumid: $k = 1.0$ to 1.5, dominant natural vegetation being forest-meadow, occupying 14.5 per cent of the total land area;

3) Semiarid: $k = 1.5$ to 2.0, dominant natural vegetation being steppe that occupies 21.7 per cent of the total land area;

4) Arid: $k > 2.0$, dominant vegetation being desert-steppe (when $k = 2.0$ to 4.0) and desert ($k > 4.0$) that

occupies 30.8 per cent of total land area in China (Figure 1-3, Table 7-1).

Again, an east coastal location leads to a greater continentality in climate, even in Eastern Monsoon China. Influence of the westerlies is usually negligible, and melioration of the Kuroshio during winter, owing to offshore winds and the Coriolis force, which deflects the warm current away from the Chinese coast, is also insignificant. Hence, a continental climate is well developed in China, generally with warmer summers and much colder winters when compared with west coastal areas at a similar latitude. For example, Huma in Heilongjiang Province has about the same latitude as London (51.5°N); Huma has a mean January temperature as low as –27.8°C, whereas London, with 3.7°C, is similar to Shanghai, which is located at about 31.2°N. Again, Tianjin and Lisbon both lie at about 39°N, the mean January temperature of Tianjin is –4.1°C and the absolute minimum –22.9°C, whereas readings of Lisbon are 9.2°C and –1.7°C respectively.

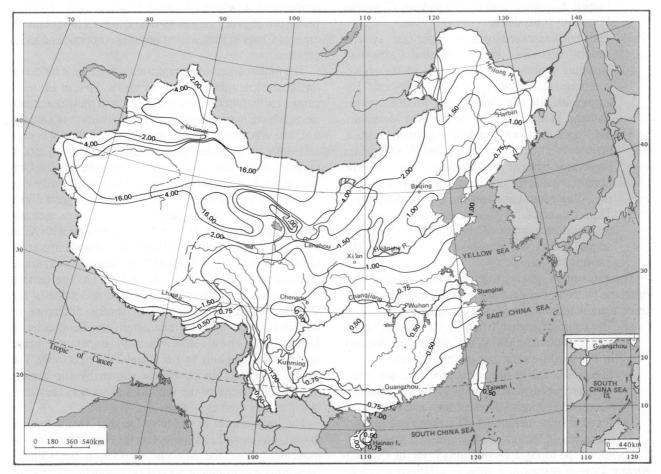

Figure 1-3 Distribution of aridity in China

MOUNTAINOUS TOPOGRAPHY

China is a mountainous country, with mountains, hills, and plateaus occupying 65 per cent of the total land area. According to an estimate, land of less than 500 m in elevation occupies only 25.2 per cent of the total land area, while that above 3000 m occupies 25.9 per cent. In the world, there are 12 high peaks of more than 8000 m above sea level (asl); 7 are located in China. The highest peak in the world, Mount Qomolangma (8848 m), stands majestically on the border between China and Nepal (see Topography of China).

China is remarkable not only in absolute relief, but also in relative relief; consequently, vertical zonation is often conspicuous, sometimes overshadowing horizontal zonation. The second lowest landmark in the world (–155 m), lies in the Ayding Lake of the Turpan Basin, which is located at the southern slope of the snow capped Mount Bogda of the Tianshan Mountains (5445 m). In the Mêdog

area of the southern Qinghai-Xizang Plateau, there are differences in elevation of more than 7000 m in a horizontal distance of less than 40 km; tropical, humid forest landscape dominates in the Yarlung Zangbo R.,whereas on top of nearby Mount Namjagbarwa (7756 m) there is continual snow. Generally speaking, changes of 100 m in vertical relief correspond to horizontal changes of about 100 km in the geographical environment.

There are five major mountain systems in China:

1) The east-westward trending mountain system, mainly controlled by the so-called Giant Latitudinal Structural System. From north to south, there are three major subsystems:the Tianshan-Yinshan-Yanshan mountain systems, the Kunlun-Qinling-Dabie mountain systems; and the Nanling mountain system. All of them are important geographical barriers and divides.

2) The north-southward-trending mountain system, mainly controlled by the so-called Longitudinal Structure System, includes the Helan Mountain, the

Liüpan Mountain and the Hengduan (Traverse) Mountain subsystems.

3) The northeastward-trending mountain system, mainly determined by the so-called Cathaysian Structure System and located east of the above-mentioned Longitudinal Structure System. Examples include the Da Hinggan Mountains, the Taihang Mountains, and the Wushan Mountain.

4) The northwestward-trending mountain system, mainly determined by the so-called West-Domain Structure System and located west of the Longitudinal Structure System. Examples include the Altay Mountains and the Qilian Mountains.

5) The arc mountain system, located mainly along the southeastern margin of the Eurasian Plate. Examples include the Himalaya Mountains and the Taiwan Mountains.

These mountain systems, together with numerous intermontane plateaus, basins, and plains are interwoven into three macrolandform complexes in China:

1) The gigantic Qinghai-Xizang Plateau, which has an area of about 2500000 sq km, uplifted and located in the southwestern part of the country.

2) The land areas in most part of China, are interwoven checkerboards of mountain ranges, plateaus, basins, and plains of different sizes and elevations.

3) A continental margin, which is composed of hills, plains,coasts, islands, and continental shelves.

Therefore, the topography of China from the Qinghai-Xizang Plateau eastward, is broadly arranged into four great steps.

1) The Qinghai-Xizang Plateau, with mean elevation above 4000 m.

2) From the eastern margin of the Qinghai-Xizang Plateau eastward up to the Da Hinggan-Taihang-Wushan mountains line, composed mainly of plateaus and basins with elevations from 2000 to 1000 m.

3) From the above-mentioned line eastward up to the coast, in which are distributed the largest plains of China — the Northeast China Plain, North China Plain and the middle and lower Changjiang Plain, also interspersed with hills generally below 500m in elevation.

4) The continental shelf, with water depth generally at less than 200 m.

Such a mountainous topography makes China quite rich in mineral resources but comparatively poor in arable land resources. China has about 10.5 per cent of its total land area in cropland, whereas the United States has more than 20 per cent. With a population of more than 1 billion and a percentage of rural population exceeding 84 per cent, China has an average of only 0.11 ha of farmland per capita and 0.15 ha per farmer; whereas in the United States the proportions are about 1.6 and 40 ha respectively. To make better use of every inch of farmland and all mountainous areas of China is, thus, a most pressing economic and social problem.

The mountainous topography also exerts great influence on climate and other physical elements. First of all, relief causes redistribution of temperature and moisture conditions, which, in turn, leads to vertical zonation of climate, soil, and vegetation. Besides, in terms of the flow of energy and matter in a physical geographical environment, mountains act as barriers and detainers. For example, the Qinling Mountains acts as a barrier to north-southward blowing cold waves and can be distinguished as the great divide between North and Central China. In January, Ankang (about 33°N) on its southern flank compared with Xián (about 34°N) on its northern flank has a mean temperature that is higher by 4.2°C and an absolute minimum temperature that is higher by 11.1°C. On the eastern coastal plain which does not enjoy the protection of the Qinling Mountains cities with a similar difference of latitudes display different characteristics; Bengbu has a January mean and absolute minimum temperature higher than Xuzhou, but only by 1.6°C and 3°C respectively. The barrier function can be clearly shown by the rapidly decreasing velocity of the typhoons as soon as they contact the hilly Chinese coasts.

The huge impact of the lofty Qinghai-Xizang (which, in turn, was caused and uplifted by the Himalayan tectonic movement since the Miocene Epoch) on the Chinese geographic environment should be emphatically pointed out. Based on recent research, the formation of the modern Chinese monsoon climate and the accelerating desiccation of Northwest China have been closely correlated with the uplifting of the Qinghai-Xizang Plateau. In the late Pliocene Epoch, when the Qinghai-Xizang Plateau attained an elevation of 1000 m, above sea level, the Chinese monsoon system was not yet formed and there existed only a weak high-pressure belt near Lhasa (about 30°N). At the end of the Tertiary period, the Tibetan Plateau together with its neighboring regions was violently uplifted, the plateau surface attained an elevation of 3000 m and the weak highpressure belt near Lhasa was strengthened and pushed northward to the southern rim of the Tarim Basin about 40°N. Yet, it was not until late Pleistocene to early Holocene, when the Qinghai-Xizang Plateau and its neighboring regions underwent violent mass uplifting again and the plateau surface attained its present elevation of more than 4000 m, that the modern Chinese monsoon system and vast desert areas in Northwest China fully developed. Since then, the Siberia-Mongolian High Pressure System has been pushed northward to its present position (about 55°N). The numerical experiments recently performed by Manabe and Terptre arrive at a similar conclusion. The January geopotential map of the northern hemisphere at 1000 millibar (mb) (with the effects of the Qinghai-Xizang Plateau) identifies clearly the Siberia-Mongolian High-Pressure system at about 55°N, whereas the model without the effects of the Qinghai-Xizang Plateau has its weak high pressure at about 30°N. Today, the Tibetan Plateau still exerts a great influence on the climate of surrounding regions. By its thermodynamic effects, there occur the peculiar plateau monsoons, with a northeastern monsoon during winter and a southwestern monsoon during summer over the eastern part of the Qinghai-Xizang Plateau. The west wind currents are also bifurcated into southern and northern branches during winter when passing through the Qinghai-Xizang Plateau.

A COMPLEX GEOLOGICAL HISTORY

The physical geographical environment is a function of latitudinal, longitudinal, and vertical dimensions as well as time. Thus, the present Chinese physical environment is a product of a long, complicated geological history. According to the plate tectonics hypothesis, China is mainly situated on the Eurasian Plate bordering the Indo-Australian Plate on the southern side and the Pacific Plate and the Philippine Plate on the eastern side. China's geological history is essentially a long, complex process of tectonic movement and interaction both between the Chinese Platform and its surrounding folding belts inside the Eurasian Plate and between the Eurasian Plate and its neighboring plates.

The Indo-China tectonic movement during the early Mesozoic era began a new epoch for the Chinese mainland; henceforth, it turned essentially into a continuous, vast land area. The Yanshan Tectonic Movement during the late Mesozoic Era had even greater impacts on the Chinese physical environment; it determined the broad geological structures and macrogeomorphological features of China. After this tectonic movement, and up to the early Tertiary period, a relatively quiet, stable time predominated in China that resulted in most of the land areas becoming peneplained, with level and undulating topography predominating (Figure 1-4). The climate was much warmer than the present, with the northern limit of the subtropical zone pushing northward more than 7° to 10° in latitude. However, the modern monsoon system was not yet established.

The supposed thrust of the Indo-Australian Plate underneath the Eurasian Plate and the impacts of the Himalayan tectonic movement since the Miocene epoch have been dominant forces in shaping the modern Chinese physical environment at its latest stage (Figure 1-5). The Himalayas and Taiwan folding belts have thus been form-

ed; the former having been uplifted more than 8000 to 9000 m. The Qinghai-Xizang Plateau and four great topographic steps from east to west have also been uplifted. This uplift of the lofty Qinghai-Xizang Plateau has been the major cause of the formation of the modern monsoon system in East Asia. Consequently, three great natural realms have been identified in China: the humid Eastern Monsoon China, the arid Northwest China, and the frigid Tibetan Plateau. In Eastern Monsoon China north of the January mean 0°C isotherm (which runs approximately in a line extending from the mouth of the Changjiang River westward to the northern rim of the Sichuan Basin, then southwestward to northwestern Yunnan), the climate has become much cooler, changing from a subtropical to a temperate environment, with landforms exhibiting remnant reddish weathering crust distributed as far north as the southern slopes of the Da Hinggan Mountains and Altay Mountains. South of the line, the climate has shifted from semiarid trade winds to a moist monsoon climate.

THE SIGNIFICANT IMPACT OF HUMAN ACTIVITY

China has a long human history and a large population; hence, mankind has made significant impact on the Chinese physical geographical environment. As early as 1.6 million years ago, there already lived tribes of *Homo erectus* in Chinese territory. Agricultural activity in the Loess Plateau began, according to recent archeological excavation, at Banpo village near Xi'an, at about 6000 B.P. On the southeast coast, large quantities of rice and tools for rice cultivation — primitive ploughs of bone or wood — have been discovered in Zhejiang Province among the remains of the Hemudu culture, which dates back to about 7000 B. P.(Table 1-4). At the present, China has about 100 million ha of cropland and more than 1 billion inhabitants. Practically no more virgin land and vegetation still exist, and mankind has made its imprint nearly everywhere. Hence, G.B. Cressey once aptly remarked, "The most significant element in the Chinese landscape is thus not the soil or vegetation or the climate, but the people. Everywhere there are human beings. In this old, old land, one can scarcely find a spot unmodified by man and his activities."

In a country with an agricultural history of more than 70 centuries and with more than four fifths of its population of 1 billion still engaged in agriculture, agricultural development is naturally the most important human activity and exerts the largest impact on the physical geographical environment. Chinese farmers in the hundreds of millions have worked upon the Chinese physical geographical environment, making a living from it and

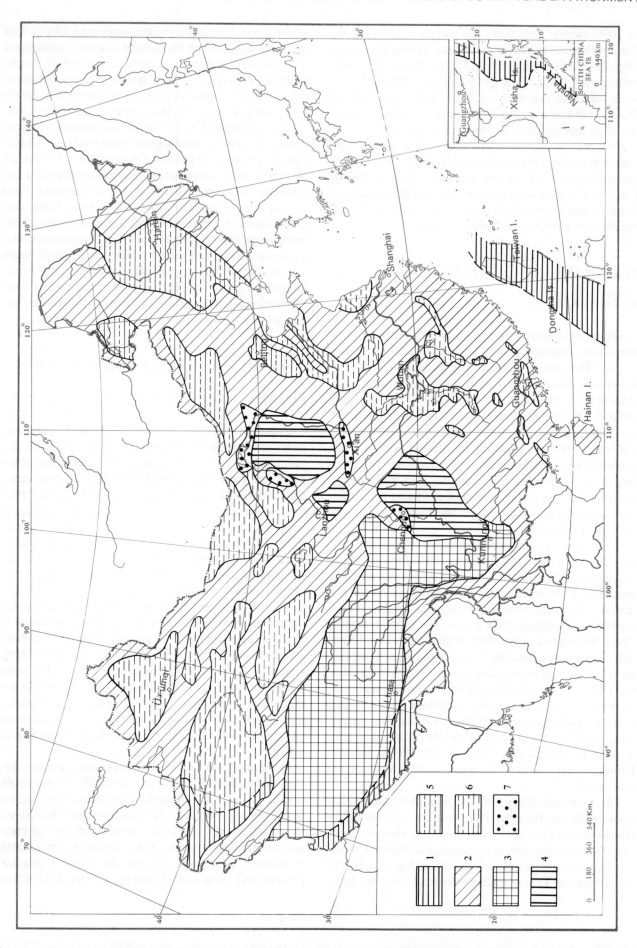

Figure 1-4 A Preliminary paleogeographical map of China in late Cretaceous and early Tertiary Periods.

1. Area of Submergence; 2. Uplifted-denundated hill and mountain; 3. Low, rolling peneplain; 4. Mesozoic era depositional basin; 5. Mesozoic-Cenozoic eras depositional basin without saline red beds; 6. Mesozoic-Cenozoic eras depositional basin with saline red beds; 7. Early Tertiary period depositional basin.

transforming it tremendously. The following outstanding agricultural successes in China might serve as examples of the advantageous side of human impact on the physical geographical environment.

China now occupies about 7 per cent of the world's total farmland and supports about 23 per cent of the world's population. Since 1949, starvation has been eliminated, and China has produced the largest food output in the world, 332.1 and 353.4 million tons in 1979 and 1982 respectively. At the same time, Chinese farmers have been improving agricultural soil in most parts of the country. One example is the widely distributed paddy soil. Another success is the irrigated oases in such ancient oases as Dunhuang of the western Hexi Corridor. The rich alluvial plains of those oases have been cultivated and irrigated for more than 2000 years; the fertile tilth has been developed to a depth of more than 2 m.

Many major crops originated in China, and many important crops have their major producing areas there. As early as Neolithic times, Chinese farmers planted sorghum, millet, wheat, rice, hemp, and mulberry trees. According to Chinese legend, Emperor Shennong in 2822 B.C. sowed five kinds of food crops, one of which was rice. China has been one of the largest rice producers in the world ever since, with an annual production of rice of 161.2 million tons in 1982. In the swampy land of South and Central China, there are, besides paddy rice, many special food crops and vegetables, including lotus, taro, calamus, water chestnut, and wild rice stem.

The Chinese have also domesticated and made use of many animals. In Paleolithic times, they domesticated the dog; since Neolithic times they have raised pigs, sheep, goats, cattle, horses, donkeys, chickens, ducks, yaks, and camels. In 1982, China raised 301 million pigs, 182 million sheep and goats, and 101 million large animals (cattle, hores, donkeys, etc.). As far as wild animals are concerned, the ancient classic *The Tribute of Yu* already discussed hunting and the domesticating of animals. Today, China has more than 70 kinds of fur-bearing wild animals, comprising about 15 per cent of the total species. Since 1949, an overall plan for the improved utilization of wild animals has been launched, with three major objects:

1) To control or eliminate economically harmful wild animals, such as rats and wolves in the Nei Mongol and Qinghai-Xizang Plateaus;

2) To make better use of, or to domesticate, useful wild animals, such as many kinds of deer, marten, and monkey;

3) To protect scarce and endangered wild animals, such as panda, swan, musk deer, red-crowned crane, white head crane, and so on.

Irrigation has long been a Chinese agricultural tradition. Irrigation in North China was described in the ancient classic *Shi Qing* (*The Classic of Poetry*) (诗经 , 781 – 771 B.C.). In 563 B.C.,the hydraulic engineers of the Zheng Kingdom made good use of river channels on the eastern slope of the Taihang Mountains to build highly efficient irrigation canals. Later, irrigation works were built in the Huaihe valley as well as in the Sichuan Basin. The famous Dujiang Dam irrigation system (see Figure 10-5, Photo IV-2) in the western Sichuan Basin, built in the 4th century B.C., was then probably the largest and most comprehensive in the world; it has been operated uninterruptedly ever since, with its irrigated area increasing from about 150000 ha initially to about 500000 ha.The Changjing (Yangtze) Delta, with its elaborate irrigation canals and transportation waterways, which were constructed in about the 4th century B.C., has now developed into one of the most highly developed and productive ecosystems in the world. Other highly developed areas, such as the middle reaches of the Changjiang River (Landsat image 3), the Zhujiang (Pearl River) Delta (Landsat image 5) and the Hexi Corridor (Landsat image 7) have all maintained and expanded their irrigation works. To date, China has about 45 per cent of her total cropland under irrigation, which comprises about one quarter of the total irrigated area in the world.

On the other hand, Chinese farmers have made mistakes in the utilization of land and other natural resources under prolonged suffering from food shortage and other socio-economic problems. Some examples follow.

During the last several thousand years, and especially over the last 100 years, many regions have been overexploited, resulting in both heavy loss of cropland, grassland, and forest resources and severe soil erosion. The most notorious example is the Loess Plateau — once the cradle of Chinese civilization. Owing to several thousand years of devastation of the natural vegetation and the misuse of land, and further handicapped by unfavorable physical conditions, this area is now subject to severe soil erosion and low agricultural productivity. The Huanghe River in this section carries 1.6 billion tons of silt per year. Another example is the old reclamation area in the pioneer settlement belt of Heilongjiang Province. Owing to extensive use of land and the positioning of most cropland on gentle slopes without conservation measures, about one half of the total cropland has suffered from soil erosion, and about a quarter of the total cropland has lost one half of its top stratum of fertile black earth after fewer than 100 years of cultiva-

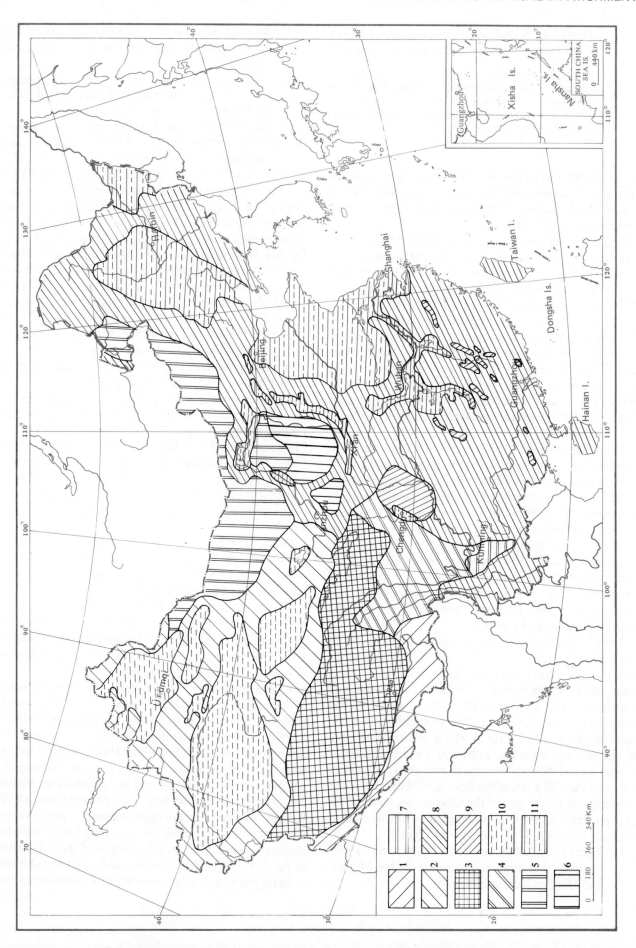

Figure 1-5 A preliminary paleogeographical map of China in the late Quaternary period.

1. Very strongly uplifted fold-fault mountains, more than 6000 m asl in elevation, with ancient and modern glaciation; 2. Strongly uplifted fault mountains, mostly more than 4000 to 5000 m asl in elevation, with ancient and modern glaciation on horst mountains and arid climate in graben basins; 3.Strongly uplifted montane plateaus more than 4000 m asl in elevation, elevated mountain ridges interspersed with depressed graben basins;   4. Strongly uplifted plateaus, 3000 to 4000 m asl in elevation, high ridges sandwiched with deep gorges; 5. Moderately uplifted inland plateaus, asl 1000 to 2000 m in elevation, with rolling topography and denundational gobi on inland plateaus and extensive fluvial-lacustrine deposits and shamo in inland basins;   6. Moderately uplifted loess plateaus, 1000 to 1500 m asl in elevation, with loess deposits in different stages of the Quaternary period;   7. Moderately uplifted red bed plateaus, 1000 to 2000 m asl in elevation, with rolling topography and extensive fluvial-lacustrine deposits, also with small-scale graben basins;   8. Moderately uplifted fold and fault mountains, 2000 to 3500 m asl in elevation, with humid climate and severe erosion ever since the early Quaternary period;   9. Slightly uplifted red bed hills and basins, with basins of the Cretaceous period-early Tertiary period strata eroded into innumerable hills since late Tertiary period;   10. Inland basins dominated by fault-sinking process, with arid climate and extensively distributed shamo, gobi, and salty lakes ever since the early Tertiary period;   11. Silting up plains, dominated by fault-sinking process, with large-scale sinking and extensively distributed fluvial networks and lakes ever since the late Tertiary period.

Table 1-4 A Brief Chronicle of Chinese History

Yuanmou Man	About 1.6 million years ago
Lantian Man (Peking Man)	About 0.7–0.3 million years ago
Upper Cave Man	About 20000 years ago (Paleolithic)
Yangshao Culture (Banpo) Hemudu Culture	About 5000–4000 B.C. (Early Neolithic)
Longshan Culture	About 2500 B.C. (Late Neolithic)
Three Rulers and Five Emperors	?2852–?2205B.C. (Traditional dates)
Xia Dynasty	?2205–?1766 B.C. (Traditional dates)
Shang Dynasty	1766–1126 B.C.
Zhou Dynasty Warring States	1126–403 B.C. 403–221 B.C.
Qin Dynasty	221–206 B.C.
Han Dynasty Eastern Han Dynasty	206 B.C.–A.D.9 A.D. 25–220
Three Kingdoms	A.D. 220–280
Western Jin Dynasty Southern-and Northern Dynasties	A.D. 280–316 A.D. 317–589
Sui Dynasty	A.D. 589–618
Tang Dynasty Five Dynasties	A.D. 618–907 A.D. 907–960
Song Dynasty Southern Song Dynasty	A.D. 960–1126 A.D. 1126–1279
Yuan Dynasty	A.D.1279–1368
Ming Dynasty	A.D. 1368–1644
Qing Dynasty	A.D. 1644–1911
Republic of China	A.D. 1911–1949
People's Republic of China	A.D. 1949–Present

tion. Some of these areas have lost all their topsoil, leaving yellowish loess parent materials exposed.

In arid and semiarid Northwest China, chiefly owing to the misuse of land, the so-called desertification problem is quite serious. The Mu-us Sandy Land, located at the semiarid southern margin of the Ordos Plateau, is probably the most notorious example, Since the Tang Dynasty, for more than 1000 years, it has been subjected to the "southward moving of a sandy desert", with shifting sands drifting more than 100 km southward. Especially during the last 300 years, owing to the accelerated removal of natural vegetation, a broad belt of shifting sands about 60-km wide has occurred along the Great Wall.

The hazard of salinization inflicts great havoc in Northwest China as well as along the coast. According to one estimate, about one fifth of the total cropland in China has suffered more or less from the salinization hazard. Major causes of salinization are the overuse of irrigation water in arid areas with inadequate drainage. For example, in oases of the middle and lower reaches of the Tarim River — the largest inland river of China — usually more than 15000 to 22500 cu m per ha of irrigation water (even 45000 to 60000 cu m for paddy rice) are used annually without any drainage. Thus, not only are huge amounts of precious water wasted, but also huge amounts of salt are deposited in the cropland every year. After such a saline buildup,

the cropland has to be abandoned, resulting in a barren wasteland. The Chinese farmers have now begun ambitious campaigns to combat this hazard.

Many wild animals and birds have been, or have nearly been, exterminated in China. One example is David's deer (*Elaphurus davidianus*), which were widely distributed in the North China Plain and the lower Changjiang Valley during the Pleistocene and early Holocene. As the forests were cleared and the swamps dried up, the deer were gradually eliminated and, in A.D. 1900, they disappeared entirely from China. Another example is the wild cattle of the Qilian Mountains. One scholar in the Qing dynasty (A.D. 1644–1911) reported that "wild cattle roam in thousands in the Qilian Mountains".But now, all the wild cattle have been exterminated, leaving behind only such relict names as Yeniu(Wild Cattle) Mountain, and Yeniu

(Wild Cattle) Ravine. Even more animal species have had their distribution areas much restricted or isolated. For example, in Eastern Monsoon China, the panda (*Ailuropoda melanoleuca*), squirrel (*Sciurus vulgaris*), monkeys (*Macaca mulatta, Rhinopithecus* spp.), river deer (*Hydropotes inermis*) and other forest-living animals have had their habitats greatly diminished as forest areas have been dramatically reduced. In Northwest Arid China and the Qinghai-Xizang Plateau, thousands of wild goats (*Procapra gutterosa, P. picticaudata, Gazella subgutterosa*, etc.) wild deer (*Moschus* sp.) wild ass (*Equus hemonius*), wild horse (*E. przewalski*) and wild camels (*Camelus bactrianus*) have been killed during the last few hundred years, some of them being nearly exterminated. The author remembers quite well when he first visited the eastern part of the Nei Mongol Plateau in 1951, and encountered many herds of hundreds, or even thousands, of "Mongolian gazelle" (*P. gutturosa*) that ran at a speed more than 80 km per hour. Now, it is most unusual to meet a large flock of these Mongolian gazelle on the Nei Mongol Plateau.

References

[1]   *Physical Geography of China* Compilation Committee, Chinese Academy of Sciences, 1985, *Physical Geography of China: General Survey,* Science Press, Beijing. (In Chinese)

[2]   *Physical Geography of China* Compilation Committee, Chinese Academy of Sciences, 1980, *Physical Geography of China: Geomormology,* Science Press, Beijing. (In Chinese)

[3]   *Physical Geography of China* Compilation Committee, Chinese Academy of Sciences, 1984, *Physical Geography of China: Climatology,* Science Press, Beijing. (In Chinese)

[4]   *Physical Geography of China* Compilation Committee, Chinese Academy of Sciences, 1982, *Physical Geography of China: Historical Physical Geography,* Science Press, Beijing. (In Chinese)

[5]   *Physical Regionalization of China* Working Committee, Chinese Academy of Sciences, 1959, *Physical Regionalization of China: Comprehensive Physical Regionalization,* Science Press, Beijing. (In Chinese)

[6]   Zhao Songqiao, 1984, *The Arable Land Resources and Their Development in China, Natural Resource,* 1984, No. 1, pp.13 – 20. (In Chinese)

[7]   Zhao Songqiao, 1981, "The Sandy Deserts and the Gobi: A preliminary Study of Their Origin and Evolution," *Desert lands of China* (ICASALS Publ., No. 81 – 1), Texas Tech University, Lubbock, Tex.

[8]   Zhao Songqiao, (1981), "Transforming Wilderness into Farmland; An Evaluation of Natural Conditions for Agricultural Development in Heilongjiang Province," *China Geographer,* No. 11.

[9]   Syukura Manabe & Theodore B. Terptre, 1974. "The effects of mountains on the general circulation of the atmosphere as identified by numerical experiments". J. of the Atmospheric Science, Vol. 31, No. 1.

[10]   Academy of Geological Sciences, 1978, *Main Features of Geological Structure in China,* Geological Press, Beijing. (In Chinese)

[11]   Xia Nai (Hsia Nai), "Carbon-14 and Chinese Prehistory." *Kaogu* (Archaeology), 1977, No. 4, pp.217 – 232. (In Chinese)

[12]   Wu Weidan, 1983, "Holocene palaeo-geography along the Hangzhou Estuary as viewed from archaeological excavations of the Hemudu Culture", *Acta Geographica Sinica,* Vol. 38, No. 2, pp.113 – 127. (In Chinese, with English abstract)

[13]   Cressey, G.B., 1934, *China's Geographic Foundations: A Survey of the Land and its People,* McGraw-Hill book Co., N.Y.

Chapter 2

Climatic Features of China

The Chinese climate is chiefly characterized by three features. First, the monsoon climate is dominant, with significant changes or even reversal of wind direction between winter and summer as well as seasonal variation of precipitation according to whether the maritime monsoon advances or retreats (Unlike the situation in India, China's winter monsoons are about twice as strong as its summer monsoons, whereas India's summer monsoons are about twice as strong as its winter monsoons). Second, climatic continentality is rather conspicuous, with higher summer temperature, lower winter temperature, and a greater annual range than other parts of the world with similar latitudes. Third, there are many climatic types in China. All these climatic features are caused by the interaction of major physical factors, including the vast area, a midlatitudinal and east coastal location, mountainous topography, and China's complicated natural history.

FEATURES OF THE MONSOON CLIMATE

In the planetary wind system, China is characterized generally as having the westerlies north of 30°N and the northeastern trades and subtropical belt of variable winds and calms south of 30°N. Because the angle of the sun's noon rays shifts through a total range of 47° from the solstice of Cancer to the solstice of Capricorn in a year, the planetary wind system migrates correspondingly: northward during summer, southward during winter. In the interface between the westerlies and the easterlies (about 25 to 35°N), the seasonal variation of prevailing winds is most conspicuous, with the westerlies dominating during winter and the easterlies dominating during summer. Such a seasonal variation of the planetary wind system combined with the seasonal variation of monsoons, which are chiefly caused by the distribution of land and sea, results in China's being one of the most famous monsoon countries in the world. Climatic features of the Chinese monsoons might be summed up as follows.

The air masses and basic air currents that control the climates of China during winter and summer are entirely different. During winter, the Polar air mass controls the Chinese climate. There is a high-pressure ridge over the upper troposphere between 90 and 100°E. The cold air mass behind it moves incessantly southward and deepens the cold high pressure near the ground — the Mongolian High Pressure. Consequently, the cold and dry northern monsoons dominate the lower troposphere in China. On the other hand, during summer, the Chinese mainland is chiefly controlled by the Tropical and Subtropical maritime air mass (Tropical continental air mass in the Qinghai-Xizang Plateau). There is a low-pressure trough in the upper troposphere between 70 and 80°E and a shallow high-pressure ridge along the coast; the pressure systems near the ground are represented by the Indian warm low pressure over most parts of the Eurasia, coupled with the maritime high pressure over the Pacific and Indian oceans. Hence, the warm and moist southern and southeastern monsoons dominate the lower troposphere in China.

The location of the main rain belt is closely related to the advance and retreat of the summer monsoon. There are two types of summer monsoon in China, the southeastern monsoon and the southwestern monsoon; the beginning and ending of the rainy season is closely correlated with these two monsoon systems. Whenever the front of the summer monsoon arrives, the rainy season begins. Generally speaking, the dividing line between these monsoon systems lies from 105 to 110°E. To the east, the southeastern monsoon begins in South China as early as March, moves to Central China in June, to North and Northeast China in July, and then retreats southward rapidly during late August and early September. It takes only one month to retreat entirely out of the Chinese mainland. In close correlation with the advance and retreat of the summer monsoon, the rainy season starts in early April in South China, early June in Central China, and early July in North and Northeast China. In Southwest China, the southwestern monsoon dominates; it bursts northward in late May when the rainy season in Yunnan and western Sichuan begins. It does not stop until October when the southwestern monsoon retreats rapidly southward.

Plentiful precipitation and high temperatures that coincide in the same season are other features of the Chinese monsoon climate. Most of the annual precipitation comes in the warm season, making agriculture quite productive. This is an important geographical factor in China's emergence as an important paddy rice producer for more than 7000 years. The northern limit for growing paddy rice reaches 53° 30'N (Mohe) — certainly the northernmost in the world. Because of these favorable climatic conditions, three agricultural crops are produced every two years in North China, two or three crops are produced per year in Central China, and a yield of three crops of paddy rice per year is possible in South China.

CHIEF CLIMATIC ELEMENTS

Solar Radiation

Figure 2-1 shows the distribution of annual total solar radiation in China, ranging from 80 to 240 kcal/cm^2 · year. The lofty Qinghai- Xizang Plateau, with 180 to 240 kcal/cm^2 · year, has the highest. Arid Northwest China, with 140 to 180 kcal/cm^2· year, ranks second, whereas, the humid Eastern Monsoon China is much lower, generally with 120 to 140 kcal/cm^2 · year in Northeast and North China and less than 120 kcal/cm^2· year in Central and South China. The lowest measurements are in the Sichuan Basin and the Guizhou Plateau, with less than 100 kcal/cm^2· year. These are also the cloudiest areas of China.

The composition of solar radiation is quite different in different regions. It is chiefly determined by the amount of water vapor and aerosols in the atmosphere. In North China or on high plateaus where the atmosphere is dry, clean, or thin, scattered radiation is relatively unimportant; consequently, direct radiation occupies a larger proportion of total solar radiation. In northern Nei Mongol and the

southwestern Qinghai-Xizang Plateau, direct radiation accounts for 60 to 70 per cent of the total. In Central and South China, where the atmosphere is rich in water vapor, scattered radiation occupies a larger proportion of total solar radiation; in Sichuan and Guizhou, it might account for more than 60 per cent. Another important area where scattered radiation accounts for more than 50 per cent of total solar radiation is the Tarim Basin where the largest sandy desert in China — the Taklimakan Desert — supplies plenty of aerosols to the overlying atmosphere.

The seasonal variation of solar radiation is determined by the "height" of the noon sun and is also somewhat modified by other climatic conditions. The monthly minimum of total solar radiation all over China occurs in December when the noon sun is "lowest"; in Sichuan, Guizhou, and the Altay Mountains and Da Hinggan Mountains, the value is less than 4 kcal/cm^2. The monthly maximum of total solar radiation generally occurs in July in South and Central China, June in North China, and in April in Southwest China, where the southwestern monsoon predominates. Most parts of China have a maximum monthly solar radiation level of between 14 to 16 kcal/cm^2 — about 20 kcal/cm^2 in Northwest China, as high as 25 kcal/cm^2 in the western Tibetan Plateau, and only 11 to 13 kcal/cm^2 in Sichuan and Guizhou.

Temperature

The distribution of temperature in China is chiefly determined by latitude and topography. China has a vast area, stretching from 3° 50' to 53° 31' N, hence, there is a great difference in temperature between the north and the south. Annual mean temperature in the Nansha Islands is about 25°C, whereas in northern Heilongjiang Province it is below 5°C, a difference of more than 30°. In Eastern Monsoon China, following the sequence of temperature belts from south to north, annual mean temperature is above 20°C south of the Nanling Mountains, about 10°C in the Changjiang River Valley, between 12 to 14°C in the Huanghe River basin, below 10°C north of the Great Wall, and below 0°C in the Da and Xiao Hinggan mountains. In Northwestern Arid China, owing to higher latitude and elevation, the annual mean temperature ranges from 0 to 10°C. The lofty Qinghai-Xizang Plateau, however, has an annual mean temperature below 0°C; its northwestern part, the Qingzang Plateau (elevation 4500 to 4800m) has annual mean temperature below −8°C.

As China is essentially a monsoon country, seasonal change in temperature is excessive and annual mean temperature has often little significance for illustrating the true climatic condition of an area. Therefore, it is preferable to discuss climate in terms of the seasons, especially conditions during January and July.

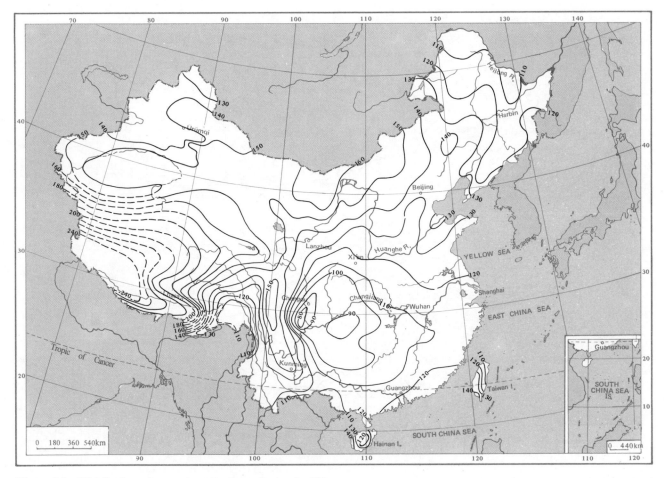

Figure 2-1 Distribution of annual total solar radiation in China

Winter

From October, strong cold air masses invade China frequently, being strongest and most frequent in January. Figure 2 - 2 shows the mean January temperature in China. The isotherms run generally latitudinal, closely paralleling each other. Latitudinal difference in temperature is great. From Heilongjiang Province to Hainan Island, the temperature differs about 1.5°C for each degree of latitude. The mean January temperature is below −30°C in the northern part of the Da Hinggan Mountains, below −10°C in most parts of Northeast China, Nei Mongol and the Junggar Basin, and below 0°C in most parts of North China and the Tarim Basin. The January −6°C isotherm (the boundary between spring and winter wheat) runs roughly along the Great Wall, and the January 0°C isotherm is at approximately 34°N latitude, or the Qin-

ling Mountains - Huaihe River line. Further southward up to the Nanling Mountains, the January temperature ranges from 0 to 10°C, while in the area south of the Nanling Mountains it is above 10°C. The Qinghai-Xizang Plateau has generally a mean January temperature between −10 to −20°C, and in its southern part, 0 to −10°C. As a whole, China has a much colder winter than other areas in the world with similar latitudes (as shown in Chapter 1), which illustrates its greater climatic continentality.

The minimum January temperature at 10 typical stations is shown in Table 2-1. At Mohe (Heilongjiang), Genhe (Nei Mongol), Qinghe (Xinjiang), and several other stations, an absolute minimum temperature of below −50°C has been recorded; Mohe has the lowest recorded temperature in China (−52.3°C). In South China where the mean January temperature is generally above 0°C, the absolute minimum temperature might also drop below 0°C. Even

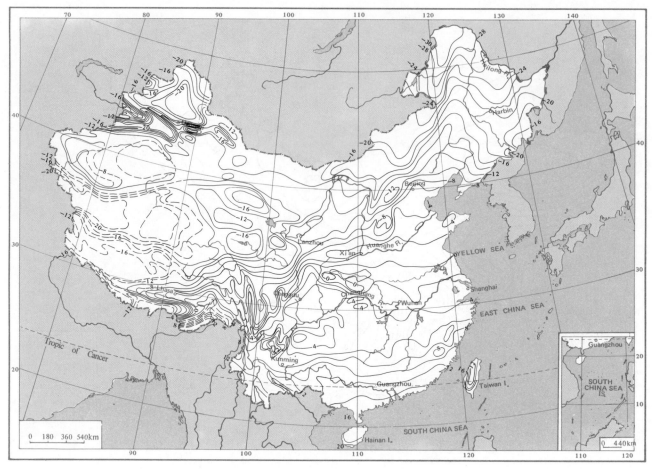

Figure 2-2 Distribution of mean January temperature in China

in northern Hainan Island and southern Taiwan Island, freezing temperatures might occasionally occur.

Spring

After January, the air gradually warms. For each month, the temperature increases by 6 to 10°C in Northeast China, 5 to 6°C in North China, 3 to 5°C in Central China and 2 to 3°C in South China. The coastal areas are usually colder. In April, China records a mean temperature above 0°C, except for the northern Da Hinggan Mountains and other high mountains and plateaus. From Heilongjiang Province southward to the Hainan Island, the temperature increases about 0.6°C for each degree of latitude.

Summer

The influence of latitude on temperature is reduced to its minimum in July, with only a 0.2°C difference of

Table 2-1 Minimum January Temperature (1951 – 1970) and Absolute Minimum Temperature in China

Station	Temperature (°C)	
	Mean Minimum in January	Absolute minimum
Nenjiang (Heilongjiang Province)	− 31.7	− 47.3
Changchun (Jilin Province)	− 22.0	− 36.5
Beijing	− 10.1	− 27.4
Nanjing	− 2.2	− 14.0
Guangzhou	9.3	− 0.3
Danxian (northern Hainan Island)	13.1	0.4
Yinchuan (Hui Autonomous Region of Ningxia)	− 15.0	− 30.6
Altay (northern Xinjiang)	− 23.1	− 43.5
Hotan (southern Xinjiang)	− 10.0	− 21.6
Lhasa	− 10.4	− 16.5

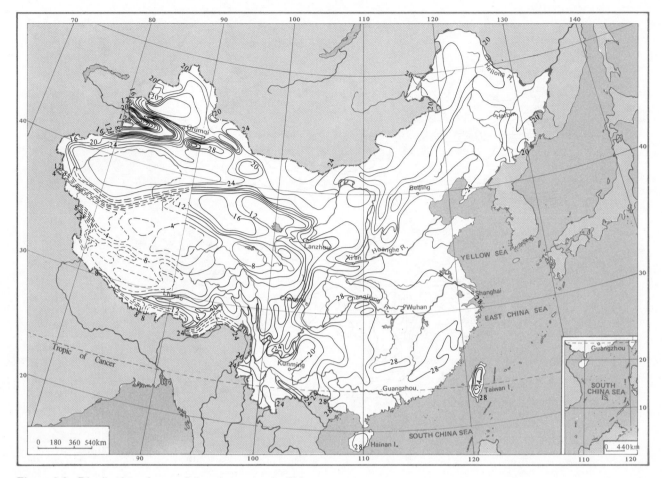

Figure 2-3 Distribution of mean July temperature in China

temperature for each degree of latitude, (Figure 2-3). Isotherms are widely spaced. They are arranged in a longitudinal pattern, generally paralleling the coast. Most parts of China have a mean July temperature of between 20 to 28°C. There are two high temperature centers: one around Poyang Lake and another in the Turpan Basin, both with mean July temperatures above 30°C. Hence, four big cities in the middle and lower Changjiang Valley — Wuhan, Changsha, Chongqing, and Nanjing are termed the "four ovens" in summer, whereas the Turpan Basin with the highest recorded temperature 47.6°C in China, was once called "Fire Prefecture."

The absolute maximum July temperature rises to above 40°C in certain parts of Northwest China and some low valleys of Eastern Monsoon China. Table 2-2 shows the mean July temperature and absolute maximum temperature of 10 typical stations.

Table 2-2 Mean July Temperature (1951 – 1970) and Absolute Maximum Temperature in China

Station	Temperature (°C)	
	Mean July	Absolute maximum
Nenjiang	20.4	38.1
Dalian	24.7	36.1
Beijing	26.0	42.6
Nanjing	28.2	43.0
Guangzhou	28.3	38.7
Haikou	28.4	40.5
Yinchuan	23.5	39.3
Turpan	33.0	47.6
Hotan	25.5	40.5
Lhasa	15.5	30.5

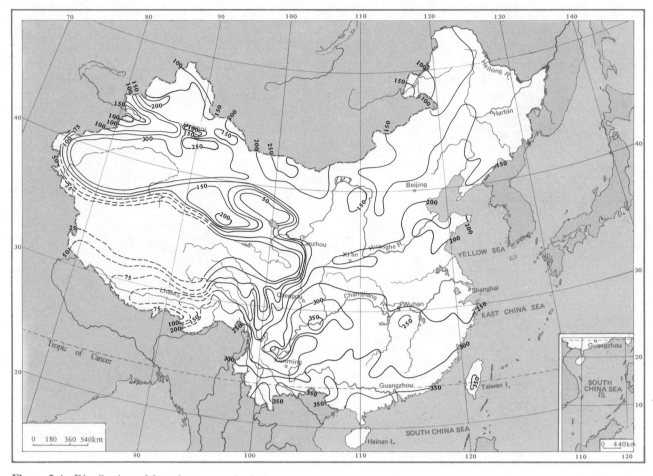

Figure 2-4 Distribution of frost-free season in China

Autumn

In October the isotherms are distributed very similarily to those in April, and most parts of China have a temperature above 0°C. Yet, the coastal areas turn comparatively warmer instead of colder, and areas north of the Qinling Mountains-Huaihe River line have autumn temperatures higher than that of spring, whereas areas south of that line have an autumn temperature that is lower. Areas north of the Great Wall drop in temperature rapidly from early September, so that there might be heavy snow in September, and the inhabitants find themselves "eating watermelon while wearing fur coats"; by October, the ground has been mostly frozen over.

Low temperature and frost are frequent menacing hazards to agriculture in China. In the Northeast and parts of North China, freezing temperatures appear in early September and last for as long as eight to nine months.

In the Qinghai-Xizang Plateau, the frost season lasts even longer, and wherever the elevation exceeds 4000 m there is practically no frost-free season at all. In North China, frost usually appears from mid-October through mid-April; in Central China, from November through March. In the Sichuan Basin, thanks to the protection of the Qinling Mountains, frost only appears from December through February. South of the Nanling Mountains, there is no frost season, although there may be frost for a few days in January. (Figure 2-4 shows the distribution of the frost-free season in China). In close correlation to the duration of the frost-free season, the crop yield is generally restricted to one crop per year in Northeast and Northwest China, three crops in two years in North China, two or three crops per year in Central China, and more than three crops per year in the South.

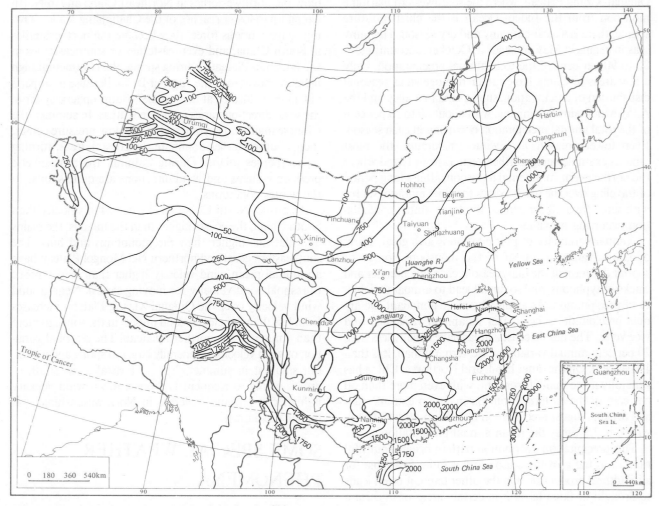

Figure 2-5 Distribution of annual precipitation in China

Precipitation

The distribution of annual precipitation in China is generally determined by distance from the sea; the farther away from the sea, the less abundant the precipitation (Figure 2-5). Therefore, the isohyets run generally parallel with the coast, and annual precipitation decreases from southeast to northwest. The 500 millimeter (mm) isohyet is roughly a dividing line: northwest of it, precipitation is scarce, resulting in chiefly pastoral areas; southeast of it, precipitation is abundant, giving rise to the chiefly agricultural areas of China. The 750 mm isohyet follows roughly the Qinling Mountains-Huaihe River line; it is the divide between the paddy rice areas and the dry farming lands. The heaviest precipitation occurs along coastal hills and mountains with more than 1500 to 2000 mm annual-

ly and with even more than 3000 to 4000 mm in some high mountains. The driest area is the mountain-enclosed heart of Eurasia — the eastern Tarim Basin with a mean annual precipitation of less than 50 mm, the lowest recorded in China is 3.9 mm (only 0.5 mm in 1968) at Toksun in the Turpan Basin.

Seasonal distribution of precipitation is uneven. Most areas have their annual precipitation concentrated in summer when the warm, moist maritime monsoon dominates. Precipitation in one month usually accounts for more than one quarter or even one half of the total annual precipitation. In Central China, because of the earlier arrival of the maritime monsoon, spring is the most important rainy season, followed by summer. In North and Northeast China, however, summer rain comprises more than one half of the total annual precipitation, and there is pronounced spring drought. In Southwest China and the southern

Qinghai-Xizang Plateau, where the southwestern maritime monsoon from the Indian Ocean is the chief moisture source, there is a clear-cut rainy and dry season. The rainy season begins in May and ends in October, accounting for 80 to 90 per cent of the total annual precipitation. Only a few areas in China have rather homogeneous seasonal distribution of precipitation, such as the Ili Valley and the Altay Mountains in Xinjiang, with about 20 to 30 per cent of the total annual precipitation occurring in each season. Even fewer areas have a winter maximum, the most famous example being northeastern Taiwan Island where the northeastern continental monsoon, after a long period of traveling over the sea, becomes the chief moisture source with more than 30 per cent of the total annual precipitation occurring in winter.

Annual variability of precipitation is also great. Early or late advance or retreat of the monsoons, longer or shorter duration of the rainy season, changing number and tracks of typhoons, and so on, all lead to annual variability of precipitation. As a rule, the more abundant the annual precipitation, the smaller its annual variability and vice versa. The humid coastal areas have usually less than 20 per cent annual variability, western Yunnan, less than 10 per cent. On the other hand, arid Northwest China has generally more than 30 per cent, with even more than 50 per cent in the eastern Tarim Basin. In an extreme case at Turpan, August precipitation in 1958 totaled 42.1 mm, which is 2.5 times the mean annual precipitation.

This precipitation distribution pattern brings bountiful rainfall in the growing season and is quite favorable for agricultural production. On the other hand, it often leads to drought or flood hazards in certain areas or in certain seasons; both may even occur in the same area in different seasons or in the same season in different areas. According to an analysis of annual precipitation data between 1950 to 1976 (supplemented by extensive ground observations), the worst conditions occur in the densely populated North China Plain, where a flood hazard occurs nearly every year in different areas. Spring drought and summer drought occurred in 20 and 21 years respectively of the 27-year survey, and the spring and summer drought combined occurred in 16 years. In another densely populated region, the middle and lower Changjiang Valley, the most menacing climatic hazard is summer drought which occurred in 25 out of the 27 years. The flood hazard is more menacing in the Changjiang Delta owing to frequent typhoons.

Wind

The distribution and velocity of near-ground winds are chiefly determined by pressure systems in the atmosphere. In winter, continental high pressure dominates the atmosphere over the Nei Mongol Plateau. From that sector blow the northwesterlies in Northeast China and in North China up to lower reaches of the Changjiang Valley. Owing to the Coriolis force, they become the northeasterlies in South China, and even easterlies or southeasterlies in the Guizhou Plateau. During spring, the continental high pressure moves over the Junggar Basin. Because most parts of China are located at the "saddle" of atmospheric systems, the wind direction varies in different areas. In summer, the Chinese mainland becomes a thermal low-pressure center, then, southeasterlies dominate the coastal areas, turning to southerlies inland. During autumn, the atmospheric pressure systems and wind directions are quite similar to the winter conditions.

As a rule, North China has higher wind velocity than South China, the coasts higher than the inlands, the plains and plateaus higher than the mountains and hills. The southeastern coast and northern Nei Mongol Plateau have an annual mean wind velocity higher than 4 to 5 m/sec (threshold wind velocity for moving sands). Near-ground wind velocity over the Qinghai-Xizang Plateau increases from less than 2 meters/second (m/sec) in the south to more than 4 m/sec in the northern plateau. The seasonal variation of wind velocity is generally larger in spring and winter, but smallest in summer. Yet, in coastal areas south of 30°N, owing to frequent typhoons, mean wind velocity is higher in autumn, whereas in Northwest China, it is higher in summer.

SOME SPECIAL WEATHER PHENOMENA

Following the advance and retreat of the monsoons, there occurs a series of special weather phenomena in different seasons, such as cold waves, continuous rain with low temperature, plum rain, drought, and typhoons, of which the cold waves, plum rain, and typhoons are the most outstanding and relevant to agricultural production.

Cold Wave

This is a frequent, large-scale atmospheric phenomenon, occurring on the average once every 10 days during winter, spring, and late autumn. It is defined by the Chinese meteorological stations as "a decrease of minimum air temperature within 48 hours of more than 10°C in the Changjiang Valley and more northern areas, a minimum air temperature below 4°C in the middle and lower reaches of the Changjiang Valley as well as strong 5 to 7-scale winds over the land and 6 to 8-scale winds over the sea." The cold wave front usually "bursts" together with strong northwestern winds. At first, there are sandstorms or snowstorms in North and Northeast China, heavy precipitation (rain or snow) in Central China, and continuous rain in South China. Then, when the polar continental air mass controls the whole region, the weather is fine and cold, warming up gradually until the arrival of the next cold wave.

Cold waves might also occur during summer time, but these are mainly restricted to Northwest China and bring some precipitation to that region.

There are three important tracks for cold waves in China:

1) The northwestern track: cold air that originates from the Barents Sea in the Arctic region, travels through European Russia, Siberia, and Mongolia; and then, enters China by way of Nei Mongol up to the middle and lower reaches of the Changjiang Valley. Sometimes it even reaches as far as South China. Cold waves along this track are most frequent and powerful.

2) The western track: cold air that originates from the Kara Sea in the Arctic region, moves southward to western Siberia, and, then, invades China by way of Xinjiang, Qinghai, and Gansu. Sometimes it reaches as far as South China.

3) The northern track: cold air from eastern Siberia moves southward by way of Nei Mongol and Northeast China, then southwestward over the North China Plain and the Yellow Sea. Cold waves along this track are comparatively weak. All these cold waves, after entering China, are subjected to much modification by the mountainous topography. For example, the lofty Qinghai-Xizang Plateau effectively blocks all these cold waves, and cold waves of the western track move generally eastward along the Tianshan Mountains, then southeastward by way of the Hexi Corridor. Another good example of a barrier is the Qinling Mountains, which effectively protect the Sichuan Basin, so that the latter has a much warmer winter than the areas with similar latitudes.

A cold wave together with its accompanying strong northern wind and sharp decrease in temperature is an unfavorable condition, sometimes even a great hazard, to agricultural production. It promotes frost hazard and restricts the growing season. Even as far south as Hainan Island (about 18°N), strong cold waves can break through a series of mountain barriers — the Yinshan Mountains, Qinling Mountains, and Nanling Mountains systems and inflict great havoc on the tropical crops of that island. Therefore, good weather forecasting and effective measures for protecting crops and animals are of paramount importance in China.

Plum Rain

Plum rain is a special weather phenomenon that occurs in the middle and lower Changjiang Valley and the Huaihe Basin during late spring and early summer; it is so-called because the plum fruit is ripening at this time. The climatic features are continuous rain, moderate temperatures, low wind velocity, and high relative humidity. The weather feels stifling and uncomfortable, and clothes and other belongings become moldy.

In June, the planetary wind system moves northward and the tropical maritime air mass reaches the Changjiang Valley. The polar front, which is a product of interaction between the polar and tropical air masses is located over the Huaihe Basin and the Changjiang Valley. When newly arrived northwestern cold air forms a cyclone or low-pressure trough along the polar front and its further eastward movement is blocked, then, prolonged rainy days occur in that area. Total precipitation during this period is determined by frequency and intensity of newly arrived northwestern cold air; the stronger and more frequent the cold air, the more plentiful the precipitation and vice versa.

Plum rain usually starts together with the onset of the summer monsoon in this area in early or middle June, with a total duration of about one month; it generally ends in early or middle July. Yet, the dates of its commencement and conclusion can vary greatly. For example, according to the statistical data of five cities (Shanghai, Nanjing, Wuhu, Jiujiang, Wuhan) between 1885 to 1974, the earliest starting date of the Plum rain was May 26 (1896); the latest, July 4 (1947). This is a difference of 40 days. The earliest and latest concluding dates were June 16 (1961) and August 1 (1954) respectively, a difference of 45 days.

As this is the greatest paddy rice producing area in China, Plum rain is a very significant factor in agricultural production. Plentiful precipitation in late spring and early summer is favorable for paddy rice growing; spring drought comprises a great hazard in most parts of China, especially in North China. Sometimes, if Plum rain fails to come on time, and, thus, does not bring the all-important rain, the local farmers express their complaint in a proverb "Little rain in Plum rain month means disaster for six months."

Typhoons

Typhoons mainly originate in the northwestern Pacific Ocean and partly in the South China Sea. They attack the Chinese coasts during summer and autumn.

The Chinese meteorological stations classify them into three categories:

1) Strong typhoon, with maximum mean wind velocity of more than 32.6 m/sec.

2) Typhoon, with maximum mean wind velocity in between 17.2 to 32.6 m/sec.

3) Tropical low pressure, with maximum mean velocity of 10.8 to 17.1 m/sec.

Most typhoons formed in the northwestern Pacific

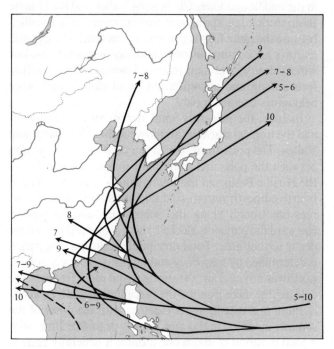

Figure 2-6 Tracks of typhoon along the Chinese Coast

Ocean do not reach the Chinese coasts. According to statistics of the 30 years between 1949 to 1979, only 276 typhoons struck China, of which 121 were the first category and 155 the second. Among the coastal provinces Guangdong ranks first with 48.2 per cent of the total landed typhoons; Taiwan is the second with 20.1 per cent; Fujian is the third with 18.1 per cent; and all other coastal provinces north of Shanghai account for the remaining 6.2 per cent. As for the seasonal distribution, 79.7 per cent of the typhoons strike China from July through September; whereas they might land in any coastal province during July and August, they are restricted to coasts south of Shanghai in September. In October and June, they are much less active, and are restricted mostly to Guangdong. In November and May, they appear only occasionally in Guangdong, Guangxi and Taiwan (Figure 2-6). The earliest typhoon has landed was on Hainan Island May 11, 1954; the latest was at Taishan (Guangdong) December 2, 1974. After landing, the wind velocity of the typhoons decreases rapidly. Nevertheless, most of them keep traveling for a considerable distance; their influence can be strongly felt in Jiangxi, Hunan, and Anhui, sometimes even in Jilin and Heilongjiang.

Typhoons bring not only high winds, but also heavy rains. In September 9 to 12, 1963, one strong typhoon landed in Taiwan; the mountain meteorological station near Taibei (elevation: 1936 m) recorded a total precipitation of 1684.0 mm in four days, of which 1172.2 mm fell in

one single day (September 11). The maximum downfall for 24 hours was 1247.9 mm. The typhoon rains play an important role in Chinese coastal provinces; they make up more than 20 per cent of the total annual precipitation south of Wenzhou (Zhejiang Province) and nearly 40 per cent in southern Hainan Island. North of Wenzhou, they account for about 10 per cent of the precipitation. From July through September, the typhoon rain is even more important, comprising more than 50 per cent of the monthly precipitation in Fujian and the southern Zhejiang coast.

High winds and heavy rains that accompany typhoons are great hazards to coastal fishing and shipping, and they can also inflict a toll on human life and produce severe flooding in the areas where they strike. Hence, reliable forecasting and effective precautions are absolutely necessary. On the other hand, the typhoons are sometimes beneficial. When the polar frontal rain belt has already moved northward to North China, the most important agricultural region in China – the middle and lower Changjiang Valley — is controlled by subtropical high pressure and is, consequently, subjected to midsummer drought. At this point, the heavy typhoon rains bring a most welcome release. In addition, hundreds of millions of Chinese who live in Central and South China and are oppressed by excessive summer heat find the typhoon with its high wind and heavy rain a great, although temporary, relief.

CLIMATIC ZONES AND CLIMATIC REGIONS

Based on a combination and integration of the climatic elements and seasonal weather conditions mentioned, China might be subdivided into several climatic regions and climatic zones. Such climatic regions and zones exert great influence on other physical factors and the total geographical environment. In turn, other physical factors and the total geographical environment affect them.

First, China might be subdivided into three great climatic realms that can be identified with the three great natural realms of China.

Eastern Monsoon China

Eastern Monsoon China spreads from Heilongjiang Province southward to the South China Sea. It is dominated by monsoons, with the southerly maritime monsoon and the northerly continental monsoon alternating in summer and winter. The temperature decreases from south to north, with sharper differences in winter. Precipitation is plentiful, with natural vegetation mostly in the form of different kinds of forest cover. Again, precipitation is concentrated in the high-sun growing season, which is quite favorable to agricultural production. Yet, owing to the variability of the advance and retreat of the monsoons as well as the invasion of typhoons and cold waves and other climatic hazards, there frequently occur droughts, floods, winds, low temperatures and other unfavorable conditions in different areas and in different seasons.

Northwestern Arid China

Northwestern Arid China is chiefly characterized by its dry climate. Summer is warm or even hot, and winter is long and severe. Agriculture depends on irrigation. Water from the melting snows and heavy rains of the surrounding high mountains are major sources of irrigation water. Thanks to bountiful sunshine, fruits and melons grow excellently. Natural vegetation consists mainly of steppes, desert-steppes, and deserts, which are primarily used for pasture when there is adequate grass cover.

The Qinghai-Xizang Frigid Plateau

The Qinghai-Xizang Frigid Plateau is chiefly characterized by high relief and, consequently, low temperatures and conspicuous vertical zonation. Natural vegetation is mostly montane desert, montane grassland, and alpine meadows and shrubs, which are mainly used as pastures. Few crops can be grown there. But owing to intense solar radiation,

crops, such as wheat, barley, potatos, and so on, usually get higher unit yields than in Eastern Monsoon China if they are planted with care.

Within these three great climatic regions, eastern Monsoon China and Northwestern Arid China might be again subdivided into six temperature zones: Equatorial, Tropical, Subtropical, Warm Temperate, Temperate, and Cool Temperate; and the whole country might be again subdivided into four aridity zones: Humid, Subhumid, Semiarid, and Arid. We have briefly outlined these zones in Chapter 1, and will discuss them in more detail in Chapter 7.

References

[1] *Physical Geography of China* Compilation Committee, Chinese Academy of Sciences, 1984, *Physical Geography of China: Climate,* Science Press, Beijing. (In Chinese)

[2] Chu Pin-hai, 1962, Climate of China, Science Press, Beijing. (In Chinese)

[3] *Physical Regionalization of China* Working Committee, Chinese Academy of Sciences 1959, *Climatic Regionalization of China,* Science Press, Beijing. (In Chinese)

[4] Coching Chu, 1964, "Some Characteristic Features of Chinese Climate and Their Effects on Crop Production", *Acta Geographica Sinica,* Vol. 30, No. 1 .(In Chinese,with English abstract)

[5] Xu Shuying & Zhang Sizhong, 1979, "Thirty Years of Climatology in China," *Acta Geographica Sinica,* Vol. 34, No. 4. (In Chinese, with English abstract)

Chapter 3

Geomorphological Features of China

In the vast domain of China, topographic and geomorphic conditions are quite varied. Not only are there numerous lofty mountains, extensive plateaus, enclosed inland basins, and fertile low plains, but also long, curved coastlines, thousands of sea islands, and broad continental shelves. Areal differentiation of landform types and their assemblages is also prominent, exerting great influence on the formation and evolution of China's physical geographical environment.

BASIC FEATURES OF CHINA'S GEOMORPHOLOGY

Landforms of China are chiefly characterized by the following three features.

Four Great Topographical Steps

In Chapter 1, we mentioned briefly four great topographic steps in China (Figure 3-1). The lofty, extensive Qinghai-Xizang Plateau (Photo I-1,Landsat image11) is the first great topographic step, with its eastern and northern borders roughly coinciding with the 3000m contour line. It generally has an elevation of 4000 to 5000 m and is, hence, sometimes called the "roof of the world". On the Plateau, there also extend many west-eastward trending or northwest-southeastward trending extremely high mountains, usually more than 5000 to 6000 m high — even more than 7000 to 8000 m high — such as the Kunlun Mountains, the Gangdisê Mountains (Photo I-4), the Nyainqêntanglha Mountains, the Tanggula Mountains and the

mighty Himalaya Mountains. The last range, with an average elevation of above 6000 m and containing the world's highest peak (Mount Qomolangma) at 8848 m (Photo I-2), is the highest mountain range in the world. Between these lofty mountain ranges, there are extensive rolling basins studded with numerous saline lakes. On the southeastern margin of the Qinghai-Xizang Plateau, however, all mountain ranges turn southeastward or directly north-southward, forming the famous Hengduan Mountains system, with deep gorges cut by the upper reaches of the Changjiang, Nujiang, Lancang rivers. The gorges are tightly sandwiched by high, steep mountain slopes.

Eastward and northward of the Qinghai-Xizang Plateau, up to the Da Hinggan Mountains (the eastern marginal mountains of the Nei Mongol Plateau (Photo III-1), the Taihang Mountains (the eastern edge of the Loess plateau), and the Wushan Mountains (the eastern margin of the Sichuan Basin), lies the second great topographic step. It is mainly composed of plateaus and basins with elevations of 1000 to 2000 m, such as the Nei Mongol, Ordos, Loess, Yunnan-Guizhou plateaus and the Tarim, Junggar, and Sichuan basins. On the borders of these plateaus and basins, there are generally uplifted mountains with elevations of more than 3000 m — such as the Tianshan Mountains, which tower between the Tarim and Junggar basins, with its highest peak (Tomul, Photo I-5) at a height of 7435 m. Yet, the Sichuan Basin has elevations of less than 1000 m, and the Turpan Basin — a faulted graben in the southern footslope of the Tianshan Mountains — is, at 155 m below sea level, the lowest land site in China.

Eastward from that line and extending to the coast is the third great topographic step. Here lie the largest plains of China, such as the Northest China Plain, the North

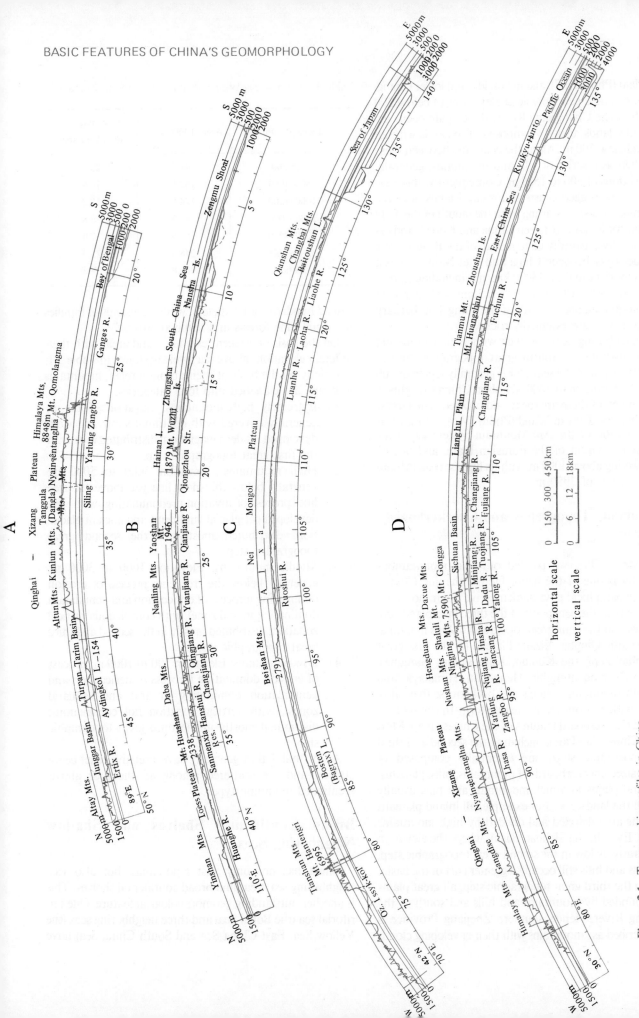

Figure 3-1 Topographic profiles in China

A. Along 89°E (radius of curvature 6367.5 km);   B. Along 110°E (radius of curvature 6367.5 km);

C. Along 42°N (radius of curvature 4730.7 km);   D. Along 30°N (radius of curvature 5514.2 km).

China Plain (Photo III-7), and the middle and lower Chang-jiang Plain. These generally lie at elevations of below 200 m. South of the Changjiang River, there again occur extensive hilly lands, usually composed of red beds with an elevation below 500 m. Along the coast and between these plains, there is a series of hills and mountains, generally with elevations of 500 to 1500 m. Consequently, there exists a series of rugged coasts and rocky islands near the coast. China's coast — starting from the mouth of the Yalu River (the border river between China and Korea), and extending southwestwardly up to mouth of the Beilun River (the border river between China and Viet Nam) — has a total length of more than 18000 km, not including islands and their coasts. Coasts along the Liaodong and Shandong peninsulas and all coasts south of the Hangzhou Estuary (Hangzhou Bay) are essentially rocky.

The neighboring seas together with their continental shelves constitute the fourth great topographic step in China. The depth of the seawater is generally less than 200 m. There are more than 5000 islands in China's neighboring seas; many of them are rocky and uplifted. The largest island in China, Taiwan Island (Photo IV-7), is located at the conjunction of the Arc Mountain System along the southeastern margin of the Eurasian Plate, and has 62 peaks towering above 3000 m, with its highest peak (Mount Yu) at a height of 3950 m.

Mountainous Topography and Checkerboard Landform

In Chapter One, we pointed out that mountainous topography is one of the major factors shaping China's physical geographical environment. Five main mountain systems serve as the backbones of the three macrolandform complexes and the four great topographic steps in China. In the gigantic Qinghai-Xizang Plateau (or the first great topographic step), the Kunlun, Kara-Kulun, Gangdies, Tanggula, Nyainqêntanglha, Hengduan, Himalaya, and other high mountain systems are backbones; they also, together with intermontane basins, form the numerous elongated checkerboard features of the Chinese map. Most parts of China's land area, including the second and third great topographic steps, are essentially composed of different-sized checkerboards of mountains, hills, plateaus, basins, and plains in which mountains and hills usually dominate the landscape. For example, all inland plateaus and basins are encircled and divided by high mountains and hills. Even in the Sichuan Basin, where the elevation is comparatively low in the second great topographic step, mountains and hills still occupy a greater part of the basin. Again, in the third great topographic step, all great plains are surrounded by mountains and hills and south of the Changjing River. Hence, southern Zhejiang Province is aptly described as "mountains with their enveloping clouds

Table 3-1 Land Areas With Different Elevation in China

Elevation (m)	Area (10000 km²)	Percentage of total land area
< 500	241.7	25.2
500–1000	162.5	16.9
1000–1500	174.6	18.2
1500–2000	65.3	6.8
2000–3000	67.6	7.0
> 3000	248.3	25.9
Total	**960.0**	**100.0**

extend hundreds of kilometers, while thousands of families live amid pine forests and bamboo groves."

On the basis of a recent estimate, land areas with their different elevations above sea level are shown in Table 3-1.

Chiefly on the basis of their absolute relief, mountains in China are classified into four categories:

1) Extremely high mountains: elevation > 5000 m; generally covered with continual snow and well-developed modern glaciation; distributed mostly in the first great topographic step.

2) High mountains: Elevation 3000 to 5000 m; generally below the snow line yet above the tree line; process of nivation predominating, with well developed periglacial landforms and ancient glaciation; distributed mostly in the second great topographic step.

3) Middle mountains: Elevation 1000 to 3000 m; generally below the tree line; process of erosion dominating (process of desiccation and denudation dominating in arid zone), with conspicuous vertical zonation; distributed mostly in the second and third great topographic steps.

4) Low mountains: Elevation 500 to 1000 m; process of erosion dominating (process of desiccation and denundation dominating in arid and semiarid zones), with vertical zonation not conspicuous; distributed mostly in the third great topographic step.

As for "hills", they generally have a relative relief below 500 m, and are scattered among all the four above-mentioned mountain types.

Broad Continental Shelves and Shallow Neighboring Seas

China has not only vast land areas, but also vast neighboring sea areas and broad continental shelves. This is another outstanding geomorphological feature. One territorial sea (the Bohai Sea) and three neighboring seas (the Yellow Sea, East China Sea and South China Sea) have

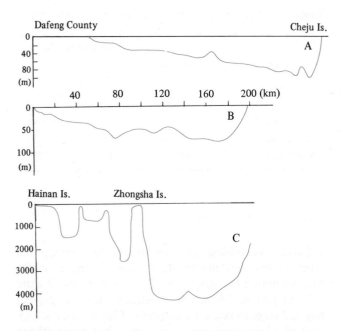

Figure 3-2 Sea-bottom topographical profiles in China's neighboring seas

A. Yellow Sea (from northern Jiangsu coast to Cheju Is.);
B. Taiwan Strait (along 25°N);
C. South China Sea (from 18°11′ N, 109°48′ E to 12°32′ N, 118°40′ E).

altogether an area of 4.7 million sq km, approximately one half of China's total land area. Except for the eastern margin of the East China Sea and the central sea basin of the South China Sea, all of them are located on broad continental shelves, with a water depth of a few dozen to little more than 100 m. The extensive Yellow Sea-East China Sea continental shelf is one of the broadest and richest of petroleum-bearing continental shelves in the world, with a width of more than 100 to 300 nautical miles. A continental shelf is a natural extension of a landmass into an ocean basin, hence, its geological and geomorphological characteristics are closely linked with the landmass. China's continental shelves have generally a level or rolling floor, gently dipping toward the southeast, and they feature many drowned deltas and ancient fluvial channels near the coast. According to recent investigations, sea level has risen nearly 100 m since the late Pleistocene and early Holocene with the present sea level formed about 3000 to 5000 years ago.

The Bohai Sea

Essentially encircled by lands, the Bohai Sea has as its only opening a narrow strait from the southern tip of the Liaodong Peninsula through the Miaodao Islands to the northeastern tip of the Shandong Peninsula facing toward the Yellow Sea. It is a territorial sea of China, with an area of about 77000 sq km and an average depth of 18 m. Geologically, it is a Mesozoic-Cenozoic sinking basin, located at the continental margin, and controlled by northeast-trending or north-northeast trending structures. The sea bottom is level or gently sloping and is mainly composed of silt and fine sand with coarse sand and gravel near the coast where the Huanghe River, Haihe River, Liaohe River, and other rivers empty into it.

The Yellow Sea

A half-closed sea, the Yellow Sea is sandwiched between the Liaodong and Shandong peninsulas on the western side and the Korean Peninsula on the eastern side; its southern boundary is roughly demarcated by a line from the mouth of the Changjiang River northeastward to Cheju Island off the southern tip of the Korean Peninsula. It has an area of about 380000 sq km and an average depth of 44 m. Geologically and geomorphologically, it is controlled by a series of NE-trending uplifts and down-warpings, with the sea bottom on the whole gently sloping southeastward. The largest river in China, the Changjiang River together with many other rivers (the Huaihe River, the Yalu River and so on) flow into the Yellow Sea. The heavily silt-burdened Huanghe River also emptied into it from A.D. 1194 to 1856, and again in A.D. 1938 to 1949, building huge deltas near its mouth.

The East China Sea

An open sea, the East China Sea faces the immense Pacific Ocean through the Ryukyu Islands. Its southern boundary with South China Sea coincides roughly with the line running from Dongshan Islands off Fujian Province eastward to the southern tip of Taiwan Island. It has an area of about 770000 sq km, and an average depth of 370 m. It is also controlled by a series of northeast-trending sea ridges and depressions, mostly located on the shallow continental shelf, with well-preserved subwater ancient deltas and river channels as well as other features of a drowned coastal plain. The subwater ancient delta of the Changjiang River extends as far eastward as 125°E. Yet, the southeastern margin of the East China Sea is occupied by a continental slope and the Ryukyu trough, with a maximum water depth of 2719 m.

The South China Sea

A deep sea basin, the South China Sea stretches extensively south of Taiwan Island up to the Indonesian Archipelago. It has an area of 3.5 million sq km and an average depth of 1212 m. Many large rivers (the Zhujiang, the Lancang, the Yuanjiang, etc.) empty into it. The sea bottom varies, although still controled by northeast-trending geological structures. Shallow continental shelves are widely distributed in the northwestern and southeastern parts but are quite narrow in the eastern and western parts. Continental slopes are stepped, and there are also deep sea troughs and trenches near the eastern and southern parts of the continental slopes and, in their northwestern part, there occurs a narrow sea depression with maximum depth of 5559 m. The central sea basin of South China Sea is scattered throughout with many isolated submarine volcanic cones. It has a water depth of generally more than 3500 m.

MAJOR FACTORS DETERMINING CHINA'S LANDFORMS

China's landforms have undergone a complicated geological history. The pattern and intensity of tectonic movements, the spatial and temporal changes of climates, the property and composition of bedrock — all have exerted great influence on the formation and development of China's landforms.

Table 3-2 Chief Features of Four Neighboring Seas

Seas	Sea area (km²)	Water depth (m)	
		Average	Maximum
Bohai Sea	77000	−18	−70
Yellow Sea	380000	−44	−140
East China Sea	770000	−370	−2719
South China Sea	3500000	−1212	−5559

Tectonic Movements

Large-scale folding, faulting, uplifting, warping, and other tectonic movements are the chief endogenic forces in determining the basic relief features in China (Figure 3-3). In Chapter 1, we demonstrated that the present physical geographical environment of China is the end product of a long, complicated geological history, involving in particular relatively recent tectonic movements during the Mesozoic and Cenozoic eras. According to recent geological investigations, macrolandforms on the Chinese mainland (particularly the above-mentioned four great topographic steps) mirror faithfully major relief forms on

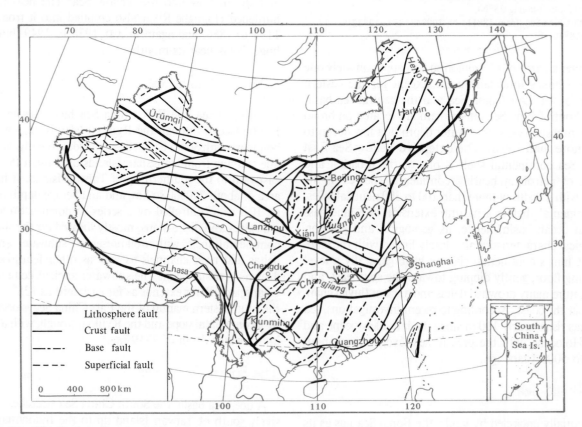

Figure 3-3 Distribution of major fault structures in China

the Moho surface[1], that is mountains and plateaus on the earth's surface are antipodes of the depressed areas on the Moho surface, whereas plains and basins correspond closely to the uplifted areas on the Moho surface. Depths of the Moho surface beneath the four great topographic steps are 50 to 70 km, around 36 to 38 km, and about 34 km going from west to east. Boundaries between these four great topographic steps correspond roughly to long-established and deep-rooted fault belts.

Another characteristic showing the close correlation of China's landforms with geological structure is that the arrangement of all macro-landform units (mountains, plateaus, plains, etc.) has a definite orientation, and their boundaries are usually identified with geological structural lines.

If we use the longitudinal Helan-Liupan-Hengduan mountains line (approximately 106°E longitude) as a boundary, China can be clearly divided into two major parts, with quite different orientation of landforms. Western China generally has its landform units arranged northwestwardly or west-eastwardly, whereas eastern China is arranged north-eastwardly or north-northeastwardly. Furthermore, in western China (using the Kunlun-Altun mountains line as a boundary), there are mainly inland basins (Tarim, Junggar, etc.) and block mountains (Tianshan Mountains, Altay Mountains etc.) distributed to the north. To the south lies the large-scale uplifted Qinghai-Xizang Plateau with its numerous folding-faulting mountains. Eastern China is essentially composed of a series of parallel, northeasterly uplifted and down-warping belts. From west to east, the first down-warped belt includes the eastern Nei Mongol Plateau, the Ordos Plateau, the Sichuan Basin, and the eastern Yunnan Plateau; the second includes the Northeast China Plain, Bohai Sea, the North China Plain, and the middle Changjiang Valley. The third is occupied mainly by the Asian marginal seas — the Sea of Japan, East China Sea, and South China Sea. Correspondingly, the first uplifted belt between the first and second down-warped belts includes the Da Hinggan, Taihang, and Wushan mountains systems. The second one is represented by the Northeastern Mountains, Shandong Peninsula, and a series of southeastern coastal mountains and hills. The Taiwan Mountains reflect the third uplifted zone east to the third down-warped zone.

Bedrock

Different kinds of bedrock, even under the action of the same erosional agent, usually produce different forms of geometric configuration. In China, there are distributed in different areas varied bedrock of different geological times. In some multicyclic folded mountainous areas in particular, the bedrock is quite divergent. Ancient crystalline rocks are generally hard and, thus, erosion resisting, and they

are usually rugged ridges and high peaks — such as the Tianshan, Kunlun, Qilian (Photo I-6), Qinling, Taishan mountains etc. The widely outcropped and divergent sedimentary rocks of the Mesozoic and Cenozoic eras usually produce low mountains and hills. The loose, unconsolidated Quaternary period deposits are the chief surface materials of China's plains. A description of some widely distributed and bedrock-oriented outstanding landforms in China follows.

Carbonate bedrocks are widely distributed in China, especially in South China and Southwest China, with a total area of about 1.3 million sq km. Under suitable climatic and structural conditions, they are liable to be turned into spectacular karst topography (Figure 3-4). A nearly continuous karst area in Guangxi, Guizhou and eastern Yunnan, with an area of more than 300000 sq km, is probably the largest and best developed karst topography in the world.

Owing to different climatic and fluvial conditions (both surface and ground water), karst topography has shown clear areal differentiation. In hot, humid Guangxi, where pure Devonian and Permian limestone and dolomite beds with a thickness of 3000 to 5000 m are distributed over about half the total area, karst topography is best developed. In Guizhou and eastern Yunnan, where limestone and other carbonates of the Carboniferous and Permian occupy more than one half of the total area, the karst-forming process is also active, forming a sharp peaked landscape known as the stone forest (Photo IV-6). Yet, owing to a higher elevation and, consequently, a cooler climate, this example of karst topography is not so well developed as in Guangxi. The Yunnan and eastern Guizhou karst was mostly formed during the Tertiary period when the climate was much warmer. In the northern subtropic zone, including Sichuan, Hubei, Hunnan, and Zhejiang, where carbonate rock beds belong mainly to the Paleozoic age, karst topography is characterized by dolines, uvala, and low hills. In the warm-temperate zone, including Shandong, Shanxi, and Hebei, there is also a considerable distribution of Cambrian and Ordovician limestone beds, although karst topography is not well developed. It is characterized only by karst springs and dry valleys; no uvala, caves, underground rivers or other karst features have been found.

Continental Mesozoic red beds are widely distributed in Central and South China. They are composed of Jurassic through early Tertiary reddish conglomerate, sandstone and shale, with a total thickness of several thousand meters. They were mostly fluvial-lacustrine deposits in intermontane basins, formed under the post-Yanshan Tectonic

---

1) Moho, a simplification of Mohorovicic, is the surface of separation between crust and mantle, named after the seismologist who discovered it.

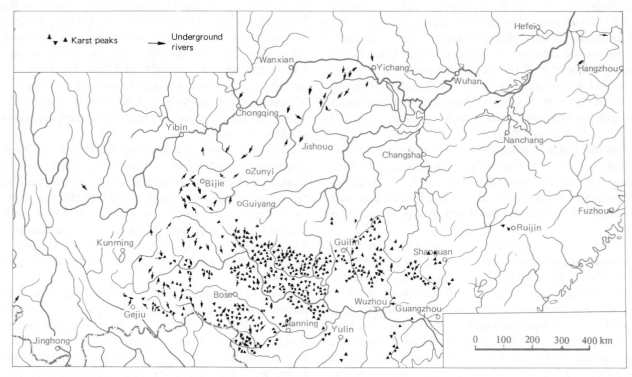

Figure 3-4 Karst topography in Central China and South China

Movement, in a hot and dry environment. During the Himalayan Tectonic Movement, the red beds were structurally uplifted and partly tilted; the climate tended to be more humid and, consequently, fluvial erosion more powerful. They were, thus, dissected into different forms of hills and terraces. These reddish, bright hilly basins are dotted conspicuously among the immense expanse of the higher and darker surrounding mountains. The Sichuan Basin, composed chiefly of nearly horizontal Cretaceous reddish sandstone and purple shale, is the largest reddish hilly basin in China. There are also hundreds of different-sized reddish hilly basins in Central and South China; sometimes, even a series or a "string" of them are distributed nearly continuously along major structural lines, such as along the Hanjiang River, Xiangjiang River, Ganjiang River and in the southeast coast region.

Landforms in these reddish hilly basins are usually arranged in concentric belts:

1) High reddish hills with a steep dip angle and cuesta along the basin's margin. In cases where conglomerate or sandstone beds have outcropped, fantastic *Danxia* landforms[1] have developed.

2) Low reddish hills with a gentle dip angle and broad intermontane basins. When sandstone and conglomerate beds are rather thick, there occur rows of mesa and hills.

3) In the centers of the basins, there is a distribution of streams, broad terraces, low hills, and sometimes also Danxia landforms.

North to the Kunlun-Qilian-Qinling mountains line, Quaternary loess and loesslike deposits are widely distributed on plateaus, mountain slopes, intermontane basins and, piedmont plains (Figure 3-5), comprising a total area of more than 600000 sq km. The deposits vary in thickness from a few meters to 100 to 200 m. The Loess Plateau, defined on its western margin by the Wushao Mountain (eastern tip of the Qilian Mountains), on its eastern margin by the Taihang Mountains, and in the south and north by the Qinling Mountains and the Great Wall respectively. It has an area of about 300000 sq km. Within it, loess and loesslike deposits are extensively and thickly distributed. It is probably the most characteristic area in the world for loess landforms. The Quaternary loess deposits in the middle reaches of the Huanghe River have their upper beds chiefly composed of loose pale yellowish Malan Loess (Q_3),lower down there are thick beds of Lishi Loess (Q_2),and the lowest beds are of Wucheng Loess

---

1) *Danxia*, literally means "reddish glow of sunset". Here, thick beds of conglomerate and sandstone are, under erosion and dissection, turned into a reddish "peak forest". It is most typically developed in the Danxia Mountain in Guangdong Province, hence, it is so-named.

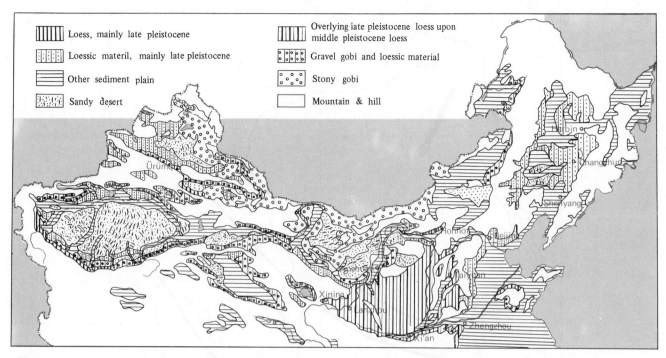

Figure 3-5 A sketch map of the distribution of loess in China

(Q_1).The last two were formerly called reddish soil.

Loess deposits are characterized as follows:

1) They are mainly composed of silts.

2) In the Ordos and Loess Plateau, following the prevailing northwestern winds, the granumetric composition of loess becomes successively finer from northwest to southeast (Figure 3-6).

3) All unstable minerals in the loess have undergone little weathering change.

4) All depositional and erosional surfaces and paleo-soil beds, although overlying unconformably on erosion surfaces of bedrock, have generally a similar geometric configuration to the bedrock.

5) There are even patches of loess distributed on mountain tops with an elevation of 3000 m asl.

All these features show that deposition of loess is a long-distance, aeolian process.

Owing to the different configuration of the underlying erosional surfaces of the bedrock, areal differentiation in landforms is quite prominent. In central Gansu, where Tertiary beds had already been dissected into hilly lands before they were mantled with loess, loessic hills dominate the landscape. In eastern Gansu, northern Shaanxi, and western Shanxi, where large-scale peneplaned landforms had already been formed before the deposition of Pliocene loess, the landforms have been turned into high, level loessic plateaus. The area between the Great Wall and the Weihe River and the area between the Liupan Mountains

and the Lülian Mountains had consisted of shallow depressions before Pliocene loessic deposition and had undergone differential uplifting after Pliocene loessic deposition, finally to be redeposited with Quaternary loess. Here the typical mosaic of liang (loessic flat ridge, Photo III-10), yuan (loessic high plain, Photo III-9) and mao (loessic gentle slope) developed. All these loessic landforms are easily subjected to erosion, especially on steep slopes where natural vegetation is very thin or absent. Therefore, the Loess Plateau is the most notorious example of a severe soil erosion hazard in China.

Different kinds of magma occurring during different tectonic movements, especially granitic intrusion and basaltic eruption, also play an important role in landform shaping. Granites of different ages are widely distributed in China, particularly in the southeast coast where colossal granitic bodies intrude many folded mountains and usually form backbones of anticlinic ranges or dome mountains. Sometimes they are eroded and dissected into lofty peaks, such as Mount Taibai (Photo I-9) of the Qinling Mountains, the Tianmu Mountains near Hangzhou, Mount Huangshan in Anhui Province, and other famous sites. Yet owing to the hot, humid climate in Central and South China, chemical weathering, exfoliation, and other erosional forces are intensive, thus most granitic mountains are rounded and dome-shaped. Furthermore, joints are well developed in granites, and surface and underground water erodes actively along these joints, resulting in many open

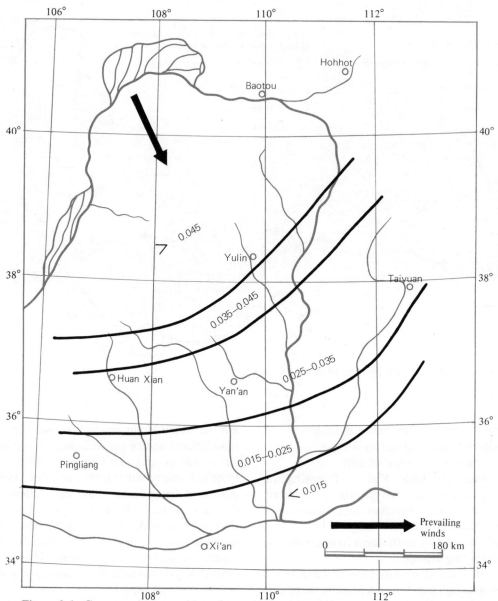

Figure 3-6 Granumetric composition of loess in the Ordos Plateau and the Loess Plateau

valleys and basins. In the granitic hills, which are distributed widely in Centra and South China, thick beds of reddish weathering crust or "red-earth hills" have been developed, usually several dozen meters in depth. When most of the granitic hills have been reduced into reddish weathering crust, the soil erosion hazard becomes severe and is even comparable to that of the Loess Plateau.

Late Cenozoic basalts appear extensively on the eastern Nei Mongol Plateau where they erupted and covered the ancient peneplaned surface to form a series of large-scale tablelands. In central Xilin Gol, these steplike basalt tablelands have a total area of more than 12000 sq km, with 204 extinct volcanic cones scattered in the region (Figure 3-7). In Northeast China there are also large tracts of basalt; on top of Mount Baitou (White Head) — the main peak of the Changbai Mountains (Photo III-5) lies a crater lake, the legendary Tianchi lake (Sky Pond). In Dedu County of Heilongjiang Province, volcanoes erupted in A.D. 1771 — the latest eruption in China. In South China, the southern part of the Leizhou Peninsula and the northern part of Hainan Island are also important basalt-distributed areas.

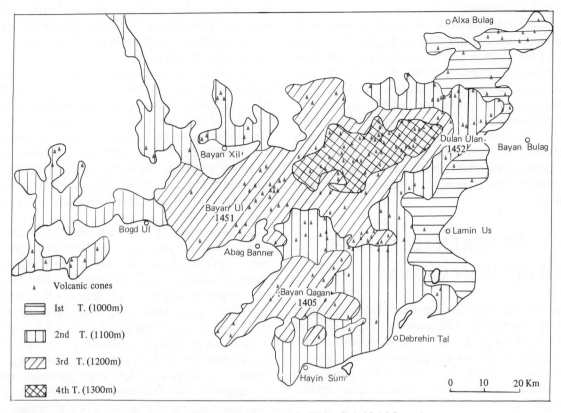

Figure 3-7 Basalt tableland and extinct volcanic cones in Xilin Gol, Nei Mongol

Climate

Different kinds of exogenic forces that work on different landforms (such as nival, fluvial, aeolian, and biological agents) are closely related with, or even determined by, climatic conditions. Temperature and moisture conditions exert especially great and direct impact on weathering, erosion, denudation, transportation, deposition, and other geomorphological processes. Different landform types are usually developed in different climatic regions, so that there is even a group of geomorphologists who refer to themselves as researchers of climatic geomorphology.

We have shown in Chapter 2 that annual precipitation in China is generally determined by distance from the sea, that is, decreasing from the southeast to the northwest (see Figure 2-5). Consequently, the density of river networks also decreases from the southeast to the northwest. In the region south to the Huaihe River and east to the Yunnan-Guizhou Plateau, the length of river networks totals more than 0.5 km/km², and even more than 1 to 2 km/km² in some delta plains. In the Yunnan-Guizhou Plateau and the Sichuan Basin, it is generally less than 0.5 km/km²; in the region north to the Qinling Mountains-Huaihe River line, it is less than 0.3 km/km², whereas in the Nei Mongol

Plateau and arid Northwest China the proportion is less than 0.05 km/km², with large tracts that have not any perennial river at all. In inland basins on the Qinghai-Xizang Plateau, owing to the lower temperature, river networks are a little bit more dense than in arid Northwest China, generally 0.1 to 0.3 km/km². These figures faithfully reflect the intensity of fluvial erosion, which also decreases from the southeast to the northwest.

Another feature of the Chinese monsoon climate is the unevenness of seasonal distribution of precipitation, with more than one half of annual precipitation concentrated in summer and a considerable part pouring down in rainstorms. The largest daily precipitation in Eastern Monsoon China generally exceeds 100 to 200 mm. Such an intensity of rain and its runoff usually results in intense erosion on slopes and in gullies, sometimes even causing rock debris flow and the overflowing of channels.

Temperature also plays an important role in geomorphological processes. For example, in tropic and subtropic zones, chemical weathering is intense, resulting in a reddish weathering crust with a thickness of several dozen meters. In addition, the karst topography is well developed, and coral reefs and mangrove forests grow along the coast. On high plateaus and mountains as well as in the northern part of Northeast China, periglacial and permafrost landforms are well developed.

Climatic areal differentiation also leads to regional differentiation of major geomorphological agents. In Eastern Monsoon China, fluvial action is undoubtedly the dominant exogenic force. In arid Northwest China, besides some high mountains, desiccation and aeolian processes are major agents, although fluvial erosion and deposition are still important along large inland rivers. In the frigid Qinghai-Xizang Plateau, glacial and periglacial processes (nivation) are important, although along the southeastern border, fluvial processes predominate. In the arid northwestern part of the plateau, the process of desiccation is of great significance.

AREAL DIFFERENTIATION OF LANDFORMS IN CHINA

Interaction and integration between different endogenic and exogenic forces result in different landforms in different areas. As landform is one of two basic elements in the physical geographical environment, such an areal differentiation certainly exerts great influence on the formation and evolution of the total physical geographical environment.

Features of Areal Differentiation of Landforms in China

Present areal differentiation of landforms in China has been chiefly governed by the intense tectonic movement and climatic change since the late Cenozoic era. The former factor not only finally determines the present distribution of land and sea, but also results in four great topographic steps, acceleration of desiccation in Northwest China, expansion of eastern coastal plains, and the uplifting of mountains and plateaus. The modern monsoon system in East Asia has also been established. Consequently, three macroclimatic realms in China have been formed, that is, the Eastern humid monsoon climatic realm, the Northwest arid climatic realm, and the Tibetan frigid climatic realm. Correspondingly, three macrogeomorphological realms have been shaped.

In Eastern humid monsoon China, the chief landform types are middle mountain, low mountain, hill, and plain. All of them are governed by NE or NNE-trending uplifting and down-warping. Under humid and subhumid climatic conditions, fluvial erosion and deposition are generally major geomorphological processes. Owing to areal differentiation in geological structure, bedrock, climate, vegetation, and other factors, areal differentiation also occurs within these fluvial processes. South of the Qinling Mountains fluvial erosion is active; landforms are heavily dissected and reduced to a mosaic of rugged mountains and hills with interspersed basins. There are also some special landforms, such as reddish hilly basins and karst topography. North of the Qinling Mountains, the landforms are less dissected; level plateaus and plains are distributed more extensively. From west to east, the Loess Plateau, the Da Hinggan-Taihang Mountains, the Northeast China-North China plains, and the Changbai Mountains-Shandong Peninsula, appear in succession. In northern Northeast China, periglacial and permafrost landforms are developed. Along the coast and around lakes in both North China and South China, there are again special landforms, such as delta plains, mud flats, sandy spits, and lagoons.

In arid Northwest China, the dominating landform types are two extensive plateaus — the Nei Mongol and the Ordos — and two vast inland basins — the Junggar and the Tarim. In addition, there are numerous lofty mountains surrounding these plateaus and basins — the Da Hinggan, the Yinshan, the Helan, the Altay, and the Tianshan mountains. Owing to the scarcity of rainfall (with annual precipitation generally less than 400 mm and less than 200 mm westward to the Helan Mountains), there are only intermittent streams and a few larger perennial rivers that originate in the high surrounding mountains. Hence, fluvial erosion and deposition processes are rather limited, whereas aeolian deflation and deposition and the process of desiccation are much stronger. On the plateaus where the elevation is generally between 1000 to 1500 m, the ground surface is usually level or undulating, and is mostly characterized by denudational, stony desert (denudational gobi) and depositional gravel desert (depositional gobi), with scattered denudational mountains and hills on the plateau proper as well as large patches of aeolian sandy deserts (such as the Badain Jaran, the Tengger, and the Ulan Buh) along their margins. In the inland basins, from the piedmont plain to the basin center, denudational hills, denudational gobi, depositional gobi, clayey flats, and extensive sandy deserts appear sucessively in concentric belts. In the surrounding mountains, following in increasing altitudes, a series of montane landforms appear in succession:

1) Denudational low mountains, generally less than 1000 m in elevation.
2) Middle mountains, 1000 to 3000 m in elevation, with both strong denudational and erosional processes.
3) High mountains, 3000 to 5000 m in elevation, mainly under the process of erosion.
4) Extremely high mountains, more than 5000 masl mainly under the process of nivation and glaciation.

In the Qinghai-Xizang frigid Plateau, with an average elevation above 4000 to 4500 m, the process of nivation dominates, many extremely high mountains are covered with continual snow and glaciers. There are two types of glaciers: (1) the so-called cold continental type, existing mainly in the northwestern arid and semiarid plateau proper, with glaciers only a few kilometers long and an elevation at their terminal moraine of about 5000 to 5500 m, and (2) is the so-called temperate maritime type, which is well developed along the southern and southeastern borders of the Qinghai-Xizang Plateau, with heavier monsoon precipitation and glaciers, usually more than 10 km long, and with terminal moraines extending as low as 2500 to 3000 m, already deep into the forest zone. Permafrost also occurs extensively on the Qinghai-Xizang Plateau, especially in the broad area between the Kunlun Mountains and the Tanggula Mountains. The continuous permafrost area extends more than 500 km. from north to south, with frozen ground more than 70 to 80 m, or even 100 m thick. On the extensive plateau surface, checkerboarded with mountains and intermontane basins, numerous lakes are also distributed. They are mostly saline, and together with their tributary rivers, they serve as important fluvial agents on the Qinghai-Xizang Plateau. Qinghai Lake and Namo Lake (north of Lhasa) are the two largest saline lakes in China. Fluvial process becomes much stronger along the southeastern border of the Qinghai-Xizang Plateau, where a series of longitudinal lofty mountains (Hengduan Mountains) and deep intermontane gorges have formed with a relative relief of more than 2000 to 3000 m, from gorge bottom to mountain top. On these mountains a sequence of vegetation — from subtropical evergreen forest, through broad-and needle-leaved mixed forest, montane coniferous forest, alpine meadow, periglacial cushion vegetation, and finally continual snow appear in succession. Other geomorphological factors, such as geological structure and bedrock, are also important in the shaping of vertical zonation and areal differentiation. In addition, temperature, moisture, and other zonal factors have a great influence on vertical zonation. In different climatic zones, there is generally a different vertical arrangement of vegetation. We might claim that every vertical zone-succession has its horizontal "brand".

Even in Eastern monsoon China, where mountains and hills are much lower and horizontal zonation, thus, plays a more important role, landform is still a basic factor in shaping areal differentiation. Every horizontal zone-succession in China also bears its geomorphological "brand". A series of west to east-trending mountain ranges — the Yinshan-Yanshan mountains (Photo I-8), the Qinling Mountains, and the Nanling Mountains, for example, exert a great effect on horizontal (latitudinal) zonation. They serve as divides and barriers between great climatic zones — the temperate (Northeast China), warm temperate (North China), subtropical (Central and Northern South China), and tropical (Southern South China). Owing to the areal differentiation of landforms on either side of these divides and barriers, diversity between these climatic zones

is enhanced. Numerous NE-SW trending uplifts and down-warping in Eastern Monsoon China also have considerable influence on horizontal zonation. Even in the same horizontal zone, owing to the diversity of landforms, areal differentiation might be quite complicated, as is the case in the extensive and rugged subtropical zone of Central and South China.

REGIONAL GEOMORPHOLOGY

As stated above, there exists conspicuous areal differentiation of landforms between three macrogeomorphological realms — the Eastern Humid Monsoon Realm, the Northwestern Arid Realm and the Qinghai-Xizang Frigid Plateau. Each of these macrogeomorphological realms is again composed of different landform assemblages or mosaics of different landforms.

Landform Assemblage in the Eastern Humid Monsoon Realm

In the Eastern Humid Monsoon Realm, where fluvial processes dominate, the chief composition from north to south is:

1) Northeastern mountains and plains in cool temperate and temperate Northeast China. These are essentially composed of down-warping plains sandwiched by block-uplifted mountains, with NNE-trending Da Hinggan Mountains on the west, NE-trending Changbai Mountains on the east, and the largest plain in China, the Northeast China Plain, is in the middle.

2) Mountains, plateaus, and plains in warm temperate North China. This extensive region is delimited by the Yinshan-Yanshan Mountain. Systems on the north, the Qinling Mountains on the south, the Helan Mountains on the west, and Bohai-Yellow Sea on the east. Structurally, it is an ancient platform, with extensive uplifting and down-warping during the Mesozoic and Cenozoic eras. It is also widely covered with Quaternary loess. From east to west, the chief landform subregions are: the Shandong low mountains and hills, the North China Plain, the Loess Plateau, the Ordos Plateau, the Yinshan-Yanshan Mountains, and the Qinling Mountains.

3) Mountains, hills and basins in subtropical and tropical Central and South China. The landforms are well dissected and sculptured and, consequently, rugged. They are characterized by dense river networks, numerous reddish hilly basins, intense

chemical weathering, and a wide distribution of karst topography and coral reef-built islands. The chief landform subregions are: the middle and lower reaches of the Changjiang Valley Plain, the Southeastern coastal mountains and hills, the mountains and plains of Taiwan Island, the hills and basins south of the Changjiang River, the Sichuan Basin, the Guangxi Basin, the Yunnan-Guizhou Plateau, and the Coral Reefs Archipelago of the South China Sea.

Landform Assemblage in the Northwestern Arid Realm

West of the Da Hinggan Mountains and north of the Kunlun-Altun-Qilian Mountain. Systems, there lies an extensive area of inland basins and plateaus, with an elevation mostly between 500 to 1500 m and generally surrounded by lofty block mountains. Here, aeolian deflation and deposition as well as the process of desiccation and denudation are dominant. Chief landform subregions are: the Nei Mongol Plateau, the Hexi Corridor, the Altay Mountains, the Junggar Basin, the Tianshan Mountains, and the Tarim Basin.

Landform Assemblage in the Qinghai-Xizang Frigid Plateau

The extensive area northwest to the Hengduan Mountains, and south of the Kunlun-Altun-Qilian Mountains. Systems is the most strongly uplifted region in China since the neo-Tertiary, with the lofty Qinghai-Xizang Plateau generally at an elevation of between 4000 to 5000 m. In addition, there are many extremely high mountains of more than 6000 to 7000 m. The process of nivation generally dominates the area; the process of desiccation is also significant in the northwestern Qinghai-Xizang Plateau and

fluvial processes are conspicuous in the southeastern border. The chief landform subregions are: the Qilian Mountains, the Qaidam Basin, the Kunlun Mountains, the Northern Qinghai-Xizang Plateau, the South Tibetan Valley, the Himalaya Mountains, the Hengduan Mountains, and so on.

Features of the above-mentioned landform subregions and their landform assemblages will be discussed in more detail as well as correlated with specific physical regions and other physical elements in subsequent regional chapters.

References

[1] *Physical Geography of China* Compilation Committee, Chinese Academy of Sciences, 1981, *Physical Geography of China: Geomorphology,* Science Press, Beijing. (In Chinese)

[2] *Physical Regionalization of China* Working Committee, Chinese Academy of Sciences, 1959, *Geomorphological Regionalization of China,* Science Press, Beijing. (In Chinese)

[3] Institute of Geography, Chinese Academy of Sciences, 1980, *Agricultural Geography of China: A General Survey,* Science Press, Beijing. (In Chinese)

[4] Academy of Geological Sciences, 1978, *Main features of geological structure in China,* Geological Press, Beijing. (In Chinese)

[5] Bi Ching-cheng, 1971, "Some Aspects of Post-Orogenic Block Tectonics," *Recent Crustal Movements,* Upper Mantle Scientific Report, No. 33, International Upper Mantle Committee.

[6] Liu Dongsheng (Liu Tong-sen) et al., 1964, *Loess in Middle Reaches of the Yellow River,* Science Press, Beijing. (In Chinese)

[7] Integrated Investigation Team of Nei Mongol and Ningxia, Chinese Academy of Sciences, 1980, *Geomorphology of Nei Mongol,* Science Press, Beijing. (In Chinese)

[8] Integrated Investigation Team of Xinjiang, Chinese Academy of Sciences, 1978, *Geomorphology of Xinjiang,* Science Press, Beijing. (In Chinese)

[9] Chao Sung-chiao (Zhao Songqiao), 1984, "Analysis of Desert Terrain in China Using Landsat Imagery" *Deserts and Arid Lands* (ed. Farouk El-Baz), pp.95 – 114, Martiuus Nijhoff Publishers, The Hague.

Chapter 4

Surface Water and Ground Water

Water is an active and movable element in the physical geographical environment. It is also a natural resource essential to mankind in their varied and intense agricultural and industrial activities. For example, total food production has been more than doubled since the founding of the People's Republic of China in 1949; probably the most important meliorating measure has been the construction of a series of hydroengineering works to improve the exploitation of water resources. Since 1952, the amount of irrigated cropland in China has been more than doubled.

According to a Chinese saying, there are three kinds of water — precipitation, surface water, and ground water. We have discussed the first in Chapter 2. Let us now examine the second and third forms of this resource.

SURFACE WATER

Surface water denotes chiefly rivers and lakes. It might also include glaciers and marshes. Glaciers are widely distributed on the Qinghai-Xizang Plateau and in the lofty mountains of Northwest China with a total area of about 57000 sq km and a total water volume of about 2964 billion cubic meters. They, thus, constitute a huge natural solid reservoir and feed numerous perennial and ephemeral rivers in Northwest China. Marshes are also extensively distributed in China, with an area of more than 110000 sq km. The Sanjiang Plain in Northeast China and the Zoigê (Aba) area of the northeastern Qinghai-Xizang Plateau are the most important marsh areas. Owing to the limited scope of the present chapter, however, a detailed discussion of glaciers and marshes will be omitted.

Rivers

China is essentially a country of great rivers. Rivers, with their beneficial freshwater and capacity for delivering fertile alluvial soil, have been feeding hundreds of millions of Chinese people for centuries. Chinese farmers are particularly well known for their close attachment to well-watered lowlands. Since very ancient times, they have skillfully employed river systems as networks for irrigation and navigation; their shallow-draft boats and bamboo rafts can carry loads upstream on less than 30 cm of river water. Now, rivers are also extensively exploited for hydroelectric power generation and urban-industrial use. According to a recent estimate, the total length of all rivers in China amounts to about 420000 km, of which more than 50000 rivers have a drainage area of more than 100 sq km. The annual discharge of all rivers totals more than 2600 billion cubic meters. The chief features of China's major rivers (each with a drainage area of more than 100000 sq km) are listed in Table 4-1.

River Systems and Drainage Basins

Rivers in China, owing chiefly to areal differentiation of climatic and geomorphological conditions, can be first of all divided into two large systems: oceanic and inland. The oceanic system comprises numerous large rivers with an abundant discharge, occupying about 64 per cent of the total land area in China. This system can be again subdivided into Pacific, Indian and Arctic drainage basins. The inland system on the other hand accounts for 36 per cent

Table 4-1 Chief Features of Major Rivers in China*

River	Area (km²)	Length (km)	Discharge	
			Annual total (10⁸m³)	Annual average (m³/sec)
Heilong	888502	3101	1181	3740
Songhua	545594	1956	706.4	2240
Nenjiang	283000	1379	240.9	764
Liaohe	219014	1390	144.8	459
Haihe	264617	1090	232.6	737
Huanghe	752443	5464	574.46	1822
Weihe	134766	818	98.0	311
Huaihe	189000	1000	459.0	1460
Changjiang	1808500	6300	9793.53	31055
Jinsha	490546	–	1546.5	4900
Hanshui	168851	1532	574.1	1820
Jialing	159638	1119	694.1	2200
Zhujiang	442585	2210	3466	11000
Lancang	164766	2354	692.9	2200
Nujiang	134882	2013	656.7	2000
Yarlung Zangbo	240280	2057	1380	4370

*All international rivers (the Heilong or the Amur, the Lancang or the Mekong, the Nujiang or the Salween, the Yarlung Zanbo or the Brahmaputra) include territory inside China only. The Songhua River (Photo II-9) is a tributary of the Heilong River, the Nenjiang River is a tributary of the Songhua River, the Weihe River is a tributary of the Huanghe River, the Jinsha River is the upper reaches of the Changjiang River, and the Hanshui River and the Jialing River are its tributaries.

of the total land area in China. It contains a few perennial rivers and has large tracts with no runoff whatsoever. All inland rivers flow into saline inland lakes or die away amid sandy deserts or salt marshes (Table 4-2).

The Pacific oceanic subsystem is primarily in humid and subhumid Eastern Monsoon China as well as a small section of arid and semiarid Northwest China and the Qinghai-Xizang Plateau. It accounts for 88.9 per cent of the total oceanic system and 56.7 per cent of the total land area in China. A majority of large rivers in China flow into the Pacific Ocean, such as the Heilong River which empties into the Sea of Okhotsk; the Liaohe, Luanhe, Haihe, and Huanghe (Photos II-7, II-8) rivers which drain into the Bohai Sea and the Yellow Sea, the Changjiang River (Photos, II-3, II-5) and the Qiantang River (Photo II-10) which flow into the East China Sea, and the Zhujiang, Yuanjiang (Red River), and Lanchang rivers which empty into the South China Sea. Furthermore, most of these large rivers, owing to the general west-eastward sloping topography in China, tend to flow eastward and are bordered by fertile and broad valley plains as well as hundreds of millions of people who live along their middle and lower reaches.

The Indian Oceanic subsystem ranks second, occupy-

ing 10.3 per cent of the total oceanic system and 6.5 per cent of the total land area in China. It is widely distributed in the southeastern part of the Qinghai-Xizang Plateau and the southwestern part of Eastern Monsoon China. Rivers flowing into the Indian Ocean have only their upper reaches located in China, such as the Nujiang, Longchuan (Irrawaddy), Yarlung Zangbo, and Shiquan (Indus) rivers. Most of them flow southward, usually with a dendritic drainage pattern that is largely determined by the longitudinal "lofty mountains-deep gorges" topography.

The Arctic Oceanic subsystem has only one large river — the Ertix River in the northern Junggar Basin. It is a tributary of the Ob River, flowing northward and emptying into the Kara Sea. It occupies only 0.8 per cent of the total oceanic system and 0.5 per cent of the total land area in China.

Trunk rivers of the oceanic system originate for the most part in three topographic belts: (1) the southeastern part of the Qinghai-Xizang Plateau, (2) the Da Hinggan Mountains–Taihang Mountains–Qinling Mountains–Yunnan Plateau belt, and (3) the Changbai Mountains–Shandong Peninsula–Southeast coast belt. These three topographic belts roughly coincide with the eastern uplifted margins of the first, second, and third great topographic

Table 4-2 Areas of Major Drainage Basins in China

River systems and subsystems	Drainage basins	Area of drainage basins (1000 km²)	Percentage of total land area in China
Oceanic system		**6120.0**	**63.76**
Pacific Ocean		5444.5	56.71
	Sea of Okhotsk	861.1	8.97
	Sea of Japan	32.6	0.34
	Bohai Sea and Yellow sea	1670.0	17.40
	East China Sea	2044.7	21.30
	South China Sea	825.0	8.59
	Directly into the Pacific	11.1	0.11
Indian Ocean		624.6	6.52
	Bay of Bengal	558.3	5.83
	Arabian Sea	66.3	0.69
Arctic Ocean	Kara Sea	50.8	0.53
Inland system		**3480.0**	**36.24**
	Nei Mongol	328.7	3.42
	Desert zone (Gansu-Xinjiang-Qaidam)	2374.1	24.73
	Northeast China	48.2	0.50
	Northesn Qinghai-Xizang Plateau	728.9	7.59
Grand Total		**9600.0**	**100.00**

steps respectively. Rivers originating from the eastern margin of the first great topographic step are the largest rivers in China; some of them, for example, the Changjiang, Huanghe, Mekong (Lancang), and Salween (Nujiang) rivers are among the leading rivers in the world. Rivers originating from the second great topographic step include the Heilong, Liaohe, Huaihe, and Zhujiang rivers. They are also important rivers in China, although not as majestic in length and in discharge as those of the first topographic step. Rivers originating from the third great topographic step, such as the Yalu and Qiantang rivers are comparatively small. However, since they are located in the heaviest rain belt in China and their upper reaches flow from mountainous areas, they are usually abundant in discharge and hydropower potential.

The inland river system is mostly located ih arid and semiarid Northwest China, the northwestern Qinghai-Xizang Plateau, and in small "islands" amid semiarid and subhumid areas, such as western Northeast China and the central Ordos Plateau. The great divide between inland and oceanic river systems in China runs essentially northeast-southwest, starting from the southern end of the Da Hinggan Mountains, then passing through the Yanshan and Helan mountains, the eastern Qilian, Bayan Har, Nyainqentanglha, and Gangdise mountains, and up to the southwestern margin of the Qinghai-Xizang Plateau. Its

annual runoff depth coincides roughly with the 50 mm isobath. The inland river system might be again subdivided into four drainage areas: (1) Nei Mongol: under semiarid climatic conditions, precipitation is still important for feeding the rivers; (2) Gansu-Xinjiang: a series of large inland rivers that originate from lofty surrounding mountains. The rivers include the Shiyang, Heihe, Shule, Tarim, Ili, and Manas rivers. On the whole, perennial rivers are infrequent, with more than one half of the total area having no rivers at all; (3) Qaidam Basin: in the northern Qinghai-Xizang Plateau, this basin has numerous short rivers, saline lakes, and salt marshes in its eastern part, whereas large tracts in its western part have no rivers at all; (4) The northern Qinghai-Xizang Plateau: with an elevation above 4500 m, this plateau is surrounded and interspersed by lofty mountains. There are numerous saline lakes together with their tributary rivers in the basins, with more than one half of the total land area without perennial rivers.

Surface Runoff

As China has a vast territory with varied and complicated natural conditions, surface runoff — as shown by annual runoff depth — is naturally quite dissimilar in different regions. Yet, as a whole, distribution of surface

Figure 4-1 Distribution of annual runoff depth in China

runoff decreases gradually from southeast to northwest (Figure 4-1). In the southeastern most part of China — the slopes of the Taiwan Mts., the annual runoff depth reaches 2000 to 4000 mm. Whereas in the northwesternmost part — the Tarim Basin the runoff is less than 5 to 10 mm and approaches zero in the center.

Distribution of surface runoff essentially corresponds to distribution of precipitation. Some isobaths of annual runoff depth are particularly meaningful.

1) An isobath of 50 mm annual runoff depth, which generally delimits the eastern boundary of the arid and semiarid areas, runs roughly from the western footslope of the Da Hinggan Mountains through the eastern part of the Northeast China Plain, the southern margin of the Nei Mongol Plateau, the eastern margin of the Ordos Plateau, and up to the northern part of the Qinghai-Xizang Plateau. It corresponds to a 400 mm isohyet of annual precipitation in its eastern part and a 200 mm isohyet in its western part. This isobath divides China into two halves: humid, agricultural eastern China with abundant surface runoff, and arid, pastoral western China with scanty surface runoff.

2) In eastern China, the 200 mm isobath is also an important divide. It corresponds roughly to the Qinling Mountains-Huai River line, dividing Eastern Monsoon China into the North and the South; the latter has abundant surface runoff (generally more than 200 mm, with the exception of some enclosed basins), whereas the former usually has less than 200 mm, except in some elevated areas.

3) In the South, the 900 mm isobath is again of great significance. It runs from the southern Hangzhou Estuary southwest through a series of coastal mountains and hills up to the Nanling Mountains. Southeast of this line lies the most abundant surface runoff belt in China.

4) The 10 mm isobath divides the arid and semiarid areas. In semiarid areas, a considerable amount of surface runoff still flows, whereas in arid areas, there is very little or no runoff at all.

Based on these specific isobaths, five major runoff belts might be identified in China.

1) Belt of abundant runoff: with annual surface runoff depth more than 900 mm and annual precipitation of more than 1600 mm; roughly coincident with belts of tropical and subtropical forests; usually three agricultural crops each year.

2) Belt of adequate runoff: with annual surface runoff depth 200 to 900 mm and annual precipitation of 800 to 1600 mm; roughly coincident with belts of subtropical evergreen broad-leaved forest and mixed broad and needle-leaved forests; usually two agricultural crops each year.

3) Transitional belt: with annual surface runoff depth 50 to 200 mm and annual precipitation of 400 to 800 mm; roughly coincident with forest-steppe vegetation zone; usually three agricultural crops every two years.

4) Belt of scarce runoff: with annual surface runoff depth 10 to 50 mm and annual precipitation 200 to 400 mm; roughly coincident with semiarid steppe zone; mostly pastoral or pastoral-agricultural mixed area, with one agricultural crop each year.

5) Belt of little runoff: with annual surface runoff depth less than 10 mm and annual precipitation less than 200 mm; roughly coincident with desert areas; mostly pastoral or empty lands, with patches of irrigated agriculture (oases).

Furthermore, surface runoff is formed by the integrated action of many physical geographical factors as well as the impact of human activities. These factors will be briefly discussed below.

Among climatic factors, the most important is precipitation which is the direct source of runoff. The quantitive correlation between precipitation and runoff varies under different climatic and geomorphological conditions. There exists distinct areal differentiation between the North and the South and between the East and the West. For example, the 200 mm isobath of runoff corresponds roughly to the 900 mm isohyet of precipitation in the central Yunnan Plateau (Southwest China), but it corresponds to the 600 mm isohyet in northern Northeast China. Again, the 50 mm isobath of runoff corresponds roughly to the 400 mm isohyet of precipitation in eastern China, but it corresponds to the 200 mm isohyet in western China. This means that there is much more evapotranspiration of precipitation in the South and in the East.

Geomorphological impact on the formation of surface runoff is great in the mountainous areas. Mountains, especially if they are perpendicular or diagonal to moisture-bearing maritime monsoons (as are numerous coastal mountains and hills in South China) usually promote abundant orographic precipitation and, consequently, rich surface runoff. Steep slopes and high elevation that lower air temperature and, thus, effectively reduce evapotranspiration, further increase surface runoff in the mountainous areas. Lofty mountains may also serve as climatic barriers that make their leeward slopes comparatively poor in precipitation and runoff. In the basins and plains, in contrast to the surrounding mountains, precipitation is usually less abundant and evapotranspiration stronger. Consequently, these areas have less surface runoff. In the extremely arid region, there is practically no surface runoff in the inland basins, and the rivers, which originate in the surrounding high mountains gradually dry and die out shortly after they flow outside the mountainous area.

Impact of surface material on the formation of runoff

is mainly realized through percolation and evapotranspiration. In karst areas, such as in Guangxi, Guizhou, and eastern Yunnan, most of the runoff goes underground, leaving few rivers on the surface. In the arid areas, sandy deserts (shamo) and gravel gobi (gebi) are widely distributed, and the scanty precipitation is mostly wasted by evapotranspiration and percolation without surface runoff being formed. A sand bed, however, is good protection against evapotranspiration; a sand bed about 30 cm thick will virtually stop further evapotranspiration of underlying soil moisture or ground water. Therefore, there is usually a good ground water reserve in the lower part of high sand dunes. Sometimes freshwater lakes are formed in the depressions between these high sand dunes.

The influence of forests on the formation of surface runoff varies according to different physical environmental conditions. For example, in the semiarid Loess Plateau where potential evapotranspiration is greater than actual evapotranspiration, the growth of forests will increase evapotranspiration, and, consequently, reduce runoff. In the humid Changjiang Valley, where potential and actual evaporation are approximately the same, growth of forest will not necessarily increase evapotranspiration. Instead, owing to conservation and regulation of surface water in the forest, runoff is generally enriched.

The impact of human activities on the formation of runoff is prominent in China, especially since 1949. On the whole, most human activities result in the increase of evapotranspiration and, thus, reduce runoff. For example, in the Haihe Basin, the average annual evaporation from artificial reservoirs in 1962–1972 equalled approximately 1.5 per cent of the total annual runoff; in the arid year of 1972, the figure was as high as 3.0 per cent. Again, China has now about 45 per cent of her 100 million hectare (ha) of croplands irrigated. Hence, a large quantity of surface runoff is being used for irrigation; in the Haihe Basin, the average annual use of irrigated water occupies about 13.2 per cent of the total annual runoff; in the arid year of 1972, the percentage was 33.8 per cent.

Fluvial Runoff Resources

Fluvial runoff is an important renewable natural resource. According to a recent estimate, annual fluvial runoff in China totals 2630 billion cu m, comprising about 6.6 per cent of the total annual fluvial runoff and ranking fifth in the world.

However, the average annual fluvial runoff depth reaches only 271 mm, whereas annual fluvial runoff per capita is less than 3000 cu m, much lower than in most countries. Nevertheless, potential hydropower resources in China, owing to the mountainous topography and many large rivers originating in the high Qinghai-Xizang Plateau, totals (theoretically) 680 million kilowatts (kW), ranking

first in the world.

Areal distribution of fluvial runoff resources is quite uneven. Most of these are concentrated in the oceanic drainage basins, accounting for 96 per cent of the total fluvial runoff resources in China. The inland drainage basins, with land areas occupying 36 per cent of the total land area in China, account for only 4 per cent of total fluvial runoff resources. There also exist conspicuous differences between the North and the South and as among the five major runoff belts. The Changjiang Basin is particularly outstanding, with an area occupying only 18.83 per cent of the total land area, yet 37.66 per cent of the total fluvial runoff resources in China. Table 4-3 shows annual runoff resources in different drainage basins in China.

Rivers of the oceanic system flow mostly into China's territoral and neighboring seas, with a total annual fluvial runoff of 1884.2 billion cu m. Their distribution is also uneven, concentrated mostly (64.9 per cent) into the East China Sea. Table 4-4 shows annual fluvial runoff resources in different sea basins.

River Regime

Distribution of fluvial runoff in China is varied and uneven not only in space, but also in time. This is clearly shown both in seasonal distribution of fluvial runoff and in fluctuation of annual runoff.

Seasonal distribution of fluvial runoff is of great significance to agricultural and industrial use of water resources. It is chiefly determined by seasonal variation of precipitation. Winter (December to February) is generally the dry season in China, with little precipitation — mainly in the form of snow, which does not feed rivers immediately. Therefore, fluvial runoff is usually low in winter, with northern Heilongjiang Province accounting for less than 1 per cent of the total annual runoff. It is only in northeastern Taiwan, where the winter coincides with the rainy season, that the winter runoff comprises about 25 per cent of total annual runoff.

In spring (March to May), fluvial runoff increases all over China, but the rate of increase differs in different regions. Two regions increase most rapidly. One is Central China, where the rainy season is in the spring, and the spring runoff comprises more than 30 per cent (up to 45 per cent in some mountainous areas) of the total annual runoff. Another is in the northern Northeast and northern Xinjiang, where there is heavy melting of snow and spring runoff or "spring flood" usually accounts for 20 to 25 per cent or even 30 to 40 per cent of total annual runoff. On the other hand, in southwestern Yunnan, where southwestern monsoons dominate, the rainy season "bursts" in early summer and spring has the lowest runoff in the year, accounting for only 6 to 8 per cent of total

Table 4-3 Annual Fluvial Runoff Resources in Different Drainage Basins in China

Drainage basins	Drainage areas		Annual fluvial runoff		Average runoff depth (mm)
	Km²	Percentage of total land area	10⁸ m³	Percentage of total runoff res.	
Rivers in Northeast China	1166028	12.15	1731.15	6.66	148
Rivers in North China	319029	3.32	283.45	1.09	89
Huanghe R.	752443	7.84	574.46	2.21	76
Huaihe R. and rivers in Shandong Peninsula	326258	3.40	597.89	2.30	183
Changjiang R.	1807199	18.83	9793.53	37.66	542
Rivers on Southeast Coast	212694	2.22	2001.33	7.70	941
Zhujiang R. and rivers in Guangdong-Guangxi Coast	553437	5.76	4466.27	17.18	807
Rivers in Taiwan and Hainan Island	68160	0.71	887.36	3.41	1302
Rivers in Southwest	408374	4.25	2160.84	8.31	529
Ocean rivers in Qinghai-Xizang Plateau	455548	4.75	2267.81	8.72	498
Arctic Basin	50860	0.53	107.85	0.41	212
Oceanic Total	**6120030**	**63.76**	**24871.94**	**95.65**	**406**
Rivers in Gansu-Xinjiang	2090162	21.77	708.62	2.73	34
Rivers in Nei Mongol	328740	3.42	27.06	0.10	8
Rivers in Qinghai-Xizang Plateau	1012848	10.55	382.97	1.47	38
Inland rivers in Northeast China	48220	0.50	12.05	0.05	25
Inland Total	**3479970**	**36.24**	**1130.70**	**4.35**	**32**
Grand Total	**9600000**	**100.00**	**26002.64**	**100.00**	**271**

Table 4-4 Annual Fluvial Runoff Resources Flowing into Different Neighboring Seas

Sea basins	Drainage areas (km²)	Annual fluvial runoff (10⁸m³)
Bohai Sea	1335910	892.28
Yellow Sea	334132	782.50
East China Sea	2044741	12222.70
South China Sea	585637	4744.83
Directly into Pacific Ocean	11112	200.02
Pacific Ocean Total	**4311532**	**18842.33**

annual runoff.

In summer (June to August), the heaviest precipitation and the richest runoff generally occur all over China. The melting of snow is also the heaviest in this season. Conse-quently, summer runoff usually accounts for more than one half of the total annual runoff. This is not only the season of luxuriant plant growth, but also of dangerous flood hazard. Unlike winter or spring, the higher the latitudes, the more concentrated is the summer runoff. In North China and the Nei Mongol Plateau, as well as the Qilian Mountains and Kunlun Mountains areas, summer runoff occupies 60 to 70 per cent of total annual runoff. In the vast areas south of the Changjiang River, where the heaviest spring precipitation and runoff occur, sum-mer precipitation and runoff are comparatively low, with summer runoff accounting for less than 35 to 40 per cent of the total annual runoff.

In autumn (September to November), runoff generally decreases all over China, with maximum autumn runoff in areas where there has been minimum spring runoff. The maximum autumn runoff happens in Hainan Island (ow-ing to heavy typhoon rain), which constitutes more than one half of the total annual runoff of that island.

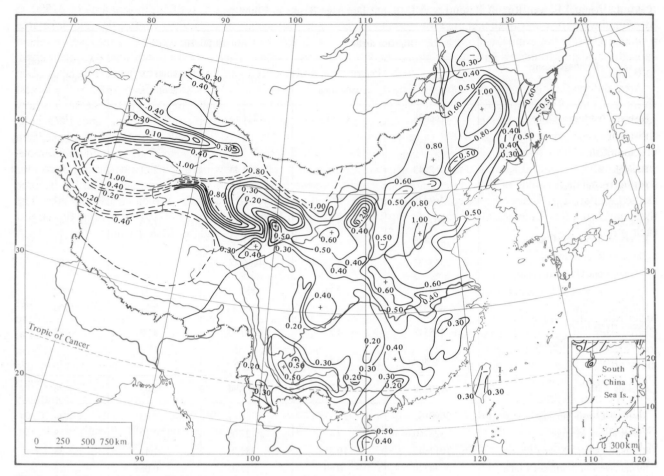

Figure 4-2 The variation coefficient (*Cv*) of annual runoff in China

On the whole, seasonal variation of fluvial runoff in China is characterized, on the one hand, by the coexistence of high temperature and abundant runoff, yet, on the other hand, by too much water in summer and autumn and too little in winter and spring. According to statistics, there occurred from 206 B.C. to A.D. 1949, 1092 large floods and 1056 severe droughts in China. Therefore, one important measure for the improvement of agricultural production is building reservoirs to store surplus flood water in late summer and autumn for irrigation use in spring and early summer. Since 1949, more than 80000 reservoirs with a total capacity more than 400 billion cu m have been constructed in China. These have greatly benefitted agricultural production and have expanded irrigated cropland. In late summer and autumn, however, great care must be taken in controlling floods. Numerous hydraulic engineering works have been undertaken since 1949 for this purpose.

Fluctuation of annual runoff is usually demonstrated by the variation coefficient (*Cv*). It also reflects the trustwor-

thiness of annual runoff. It is determined by the volume of annual runoff and the *Cv* of precipitation as well as by source area features and landforms.

The *Cv* of annual runoff is generally proportional to the *Cv* value of annual precipitation and inversely proportional to the volume of annual runoff. Hence, it reflects faithfully the volume of annual runoff, increasing generally from 0.2 to 0.3 in the southeast to 0.8 to 1.0 in the northwest (Figure 4-2). Yet, owing to the influence of landform features, it is usually smaller in the mountainous areas than in the plain, and its value is usually larger than the *Cv* value of annual precipitation. The source area also has a great impact on the *Cv* value of annual runoff. It is usually larger in unreliable rain-fed areas, such as along the Southeast coast, where the track and intensity of rain-inducing typhoons change every year and results in a larger *Cv* value of both annual precipitation and runoff. The *Cv* value of annual runoff in Hainan Island, for example, is as high as 0.5. Nonperennial, rain-fed rivers in Arid Northwest China have even an larger *Cv* value of annual runoff, reaching 0.8 to 1.0. On the other hand, rivers mainly fed by ground water or water from melting snow usually have a smaller Cv value of annual runoff. Rivers in the snow-capped Tianshan Mountains have a Cv value of only 0.1 to 0.2.

Areal differentiation of Cv value also exists among large rivers in China. The Huaihe River has the largest one, reaching 0.63 at Bengbu. This is because it is located at the transitional zone, with a great Cv value of annual precipitation. The Changjiang River, however, has the smallest value — only 0.13 at Wuhan. This is owing to its much smaller Cv value of annual precipitation as well as the fact that it stretches over several natural zones, so that annual runoff volumes in different sections supplement each other.

Lakes

China is also a country with numerous lakes. According to statistics, there are approximately 2800 natural lakes in China, each with an area greater than 1 sq km, and with total area more than 80000 sq km. In addition, there are many artificial lakes — reservoirs, mainly constructed since 1949. Distribution of major lakes in China is shown in Table 4-5 and Figure 4-3.

The origin of lake basins is quite divergent. They might be formed by tectonic, volcanic and other endogenic forces or they might result from fluvial, glacial, or a aeolian excavation. The distribution of lake basins in China depends heavily on the feeding conditions of lake water. Five major lake regions can be identified.

The Northeast Lake Region

There exist large tracts of marsh and numerous small lakes in the Northeast China Plain, with a total lake area 3722 sq km, which constitutes 4.7 per cent of the total lake area in China. On the upper reaches of the Mudang River lies the largest lava-dammed lake in China — the Jingpo Lake, with an area of about 95 sq km. In the famous volcanic area around Dedu County, Heilongjiang Province, the second largest lava-dammed lake in China — the scenic Wudalianchi Lake (Five-Linked Ponds) is found. Its total area is 18.5 sq km. On the highest peak, Mount Baitou of the Changbai Mountaius, lies the largest crater lake in China — the Tianchi Lake, with an area of 30 sq km and a maximum depth of 373 m. The Xingkai Lake located on the border between China and the USSR, has an area of 4500 sq km, of which approximately one third is on the Chinese side.

The Northwest Lake Region

In extensive arid and semiarid Northwest China, there are numerous inland lakes with a total area of 22500 sq km, occupying about 27.9 per cent of the total lake area in China. Most of them are terminal saline lakes of inland rivers. Owing to fluctuation and shifting inland river channels, many of them are described as "wandering lakes". Lop Lake (Lop Nor) with an area of 3006 sq km is the largest and most famous "wandering lake" in the region; it is now entirely dry and has become a "Great Ear" since 1972, judging from the image (Landsat image 8) . Juyan Lake (Ejin Nor), a terminal lake of the Heihe River, is another famous example. It has now been split into two lakes: Sogo Nor (eastern) and Gaxun Nor (western); the latter has become entirely dry, the former has turned into a salty lake. The Bosten Lake, with an area of 1019 sq km, is the largest freshwater lake in arid Northwest China.

The Qinghai-Xizang Lake Region

On the lofty Qinghai-Xizang Plateau, there exist numerous lakes in intermontane basins, with a total lake area of 30974 sq km, which is 38.4 per cent of the total lake area in China. The water bodies are mostly inland saline lakes, fed by melting snow, and their lake basins are mainly controlled by tectonic structure. Qinghai Lake with an area of 4583 sq km, is the largest lake in China; it is located in a structural basin without any outlet, consequently, it has turned into a highly salty (mainly NaCl) lake.

The Eastern Lake Region

In the middle and lower reaches of the Changjiang, Huaihe, Zhujiang, and other rivers, there is a distribution of numerous lakes, with a total lake area of 22161 sq km, occupying 27.5 per cent of the total lake area in China. Owing to heavy precipitation, all of them are freshwater lakes with abundant inflow and good drainage. They are characterized by low elevation and shallow water being mostly less than 4 m in depth. Poyang Lake with area of 3583 sq km, is now the largest freshwater lake in China. Dongting Lake was formerly a part of the extensive Yun-mong Swamp Area and the largest freshwater lake in China; for many years, it has been gradually silting up, a process that has been accelerated since 1949 because of increased reclamation. From 1949 to 1976, its area diminished from 4350 sq km to 1840 sq km, and the number of lakes in the ancient Yunmong Swamp Area decreased from 1066 to 326 (Landsat image 3). Taihu Lake is the largest freshwater lake in the lower reaches of the Changjiang River. It was formerly a lagoon and is now located in the center of the highly productive Changjiang Delta (Landsat image 4).

Table 4-5 Major Lakes (with area exceeding 1000 km²) in China

Lakes	Province (Autonomous region)	Location		Basin area (km²)	Elevation (m)	Maximum depth (m)	Total volume (10⁸m³)
		Latitude	Longitude				
Qinghai	Qinghai	36°40′N	100°23′E	4583	3195.0	32.8	1050.0
Poyang	Jiangxi	29°05′N	116°30′E	3583	21.0	16.0	248.9
Lop	Xinjiang	40°20′N	90°15′E	3006	778.0	–	–
Dongting	Hunan	29°20′N	112°50′E	2820	34.5	30.8	188.0
Taihu	Jiangsu	31°20′N	120°16′E	2420	3.0	4.8	48.7
Hulun	Nei Mongol	48°57′N	117°23′E	2315	545.5	8.0	131.3
Hongze	Jiangsu	33°20′N	118°40′E	2069	12.5	5.5	31.3
Nam	Xizang	30°40′N	90°30′E	1940	4718.0	–	–
Siling	Xizang	31°50′N	89°00′E	1640	4530.0	–	–
Southern Four Lakes	Shandong	34°59′N	116°57′E	1266	35.5 – 37.0	6.0	53.6
Ebinur	Xinjiang	44°55′N	82°53′E	1070	189.0	–	–
Bosten	Xinjiang	41°59′N	86°49′E	1019	1048.0	15.7	99.0

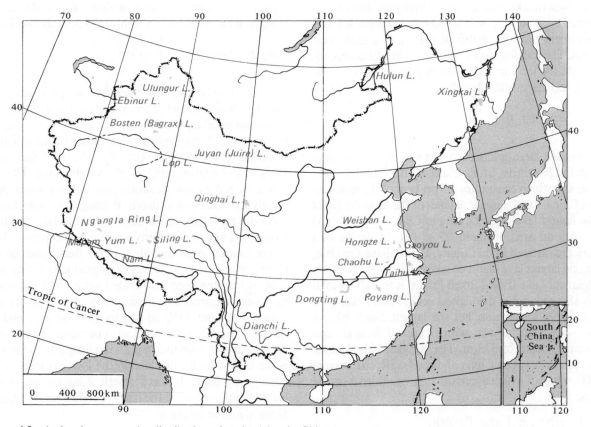

Figure 4-3   A sketch map on the distribution of major lakes in China

The Southwest Lake Region

On the Yunnan-Guizhou Plateau, especially in eastern Yunnan, many structure-controlled, longitudinally elongated freshwater lakes are distributed. They are mostly located in the karst areas, with a total lake area of 1188 sq km.

Dianchi Lake is the largest lake on the Yunnan-Guizhou Plateau, with an area of 297 sq km. Fuxian Lake is the deepest large lake in China, with an area of 217 sq km and a maximum depth of 151.5 m.

In these five lake regions, there are abundant limnological water resources — more than 57 billion cu m for the five largest lakes alone. Many lakes in mountainous areas have rich hydropower resources, such as Riyue Lake (Sun Moon Lake) in Taiwan Island, with a capacity of more than 100000 kw. There are more than 200 kinds of edible fishes in China's lakes as well as numerous water fowl, birds, shrimps, and shellfish. In addition, numerous salty and saline lakes in the Qinghai-Xizang Plateau and Northwest China contain abundant and varied salt resources; the Qarhan Salt Lake in the Qaidam Basin has a salt bed more than 50 m thick and includes not only NaCl, Na_2SO_4, $CaSO_4$, and other common salts, but also many kinds of rare elements.

GROUND WATER

The vastness of China's territory and her varied natural conditions are again reflected in the formation and distribution of ground water resources. According to an estimate, China has a total ground water resource of about 800 billion cu m, distributed and concentrated mostly in several large-scale alluvial plains (including the Northeast China Plain, the North China Plain, the Middle and Lower Changjiang Valley) and structural basins, (such as the Tarim, Junggar, and Sichuan basins). In the extensive mountainous areas, however, geological structures and topographical conditions are quite complicated, resulting in varied conditions for the formation and distribution of ground water resources.

In arid and semiarid Northwest China where both precipitation and surface water resources are scanty, exploitation of ground water resources is of great use, sometimes constituting the only source of water for irrigation as well as drinking. In humid and subhumid Eastern Monsoon China, owing to rather uneven seasonal distribution of precipitation and fluvial runoff, exploitation of ground water resources is also important as a supplement to irrigation during the dry season. In terms of industrial and municipal use of water, ground water sometimes has great advantage over surface water, for example, ground

water is less subject to pollution and, being much lower in water temperature, it might be inexpensively used for cooling. Hence, many large cities in China, including Beijing, Xi'an, Jinan, Taiyuan, Baotou and Ürümqi, take their municipal water supply mainly from ground water resources. Valuable ground hot-water and ground mineral-water resources are also widely distributed in China.

Factors in the Formation and Distribution of Ground Water

The formation and distribution of ground water are first determined by climatic conditions, of which the most important is the ratio between precipitation and evapotranspiration. The ratio determines the water quantities of the feeding source and the drainage conditions of ground water in different areas. The climatic impact on shallow ground water is particularly obvious. In China's physical geographical environment, temperate, warm-temperate, subtropical, and tropical zones appear successively from north to south; and humid, subhumid, semiarid, and arid regions appear successively from southeast to northwest. The quantity and quality of ground water resources vary correspondingly. For example, in arid regions, the continental salinization process is most active, consequently, ground water in inland basins is usually highly mineralized, and it is mostly fed by percolation of surface runoff, especially when rivers flow from surrounding mountains into gravel gobi. In humid regions, where precipitation is plentiful, there is no continental salinization process at all, and ground water is richly fed both by precipitation and surface runoff. The dynamics of ground water are also greatly influenced by seasonal distribution of precipitation; curves of the annual ground water regime usually appear in multi-peaks in hot, humid South China and Central China and in double or single-peak in North China and Northeast China; there is also a gradual temporal lag of the first peak from south to north.

Second, different geological and geomorphological conditions result in different feeding, moving, and drainage conditions of ground water. In mountainous areas, ground water is generally fed by precipitation and melting snow, and it is then quickly drained by drainage networks with steep slopes. Consequently, it is usually fresh bicarbonate water. In structural basins, drainage conditions are rather poor. Different kinds of salts are deposited and high mineralization of ground water results. Ground water in the permafrost area on the lofty Qinghai-Xizang Plateau is another outstanding example that shows the great impact of high relief on ground water. Furthermore, hydrologic impact on the formation and distribution of ground water is also obvious. Surface water and ground water are closely related and might interchange with each other under certain conditions. Surface water exerts great influence on shallow ground water and sometimes also on deep ground water, such as on the piedmont plain bordering a large-scale inland basin in which surface runoff feeds both shallow and deep ground water. In Central and South China, fluvial channels are numerous and precipitation is abundant. Ground water is well fed and well drained, resulting in the richness of high-quality subsurface water resources. In North and Northeast China, both surface and precipitation conditions are not so good, and the lower reaches of some rivers (such as the Huanghe River) usually flow "above" the ground and become important feeding sources for the ground water of nearby areas.

Distribution of Shallow Ground Water

Chiefly on the basis of porosity and other features of water-bearing strata, shallow ground water in China (Figure 4-4) might be classified into four types:

1) Pore-water: In loose deposits, mainly distributed in North China and the Northeast China plains, the Loess Plateau as well as piedmont plains and sandy deserts in Northwest China.
2) Fracture-water: In bedrocks, widely distributed in montane and hilly areas.
3) Fracture-cave water: In karst areas, mainly located in Southwest China and South China.
4) Pore-fracture water: In the permafrost regions distributed chiefly on the Qinghai-Xizang Plateau and northernmost Northeast China.

A brief discussion of some sample areas follows.

The Northeast China Plain

This is the largest plain in China, deposited with thick loose Cenozoic sediments and containing rich ground water in a NNE-SSW trending down-warped basin. It is composed essentially of three smaller plains.

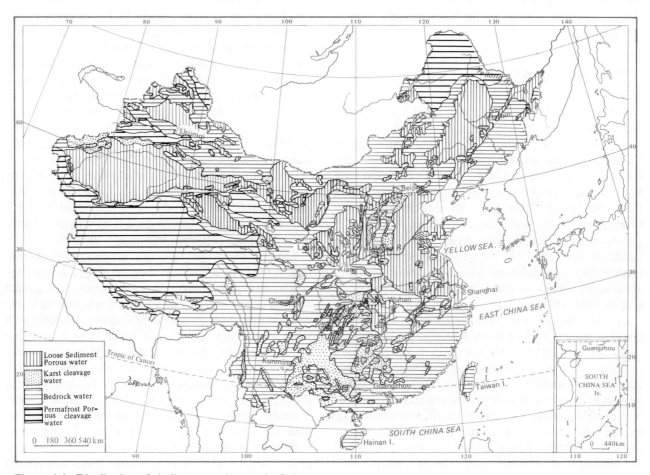

Figure 4-4 Distribution of shallow ground water in China

The Songhua-Nenjiang Plain is encircled by mountains, except on its southern side. In marginal piedmonts there are ground-water-bearing beds of sandy gravel generally 20 to 60 m thick, with the depth of the ground water table 5 to 15 m and water mineralization less than 0.5 grams/liter (g/l). In the basin center, ground water resources are rich both in quantity and in quality, with water-bearing beds mainly composed of sand and sandy gravel about 80 to 100 m thick, with the depth of the ground water table less than 5 m, and with mineralization (mainly bicarbonate) less than 1 g/l. In local depressions, owing to poor drainage, mineralization (mainly chloride) rises to 1 to 3 g/l.

The Liaohe Plain is essentially composed of sand deposits about 80 to 210 m thick, with a ground water table 1 to 5 m deep (mostly freshwater).

The Sanjiang Plain is composed of sandy deposits that overlie a thin clay layer of 5 to 17 m thick. There are extensive marshes on the ground surface. Underneath the clay layer, there exists a sandy gravel bed of more than 100 m thick that contains rich ground water under pressure.

The North China Plain

The North China Plain is also a NNE-SSW trending Cenozoic down-warped basin with Quaternary deposits 200 to 600m thick. From piedmonts along the Taihang Mountains footslope eastward up to the coastal plain, hydrogeological and geomorphological conditions change in succession.

The piedmont plain is characterized by thick (more than 100 m) ground water-bearing deposits, coarse ground surface materials (mainly sandy gravel and pebble), good drainage conditions and high-quality ground water. The depth of the ground water table exceeds 5 m, mineralization is less than 1 g/l and, hence, there is practically no salinity hazard.

Alluvial plains are extensively distributed, with varied ground surface materials. There are several water-bearing strata, (mainly composed of interbedded sand and clay), although each is rather thin. Depth of the ground water table ranges from 2 to 4 m (less than 2 m in depressions), mineralization is 1 to 3 g/l and, hence, the water is easily subjected to a salinization hazard.

The coastal plain is composed of fine-ground materials with poor drainage conditions and poor water quality; mineralization is usually greater than 5 g/l, even 30 g/l in local areas. This is not suitable for irrigation use, although the ground water table is quite near the surface.

The Hexi Corridor

The Hexi Corridor, located deep within arid Northwest China and sandwiched between the Nei Mongol Plateau and the Qinghai-Xizang Plateau, is famous for its surface water and ground water resources as well as its highly developed agriculture and industry. Its middle section — the Jiuquan Basin — is a good example. From the lofty Qilian Mountains (the marginal mountains of the Qinghai-Xizang Plateau) northward to Jashan Mountain (the marginal frontal hill of the Nei Mongol Plateau), zonal distribution of ground water resources is perfect:

1) Zone of Qilian frontal hills — depth of ground water table varies from 0 to 50 m.
2) Zone of diluvial piedmont plain (gravel gobi) — ground water table buried deeper than 50 m, yet, very rich and characterized by high-quality ground water runoff.
3) Zone of diluvial-alluvial piedmont plain (sandy gravel gobi) — depth of ground water table less than 50 m and rich ground water runoff. It is the best site for developing well irrigation.
4) Zone of alluvial plain — ground water table less than 5 m deep, yet, with ground water quality deteriorating.
5) Isolated sandy desert with varied depth of ground water table.

Northern Qinghai-Xizang Plateau

There are extensively distributed Quaternary alluvial-lacustrine loose deposits (mainly sandy gravel and clay) on northern Qinghai-Xizang Plateau (elevation: above 4500 m). Each lake basin is an independent drainage unit; from its margin to basin center, the chemical composition of ground water changes from bicarbonate to chloride. There also exists extensive permafrost; the higher the elevation, the thicker the permafrost. The permafrost attains a thickness of 70 to 190 m where the elevation is 4700 to 4950 m. The lower limit of permafrost varies in different areas; it is generally located at between 4200 to 4300 m on the northern slopes of the Kunlun Mountains at 4700 to 4900 m on the southern slopes of the Tanggula Mountains, and as high as 5700 to 5800 m in the Hengduan Mountains. There are three types of ground water in the permafrost area: water overlying frozen beds, intrawater among frozen beds, and water underlying frozen beds. The first two are rather scattered and small in quantity, whereas the third is rich both in quantity and quality. A series of nonfrozen springs along the northern footslopes of the Kunlun Mountains have become the most important water supply for local inhabitants.

South China and Southwest China

In the karst areas of South China and Southwest China,

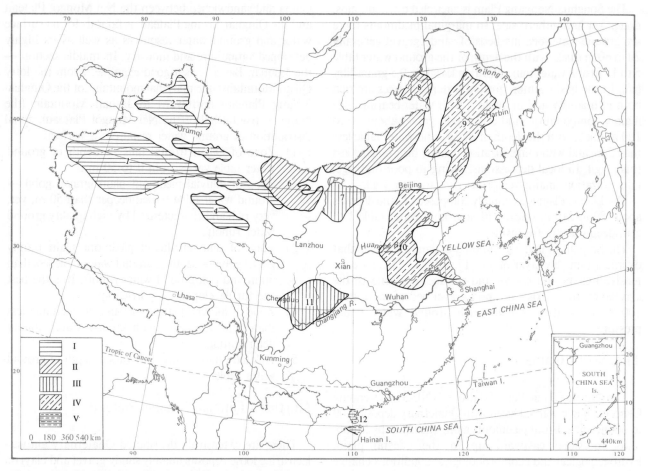

Figure 4-5 Distribution of major artesian basins in China

I. Quaternary fresh artesian water in the margins, highly mineralized water of Mesozoic and Tertiary Periods in the center; II. Artesian basins composed of Mesozoic and Tertiary beds; III. Synclinic artesian basins composed of thick Mesozoic beds, freshwater in upper layers, highly mineralized water in lower layers; IV. Quaternary fresh artesian water in upper layers, Cretaceous and Tertiary artesian water in lower layers: V. Fresh artesian water in basalt overlain coastal Cenozoic deposits.

thick beds of pure limestone and dolomite are extensively distributed (see Chapter 3). For example, in Guangxi, nearly one half of the total area is occupied by Devonian through Triassic limestones, with a total thickness more than 8000 m.

Just as local inhabitants remark, "No mountain can be without caves," hence in such karst areas, rivers frequently disappear from the surface and flow underground. In Guangxi alone there are more than 100 underground rivers with a discharge of more than 0.05 m³/sec each. They have a total low-water discharge of more than 100 m³/sec. Discharge of these underground rivers changes greatly owing to uneven seasonal distribution of precipitation. For example, one underground river in Duan County, Guangxi, has a low-water discharge of 4 m³/sec but a flood discharge as high as 390 m³/sec. Furthermore, the ground water table

is usually quite low in karst areas, and underground rivers are distributed sparsely and unevenly. All this leads to serious water-shortage problems in many localities.

Distribution of Major Artesian Basins

The formation and distribution of artesian basins are mainly determined by hydrogeological structure. Areal differentiation between northern China (north to the Qinling Mountains - Huaihe River line, including North China, Northeast China, and Northwest China) and southern China (south to the Qinling Mountains-Huaihe River line, including Central China, South China, and Southwest China) is conspicuous. In northern China, there occur many large-scale artesian basins, with artesian water chiefly located in synclines or faulted zones, intermontane basins and alluvial fans of lower Paleozoic limestone areas, and in artesian basins of Mesozoic and Tertiary bedrocks. In southern China, artesian water is distributed mainly in Mesozoic and Cenozoic as well as Tertiary intermontane basins. All artesian basins are rather small in size except the huge Sichuan Basin (Figure 4-5). The following are some typical artesian basins in China.

Tarim Artesian Basin

Jurassic and Cretaceous highly mineralized artesian water is distributed on the northern and southern margins of the Tarim Basin. Tertiary and Quaternary artesian water-bearing beds are mainly located in the basin center, with a thickness of 2000 to 3000 m, overlying directly on ancient metamorphic rocks. All consist of continental saline ground water, except along old river channels and deltas. Around the modern drainage center of the Tarim Basin, the Lop Lake, mineralization of ground water to a depth of 120 m is generally greater than 50 g/l. However, there are rich fresh ground water resources in Quaternary beds located along margins of the Tarim Basin.

Junggar Artesian Basin

This is a closed hydrogeological synclinorium structure, with Mesozoic and Cenozoic overloads several thousand meters thick, overlying directly on Paleozoic metamorphic rocks. Fresh artesian water-bearing beds more than 1300 m thick occur in the southern part of the basin, whereas in the northern part, they are only 300 m thick.

Hexi Corridor Artesian Basin

Hexi Corridor is a typical down-warped basin composed of three smaller basins — the Jiuquan, Zhangye, and Wuwei. Major ground water-bearing beds are Tertiary and Quaternary, the former with a total thickness of more than 3500 m. Ground water is characterized by conspicuous vertical zonation in hydrodynamics and hydrochemistry. For example in the Jiuquan Basin, ground water is subjected to active circulation above local base level (about 1200 m asl), with mineralization (mostly bicarbonate) less than 1 g/l. From 1200 m to sea level, ground water moves rather slowly, with mineralization (mostly sulphate or sulphate chloride) of 1 to 10 g/l. Below sea level, ground water stagnates, with mineralization (mostly chloride) increasing rapidly; it is usually greater than 10 g/l, sometimes even greater than 80 g/l.

North China Plain Artesian Basin

From the piedmont eastward, there are three artesian zones:

1) The Piedmont zone, with artesian water flowing within a large gradient.
2) The alluvial plain zone, with artesian water flowing slowly in several strata.
3) The coastal plain zone, with artesian water originating from the sea. Vertical zonation is also conspicuous. Beneath the ground surface, there usually occurs a saline ground water belt 100 to 300 m in thickness and 3 to 10 g/l in mineralization. Freshwater appears again at a depth of 250 to 400 m.

Sichuan Artesian Basin

The Sichuan Basin has been formed by large-scale downwarping since the Mesozoic and deposited with Jurassic and Cretaceous beds several thousand meters thick. After the Yanshan Tectonic Movement, these Mesozoic rocks were folded and the basin was divided into two portions, the eastern Sichuan broad folded artesian basin and the western Sichuan dome-folded artesian basin. In the latter, mineralization of artesian water increases rapidly with increase of depth; salty water occurs at a depth of 100 m beneath ground surface. Around Zigong (the famous salt-mining city), salty water with mineralization as high as 200 to 250 g/l appears at depths of 800 to 1000 m beneath ground surface.

References

[1]   *Physical Geography of China* Compilation Committee, Chinese Academy of Sciences, 1981, *Physical Geography of China: Surface Water,* Science Press, Beijing. (In Chinese)

[2]   *Physical Geography of China* Compilation Committee, Chinese Academy of Sciences, 1982, *Physical Geography of China: Ground Water,* Science Press, Beijing. (In Chinese)

[3]   *Physical Regionalization of China* Working Committee, Chinese Academy of Sciences, 1959, *Physical Regionalization of China: Regionalization of Surface Water in China,* Science Press, Beijing. (In Chinese)

[4]   *Physical Regionalization of China* Working Committee, Chinese Academy of Sciences, 1959, *Physical Regionalization of China: Regionalization of Ground Water in China,* Science Press, Beijing. (In Chinese)

[5]   Qian Ning (Chien Ning) et al., 1978, "The Regulation of Flow and Sediment Based on the Principles of Fluvial Processes for the Improvement of the Lower Huanghe River" *Acta Geographica Sinica,* Vol. 33, No. 1. (In Chinese, with English abstract)

[6]   Liu Chongming et al., 1978, "The Influence of Forest Cover Upon Annual Runoff in the Loess Plateau," *Acta Geographica Sinica,* Vol. 33, No. 2. (In Chinese, with English abstract)

[7]   Tang Qichang, 1979, "An Analysis on Main Features of Runoff in the Tianshan Mountains" *Acta Geographica Sinica,* Vol. 34, No. 2. (In Chinese, with English abstract)

[8]   Shi Chengxi, 1979, "Thirty Years of Limology in the People's Republic of China", *Acta Geographica Sinica,* Vol. 34, No. 3. (In Chinese, with English abstract)

[9]   Cheng Kezhao, et al., 1981. "The Salt Lakes on the Qinghai-Xizang

Plateau," *Acta Geographica Sinica,* Vol. 36, No. 1. (In Chinese, with English abstract)

[10] Chen Mon-yung, et al., 1959. "Types and Distribution of Artesian Basins in China." *Hydro-geology and Engineeringgeology,* 1959, No. 7. (In Chinese)

[11] Zhao Songqiao, 1983, "Evolution of the Lop Desert and the Lop Lake," *Geographical Research,* Vol. 2, No. 2. (In Chinese, with English abstract)

Chapter 5

Soil Geography

A vast area and complicated physical conditions make China rich and varied in soil resources. The Chinese people have studied, utilized and transformed these soil resources for more than 7000 years. As early as the fifth century B.C., *The Tribute of Yu* subdivided China into nine great regions and classified their soils mainly on the basis of soil color. Soils were again demarcated, according to their fertility into three levels and nine grades; this was probably the first systematic nationwide soil classification system in the world. A little later, another classic, *The Book of Master Kwan,* made a more detailed study of China's soils, classifying them into 18 types, each type with 5 fertility grades. Since then, billions of Chinese farmers have laboriously worked in close contact with China's soils, although little scientific research work on soils has ever been done.

Modern scientific studies on China's soils started in the 1930s, and the American soil scientist James Thorp was one of the distinguished pioneers. At this early stage, the American school of soil science exerted great influence on the soil classification system in China; more than 10 great soil groups and 2000 soil series were identified. Since 1953 the Russian school's genetic method of evaluation has been adopted. In 1958 to 1961, a nationwide soil survey was completed. In 1978, the current soil classification system of 11 orders, 47 groups and 139 subgroups was finally developed. Up to now, little research work has been conducted on any "new" international soil classification systems, such as the American Soil Taxonomy and the FAO-Unesco Soil Map of the World. For the sake of making necessary comparisons with soils of other parts of the world, we are now tentatively presenting Figure 5-1 and Table 5-1 as a basis for discussing the soil geography of China. The table and figure have been compiled from old soil maps and traditional soil classification systems with a view to correlating them with the FAO-Unesco system. Such an effort is still preliminary. Much revision based on solid fieldwork and laboratory analysis must be done in the future. Lacking trustworthy scientific data for a new soil classification system, we must for the time being continue to use a traditional soil classication system in most regional geographical studies, as in chapters 8 through 14 in this volume.

CHIEF SOIL-FORMING PROCESSES IN CHINA

The soil-forming process is chiefly a function of parent material, climate, landform, vegetation, and time. In such an ancient and densely populated country as China, the impact and feedback of past and present human activities are also important.

Under China's specific physical and human conditions, the following 12 soil-forming processes are most important. Each soil-forming process usually gives birth to more than one soil type and each soil type usually represents the end product of more than one soil forming process. For example, within a semiarid temperate steppe environment, the interplay of calcification and humification produces chernozems; whereas within a tropic humid forest environment, the joint action of leaching, laterization, and humification processes results in ferralsols.

The processes that give character to the soil of China include the following:

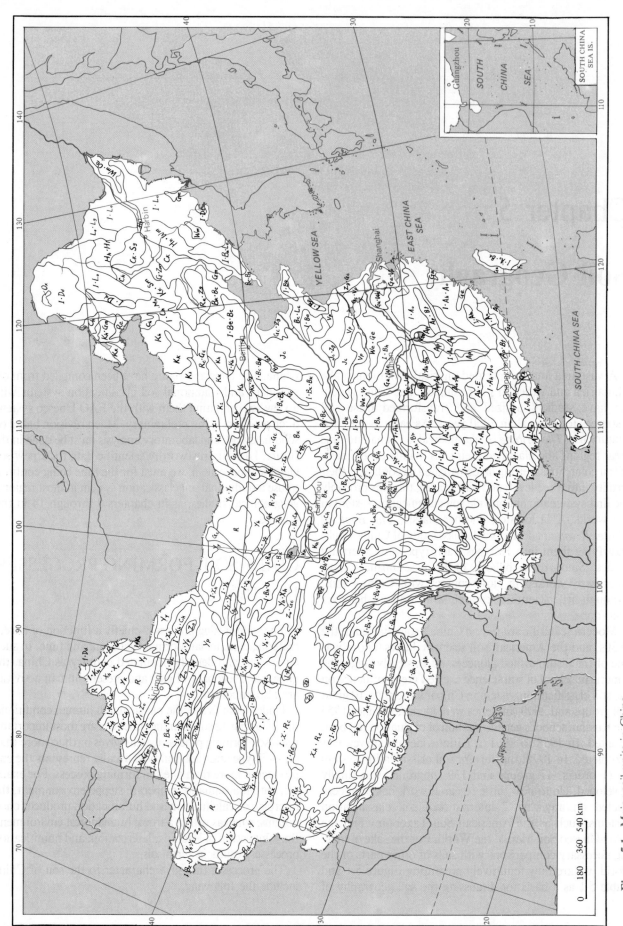

Figure 5-1 Major soil units in China
(Based on the FAO-Unesco system. legends see table 5-2)

Table 5-1 Correlation of Soil Types Between China's Traditional Soil Classification System (1978) and the FAO-Unesco *Soil Map of the World* (1977)

Soil units FAO-Unesco system	Soil groups of China's traditional classification system
1. Fluvisols (J)	Chou tu (wet soil), meadow soil
2. Gleysols (G)	Meadow soil, bog soil, paddy soil, irrigated oases soil, alpine meadow soil
3. Regosols (R)	Alpine frozen soil, aeolian sandy soil, purple soil, saga soil (alpine steppe soil)
4. Lithosols (I)	Soils of mountainous areas
5. Rendzinas (E)	Limestone soil, phosphocalcic soil
6. Rankers (U)	Alpine meadow soil, subapline meadow soil
7. Vertisols (V)	Shachiang soil, paddy soil
8. Solonchaks (Z)	Solonchak
9. Solonetz (S)	Solonetz
10. Yermosols (Y)	Grey desert soil, grey-brown desert soil, brown desert soil, takyric soil, alpine desert soil
11. Xerosols (X)	Sierozem, semidesert brown soil, irrigated oases soil
12. Kastanozems (K)	Chestnut soil
13. Chernozems (C)	Chernozem
14. Phaeozems (H)	Black earth
15. Greyzems (M)	Grey forest soil
16. Cambisols (B)	Burozem, drab soil, greydrab forest soil, mein tu (cultivated loess), lou tu (stratified old manual loess), heilu tu (dark loess), subalpine meadow soil
17. Luvisols (L)	Dark-brown forest soil, burozem, yellow-brown earth, heilu tu, limestone soil, dry red earth
18. Podzoluvisols (D)	Bleached grey soil
19. Planosols (W)	Baijiang tu, yellow-brown earth, burozem
20. Acrisols (A)	Lateritic soil, red earth, yellow earth
21. Nitosols (N)	Laterite, red earth, dry red earth
22. Ferralsols (F)	Laterite
23. Histosols (O)	Peat soil, bog soil

1) Weathering is a soil-forming process under which rocks and other parent materials become decayed and disintegrated into clay, silt, and sand particles. Thus, the soil profile starts to be developed. Weathering occurs all over China and is particularly strong in humid tropic zones. Its typical end products are lithosols and regosols.

2) Leaching is a process that involves soluble salts and other movable materials in a soil body being removed or leached from the upper parts to the lower parts. It occurs mainly in humid and subhumid regions where water is plentiful.

3) Podzolization predominates in climates that have sufficient cold to inhibit bacterial action but sufficient moisture to permit larger green plants to thrive. In its extreme development, podzolization is associated with coniferous trees (pine, fir, spruce, etc.). Humic acids, produced from the abundant leaf mold and humus strongly leach the upper soil of bases, colloids, and the oxides of iron and aluminum, leaving a characteristic ash grey A2 soil horizon composed largely of silica (SiO_2). Typical soils are podzoluvisols.

4) Argillation includes the disintegration of primary minerals into secondary minerals and clays in upper layers that are then deposited in lower horizons. Typical soils are luvisols.

5) Laterization takes place in a tropical or subtropical climate that has copious rainfall that permits sustained bacterial action to destroy dead vegetation as rapidly as it is produced. In the absence of humic acids, the sesquioxides of iron (Fe_2O_3) are insoluble and accumulate in the soil as red clays, nodules, and rocklike strata (laterite). Silica, on the other hand, is leached out of the soil. The soil tends to be firm and porous rather than sticky and plastic. Its typical end products are ferralsols.

6) Calcification occurs mainly in semiarid and arid regions where evaporation on the average exceeds precipitation. Rainfall is not sufficient to leach

out the bases and colloids. Calcium carbonate is precipitated in the B horizon in the form of nodules, slabs, and even dense stony layers (caliche). Its typical end products are kastanozems and chernozems.

7) Salinization is chiefly associated with the steppe and the desert environments as well as with poorly drained locations. Sulphates and chlorides of calcium and sodium accumulate in the soil. According to an estimate, about one fifth of China's total farmlands, or approximately 20 million ha, are more or less affected by the salinization process. Typical soils are solonchaks.

8) Alkalinization is similar to salinization but with a higher proportion of sodium content. Typical soil are solonetz.

9) Gleization is characteristic of poorly drained (but not saline) environments under a moist and cool or cold climate. Low temperatures permit heavy accumulations of organic matter to form a surface layer of peaty material; beneath this is the "glei" horizon, a thick layer of compact, sticky, structureless clay of a bluish-grey color. Its typical end products are gleysols and histosols.

10) Humification is essentially the slow oxidation of organic matter under cold humid climates. The organic acids help in the decomposition of the minerals of the parent materials. The hydrogen ions of the acid solution tend to replace the ions of potassium, calcium, magnesium, and sodium. The typical soil are phaeozems and histosols.

11) Leucinization is essentially a bleaching action of upper soil horizons under seasonal flooding conditions. Iron and other colorful materials of upper soil horizons are reduced and removed either by laterally flowing water or by forming concretions in situ, with a resultant whitish layer. Its end products are planosols, the wide spread *Baijiang tu* (literally whitish soup earth) in humid temperate Northeast China being the most famous.

12) Cultivation in China should be considered as one of the most important soil-forming processes. All cultivated soils have been more or less changed from their original conditions. For example, in the Hexi Corridor, after more than 2000 years of cultivation, the irrigated farmlands have been practically transformed into entirely man-made soils, with a fertile tilth layer generally more than 1 to 2 m deep. The widely distributed paddy soils in South China and Central China are another excellent example of man-made fertile and productive soil. On the other hand, if farmlands are poorly managed or misused, orginally excellent soils might deteriorate or even be destroyed.

Again, soil is a dynamic natural body in which, many complex chemical, physical, and biological (including human) activities are constantly in progress. It is always changing and evolving. In different regions, at different stages, and under the different impacts of varied environmental factors, one soil type may be transformed into another. According to recent studies on paleoweathering crusts and paleosoil horizons, in China many soil types have greatly evolved since the late Tertiary and early Quaternary periods. For example, the origin and formation of acrisols in South China might be traced back to the late Tertiary period. During the early and middle Pleistocene epoch, when the paleoclimate was much warmer, the subtropical climate together with acrisols pushed northward, up to about 34°N; their northern boundary retreated southward to about 30°N during the late Pleistocene epoch, then moved northward again to their present position (about 32°N) during the early Holocene epoch. In the same natural division, the composition of major soils has also been changed since the Quaternary period as a consequence of environmental changes. For example, in the lower Changjiang Plain, the major soil types were acrisols during the early Pleistocene epoch, but became luvisols and planosols during the late Pleistocene epoch. In the southeastern Loess Plateau, the major soil types changed from luvisols during the early and middle Pleistocene epoch to cambisols (developed on the Malan loess) during the late Pleistocene epoch.

MAJOR SOIL TYPES AND THEIR DISTRIBUTION

Under the action and interplay of the above-mentioned soil-forming processes, varied soil types have been formed. Figure 5-1 and Table 5-2, based on the FAO-Unesco soil classification system, show the identification of soil types. It is far from being complete. The all important agricultural soils lack representation and the classification, and the association of major soil types on the Qinhai-Xizang Plateau seems quite problematic. However, this categorization is a good beginning, and it also offers a common denominator with soil classification systems of other parts of the world.

The "new" soil classification system, unlike the "traditional" one, does not emphasize the principles of genetics and zonation. Yet, major soil units classified after the FAO-Unesco system beautifully reflect zonal distribution, both horizontally and vertically. Figure 5-2 shows the model of horizontal zonal distribution of China's major soil units. From south to north, as the temperature decreases soils change from ferralsols to acrisols, cambisols, luvisols, Phaeozems, and podzoluvisols successively; this illustrates

Table 5-2 Major Soil Units in China

A. Acrisols	Fx Xanthic ferralsols	Lc Chromic luvisols	Vp Pellic vertisols
Ao Orthic acrisols	Fr Rhodic ferralsols	Lk Calcic luvisols	
Af Ferric acrisols	Fp Plinthic ferralsols	Lf Ferric luvisols	**W. Planosols**
Ah Humic acrisols		La Albic luvisols	We Eutric planosols
Ap Plinthic acrisols	**G. Gleysols**	Lg Gleyic luvisols	Wm Mollic planosols
Ag Gleyic acrisols	Ge Eutric gleysols		
	Gc Calcaric gleysols	**M. Greyzems**	**X. Xerosols**
B. Cambisols	Gm Mollic gleysols	Mo Orthic greyzems	Xh Haplic xerosols
Be Eutric cambisols	Gx Gelic gleysols		Xk Calcic xerosols
Bh Humic cambisols		**N. Nitosols**	Xy Gypsic xerosols
Bg Gleyic cambisols	**H. Phaeozems**	Nh Humic nitosols	Xl Luvic xerosols
Bk Calcic cambisols	Hh Haplic phaeozems		
Bx Gelic cambisols	Hl Luvic phaeozems	**O. Histosols**	**Y. Yermosols**
Bc Chromic cambisols	Hg Gleyic phaeozems	Oe Eutric histosols	Yh Haplic yermosols
Bf Ferric cambisols		Ox Gelic histosols	Yk Calcic yermosols
	I. Lithosols		Yy Gypsic yermosols
C. Chernozems		**R. Regosols**	Yl Luvic yermosols
Ch Haplic chernozems	**J. Fluvisols**	Re Eutric regosols	Yt Takyric yermosols
Ck Calcic chernozems	Je Eutric fluvisols	Rc Calcaric regosols	Yp Plinthic Yermosols
Cl Luvic chernozems	Jc Calcaric fluvisols	Rx Gelic regosols	
			Z. Solonchaks
D. Podzoluvisols	**K. Kastanozems**		Zo Orthic solonchaks
Dd Dystric podzoluvisols	Kh Haplic kastanozems	**S. Solonetz**	Zm Mollic solonchaks
	Kk Calcic kastanozems	Sm Mollic solonetz	Zt Takyric solonchaks
E. Rendzinas	Kl Luvic kastanozems	Sg Gleyic solonetz	Zg Gleyic solonchaks
F. Ferralsols	**L. Luvisols**	**U. Rankers**	
Fo Orthic ferralsols	Lo Orthic luvisols		
		V. Vertisols	

the socalled latitudinal zonation. From east to west, according to decreasing precipitation and increasing distance from the sea, luvisols, phaeozems, podzoluvisols, greyzems, chernozems, kastanozems, xerosols, and yermosols appear in succession. Figure 5-3 considers the Tianshan Mountains as an example and shows vertical zonal distribution. On the southern slope of the Tianshan Mountains, from lower to higher elevations, yermosols, xerosols, kastanozems, cambisols, regosols, and continual snow appear successively. And, on its northern slope, xerosols, chernozems, cambisols, regosols, and continual snow are distributed in an altitudinal sequence.

Chief Features of China's Major Soil Units

Fluvisols (J) These are the soils of river valley plains, formed mainly under seasonal flood and continual alluvial deposition as well as various soil-forming processes, such as leaching, gleization, and humification. These soils are widely distributed along the middle and lower reaches of larger rivers, especially in the North China Plain. In general, they are naturally fertile soils with good soil moisture conditions, but sometimes they are liable to flood hazards. Vegetation cover is luxuriant, resulting in a high content of humus (about 1 per cent in the top soil layer of the North China Plain and as high as 4 per cent in some parts of the Zhujiang Delta). Practically all of these soils have been cultivated with a tilth layer more than 30 to 40 cm deep. These soils are especially valuable in extreme environmental conditions (e.g., desert, high mountain), where they constitute the only agriculturally important land.

There are two important subunits of fluvisols in China: (1) eutric fluvisols (Je) are mainly distributed in the middle and lower Changjiang Plain, often in association with various gleysols, with a humus content of the top soil layer of about 1.5 to 2.0 per cent; (2) calcaric fluvisols (Jc) are distributed mainly in North China and Northwest China, often associated with solonchaks, with high lime content

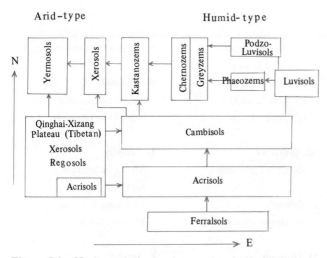

Figure 5-2 Horizontal distribution model of soils in China

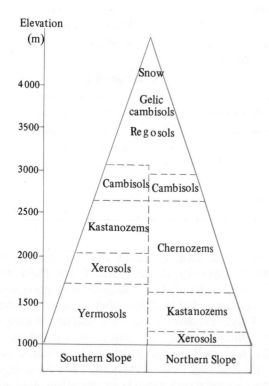

Figure 5-3 Vertical distribution of soils in the Tianshan Mountains.

cultivated and transformed into paddy soils and irrigated oasis soils with the exception of a small portion of gleysols that are waiting for reclamation in Northeast China and Northwest China. They are excellent agricultural soils, with a humus content in the top layer ranging from 1 to 10 per cent and a nitrogen content of 0.1 to 0.5 per cent.

Four subunits are identified: (1) eutric gleysols (Ge) are mainly distributed in alluvial-lacustrine plains in Central China and South China, with a humus content in the top layer of 2 to 3 per cent; (2) calcaric gleysols (Gc) are distributed mostly in alluvial plains in North China and Northwest China, with a high lime content; (3) mollic gleysols (Gm) are mostly distributed in alluvial plains in Northwest China, with a high humus content in the top soil layer (usually 3 to 6 per cent, sometimes even 10 per cent); (4) gelic gleysols (Gx) are mainly located on lofty mountains in the Qinghai-Xizang Plateau and the Altay Mountains with permafrost about 1 to 2 m beneath the ground surface.

Regosols (R) Widely distributed in sandy and mountainous areas, with the soil profile poorly developed. Most of them lay as waste or are slightly used as pasture. Three subunits are identified: (1) eutric regosols (Re): Distributed in sandy areas of the eastern Nei Mongol Plateau and rocky mountains of southern China. The chief feature of the soil profile is the absence of lime content; (2) calcaric regosols (Rc) are distributed in sandy deserts and rocky mountains of the arid zone, with a lime content ranging from 1 to 15 per cent in the soil profile; (3) gelic regosols (Rx) are located on the tops of the Da Hinggan Mountains and the Altay Shan as well as on alpine slopes above 4600 to 5200 m in the Qinghai-Xizang Plateau. They have a very coarse soil texture, with gravels usually occupying more than 40 per cent of the solum.

Lithsols (I) Widely distributed on steep slopes, especially in Northwest China. With a soil profile of less than 10 cm and a humus content approaching zero, they generally lie waste.

Rendzinas (E) These are the soils of karst topography in humid subtropic Central China and South China, with a high content of calcium carbonate (ranging from 1 to 30 per cent). On the South China Sea Islands, which are essentially built up by coral reefs, a special kind of rendzinas (phosphocalcic soil) is developed. As a result of the action of thick guano deposits, a layer of phosphorous 20 to 30 cm thick lies immediately under the humus top layer (see Chapter 11).

Rankers (U) These soils are chiefly distributed in the southeastern border of the lofty Tibetan Plateau and on alpine meadows of arid Northwest China. They usually coexist with lithsols and might be regarded as a further development of such a soil. They differ from lithsols, however, by the existence of a humus layer and a developed profile generally 10 to 25 cm thick. They are used mainly

throughout soil profile (about 6 to 10 per cent).

Gleysols (G) These are the soils of low depressions, usually with a high water table or impeded drainage. Gleization and humification are dominant soil-forming processes. Gleysols are widely distributed in northern China as well as in southern China. They are now mostly as summer-autumn pasture.

Vertisols (V) These special soils are mostly restricted to the lower reaches of the Huaihe River with clayey alluvial-lacustrine deposits and low, depressed landforms. Ground water tables are usually less than 1 m deep. The chief feature of the soil profile is the existence of a blackish, clayey layer 40 to 60 cm thick immediately under the grey-brownish cultivated layer, which, in turn, merges into a conspicuous fluvial depositional parent material strata. This is the so-called Shachiang layer, which is composed mainly of lime concretions. On the whole, the soil profile is clayey and sticky, with clay occupying about 34 to 38 per cent of the solum. Humus content of the top soil layer amounts to about 1.3 per cent.

Solonchaks (Z) These soils occur widely in steppe and desert zones where surface evaporation is higher than precipitation and, thus, salts are brought to the surface from the subsoil by the ascending capillary water. They are also distributed in small patches along the coast. The dominant soilforming process is, of course, salinization, yet, when salt content is not too high, gleization and humification are also important. The salt content varies with different environments, generally increasing with increasing aridity. In the extremely arid environments, salt crusts more than 3 cm thick are usually formed on the ground surface; and in the Turpan Basin and Hami Basin where annual precipitation drops below 10 mm, nitrates with an NO_3 content of 0.4 to 1.0 per cent are included in the soil profile. Solonchaks are unsuitable for traditional agriculture, being too saline for the growth of ordinary crops. All agricultural utilization of these soils depends on the leaching of toxic salts through physical, biological, and technical reclamation measures. Irrigation and leaching of solonchaks are successful only if drainage facilities are good. Thus, when natural drainage is impeded or insufficient, an artificial drainage system is always a prerequisite for the successful reclamation of these soils.

Four subunits of solonchaks can be identified: (1) orthic solonchaks (Zo) are widely distributed in inland basins of the desert zone, with salt crust (salt content from 10 to 60 per cent) 5 to 15 cm thick and a humus content of less than 1 per cent, making them entirely useless for agriculture; (2) mollic solonchaks (Zm) are mainly distributed in the Northeast China Plain and eastern Nei Mongol Plateau, with thin salt crust (1 to 3 cm thick) containing 1 to 3 per cent salt and 4 to 6 per cent humus, and, thus, are usable for pasture; (3) takyric solonchaks (Zt) are distributed chiefly in the clay desert flats, forming polyhedron cracks when dried, with a salt content of 2 to 4 per cent on top layers, which are usually entirely barren; (4) gleyic solonchaks (Zg) are distributed widely from the coast to the extremely arid inland basins, usually with a very shallow ground water table and a salt content on the top layer ranging from 1 to 2 per cent.

Solonetz (S) Small patches of mollic solonetz (Sm) and gleyic solonetz (Sg) are scattered in northern China, usually in association with chernozems, kastanozems, xerosols, and yermosols. The most notable property of solonetz soils is their strong alkalinity (pH value of the top layer is usually higher than 9), frequently combined with salinization of the subsoil. Chemical improvement is always needed to make them productive.

Yermosols (Y) These soils are widely distributed in the desert zone of China. Their profiles have been only weakly developed and are usually less than 1 m thick. Humus content is low, usually less than 1 per cent in the topsoil layer. Soil moisture is also low and farming is not possible without irrigation. On the other hand, the salt content is high and the soil texture is coarse. With irrigation and other meliorating measures, these soils can be used for different agricultural purposes.

Six subunits of yermosols are identified in the extensive desert zone of China, which accounts for about 22 per cent of China's total land area: (1) haplic yermosols (Yh) are mostly distributed on the upper parts of piedmont plains (depositional gobi), with a humus content of about 0.5 per cent and a salt content of about 0.5 to 1.0 per cent in the topsoil layer; (2) calcic yermosols (Yk) are formed chiefly on the denundational gobi or peneplaned plateau and are characterized by a calcified layer about 10 to 30 cm thick in the subsoil between depths of 40 to 80 cm; (3) gypsic yermosols (Yy) are restricted to extremely arid areas, such as the Nomin Gobi and the Hoshan Gobi in eastern Xinjiang. They feature a conspicuous gypsum layer in the subsoil between depths of 10 to 40 cm; (4) plinthic yermosols (Yp) are also restricted to extremely arid areas, with a salt pan (containing soluble salts 30 to 50 per cent) in the subsoil between depths of 20 to 40 cm; (5) luvic yermosols (Yl) are developed in the silt flats of the desert zone and feature an argillated, sodic B horizon about 10 to 15 cm thick; (6) takyric yermosols (Yt) are restricted to clayey flats of the desert zone. The ground surface is cracked into polyhedrons about 5 to 13 cm in diameter when dry.

Xerosols (X) These soils are developed in the transitional zone between semiarid and arid regions mainly through calcification and partly by humification and salinization. The humus content in topsoil layers ranges from 1.0 to 2.0 per cent. Four subunits are identified: haplic xerosols (Xh), calcic xerosols (Xk), gypsic xerosols (Xy), and luvic xerosols (Xl). The last one is the major soil of the vast desert-steppe zone.

Kastanozems (K) These soils are also called chestnut soils. They are the major soils of the extensive steppe zone in China. They are developed under the interplay of calcification and humification, with a humus layer 15 to 25 cm thick, containing 1.5 to 4.0 per cent humus. Three sub-units are identified (1)haplic Kastanozems (Kh) are non-sodic and slightly calcareous and are mainly distributed in the best grassland in China, such as the *Hulun Buir Steppe* and the *Xilingol Steppe;* (2) calcic kastanozems (Kk) have a calcareous layer in the subsoil that is more than 30 to 50 cm deep; (3) luvic kastanozems (Kl) have an argillated B-horizon, usually in association with xerosols in the desert-steppe zone.

Chernozems (C)   These soils develop under temperate subhumid forest-steppe and semiarid steppe environments and are excellent agricultural soils. The dominant soil-forming processes are similar to those of the kastanozems, except for stronger humification and weaker calcification. The humus layer generally has a thickness of 20 to 40 cm, with a humus content of 3 to 15 per cent. The calcareous layer usually starts 40 to 60 cm from the ground surface. Three subunits are identified: (1) haplic chernozems (Ch) have a very thick humus layer; (2) calcic chernozems (Ck) have a subsoil calcareous layer of 20 to 60 cm; (3) luvic chernozems (Cl) have a conspicuous argillated B-horizon and a calcareous layer deeper than 100 cm.

Phaeozems (H)   These soils develop under a temperate subhumid forest-meadow environment and are probably the most fertile soils in China. Yet, their distribution is limited to Heilongjiang and Jilin provinces. The humus layer generally has a thickness of 30 to 70 cm (sometimes more than 100 cm), with a humus content of 3 to 15 per cent. The calcareous layer is practically nonexistent. Three subunits can be identified: (1) haplic phaeozems (Hh) are distributed on well-drained gentle slopes, with a humus layer of 30 to 50 cm; (2) luvic phaeozems (Hl) have impeded drainage and an argillated layer immediately below the humus layer; (3) gleyic phaeozems (Hg) are restricted to low depressions, with a thick humus layer as well as an argillated layer.

Greyzems (M)   These soils, also known as grey forest soils, are mainly restricted to the tree-clad western slopes of the Da Hinggan Mountains and the southern slopes of the Altay Mountains. There is a litter layer of 2 to 7 cm on the ground surface, then a drab-greyish humus layer of 30 to 50 cm, with a humus content of 5 to 15 per cent. Only one subunit (orthic greyzems [Mo]) is identified in China.

Cambisols (B)   These soils are at an early stage in the soilforming process. They are widely scattered in China, especially in the Liaodong-Shandong peninsulas, the southeastern Loess Plateau, and the South Changjiang hills and basins. The soil profile is characterized by a top humus layer overlying a brownish transitional B-horizon. Areal differentiation of physical-chemical features of cambisols can be quite divergent according to different conditions of climate, topography, and parent materials. Hence, seven subunits can be identified in China: (1) eutric cambisols (Be), mainly distributed in the Liaodong-Shandong peninsulas and other hilly areas in North China, are characterized by a thin humus layer (with a humus content of 1 to

3 per cent) overlying a drab to brown transitional B-horizon; (2) humic cambisols (Bh), mainly developed in the mountainous areas of southern China, are characterized by a humus layer of 15 to 20 cm and have a humus content of more than 5 per cent; (3) gleyic cambisols (Bg) are located on lower slopes and terraces, with iron mottles in the lower soil profile; (4) calcic cambisols (Bk) are distributed mainly in the Loess Plateau and intermontane basins of North China, with humus content only 1 per cent in the top layer and with a calcareous layer in the subsoil of between 50 to 100 cm deep; (5) chromic cambisols (Bc), located mainly on the low mountains and terraces of North China, are characterized by a brownish to reddish B-horizon; (6) ferralic cambisols (Bf) are widely distributed in the reddish B-horizon; (7) gelic cambisols (Bx) are distributed on the northern Qinghai-Xizang Plateau and its surrounding high mountains, with permafrost between 1.3 to 2.0 m from the ground surface.

Luvisols (L)   These soils are mainly formed within the mixed broad and needle-leaved forest environment and are widely distributed in the Da and Xiao Hinggan mountains and Changbai Mountains. The typical soil profile is composed of a litter layer of 3 to 5 cm, then there is a dark brown or drab argillated B-horizon followed by the parent material layer. Six subunits of luvisols can be identified in China: (1) orthic luvisols (Lo) are characterized by a well-developed argillated B-horizon; (2) chromic luvisols (Lc): possessing a colorful agrillated B-horizon; (3) calcic luvisols (Lk) feature a calcareous layer about 60 to 70 cm from the ground surface; (4)ferric luvisols (Lf) have a reddish, ferric argillated B-horizon; (5) albic luvisols (La), mainly distributed in the mountainous areas of Northeast China and the western Sichuan Basin, and possess a conspicuous bleached E-horizon; (6) gleyic luvisols (Lg) show conspicuous gleization action in the upper soil profile.

Podzoluvisols (D)   These soils characterize the taiga forest environment in the northern Da Hinggan Mountains and the northwestern Altay Mountains. The dominant soil-forming processes are leaching, podzolization, and leucinization. Their profiles include a litter layer on the ground surface, then a dark grey humus layer of 5 to 20 cm thick. This is followed by the bleached E-horizon of 5 to 15 cm, then the brown or drab argilated B-horizon. Finally, the half-weathered parent materials appear. Podzoluvisols can be divided into three subunits: (1) eutric (saturated), (2) dystric (Dd, unsaturated), and (3) gleyic, of which dystric podzoluvisols are most widely distributed.

Planosols (W)   These soils are the end products of intense leaching and leucinization and are widely distributed in the northern parts of Eastern Monsoon China, particularly in temperate humid Northeast China where the humification process is also strong. Their formation is closely related to heavy monsoon rainfall and clayey parent materials. The most important feature in their profile is the whitish bleached E-horizon overlying the argillated B-horizon. Two subunits of planosols can be identified in China: (1) eutric planosols (We) are distributed on mountainous and hilly areas of North China and Central China, with a humus content of less than 1 per cent in the humus layer, a bleached horizon that contains plentiful iron concretions, and an argillated B-horizon composed mainly of clay; (2) mollic planosols (Wm) are distributed mainly in Northeast China and feature a dark grey, soft humus layer, with a humus content of 8 to 10 per cent, a clayey, bleached horizon about 20 cm thick, and a dark brown, argillated B-horizon.

Acrisols (A)   These soils are mainly distributed in humid, subtropic zones, under intense actions of leaching, argillation, and laterization. Their profiles include a litter layer; a greyish brown humus layer 10 to 20 cm thick, with a humus content of less than 1 per cent; a reddish or yellowish argillated B-horizon that ranges from 15 cm to more than 1 m in thickness; and, finally, parent materials derived from different rock outcrops. Owing to intense leaching action, the pH value usually drops below 5.0 to 6.0. Five subunits of acrisols can be identified in China: (1) orthic acrisols (Ao) are widely distributed in the low mountaion areas of southern China; (2) ferric acrisols (Af) are formed mainly in the red clay areas of the Yunnan-Guizhou Plateau, with iron concretions in the subsoil that are 30 to 40 cm from the ground surface; (3) humic acrisols (Ah) are characterized by a humus layer with a humus content of 5 to 20 per cent, largely located on the high mountains of southern China; (4) plinthic acrisols (Ap) are characterized by a subsurface net of streaks below that are 60 to 70 cm from the ground surface, mostly located along the southeastern coast; (5) gleyic acrisols (Ag), associated with the paddy-growing areas, have iron mottles in the upper soil profile.

Nitosols (N)   These soils are developed chiefly on basalt and red clay areas in humid, subtropic zones. They are characterized by the existence of an argillated B-horizon, with clay accounting for 50 to 60 per cent, and by the absence of the bleached horizon. Two subunits are identified; (1) dystric nitosols and (2) humic nitosols (Nh). The latter are restricted to better forested areas and have a humus content of 3 to 9 per cent in their darkish A-horizon.

Ferralsols (F)   These soils are distributed mainly in humid tropic South China. They are the end product of laterization and are also influenced by leaching and humification. Their profiles include a litter layer of 2 to 3 cm; a grey-brownish or dark brownish humus layer of 10 to 30 cm; a reddish, oxidized B-horizon with duripan or fragipan; and, finally, the reddish or brownish parent materials. Total profiles are usually quite thick, sometimes more than 3 m. Four subunits of ferralsols can be identified in China: (1) orthic ferralsols (Fo) are developed mainly on granite or metamorphic rocks with a reddish B-

horizon; (2) xanthic ferralsols (Fx) are developed under moist conditions and, hence, display a yellowish B-horizon; (3) rhodic ferralsols (Fr) are developed mainly on basalt and limestone beds with a clayey, reddish B-horizon; (4) plinthic ferralsols (Fp) are characterized by yellowish mottles and streaks at 80 cm below ground surface.

Histosols (O)   These soils are distributed mainly in the marshy Sanjiang Plain and the Zoigê area of the eastern Qinghai-Xizang Plateau. They are characterized by a high ground water table, usually less than 1 m from the surface. A peat layer of from 30 to 40 cm to 2 to 3 m (or even 10 m) and a humus content of 50 to 70 per cent is accumulated on the ground. Two subunits of histosols can be identified in China: (1) eutric histosols (Oe) and (2) gelic histosols (Ox); the former have a wider distribution with a pH value of about 5.5 and they are covered mainly by luxuriant Carex vegetation.

SOIL GEOGRAPHICAL REGIONS

The geographical distribution of each above-mentioned soil unit is determined by zonal factors, such as climate and vegetation, as well as by azonal factors, such as bedrock, topography, and hydrologic geology. The integration of the areal differentiation of all these zonal and azonal factors is the basis for demarcating soil regions, which, in turn, is the basis for overall planning in the use and transformation of each soil-unit according to its specific features and capability. Thus, 10 soil regions and 51 subregions are demarcated in China (Figure 5-4, Table 5-3).

1. Rendzinas Region

This soil region is restricted to the South China Sea Islands, including four archipelagos — Dongsha, Xisha, Zhongsha and Nansha. These tropical archipelagos are composed of numerous islands, beaches, reefs, and shoals — all of them being built up by limy coral skeletons. Major soil units include rendzinas in the higher parts of the islands and sandy regosols and solonchaks along the low coasts. This is a soil region for developing tropical crops and fisheries.

2. Ferralsols Region

This soil region is located in humid tropical South China. Dominant soils are ferralsols, including in part nitosols, luvisols, and gleysols. In addition, solonchaks and sandy regosols are distributed along the coast. Vertical distribution is conspicuous in the mountainous areas, with ferralsols on the foothills and acrisols, cambisols, and lithosols on the higher slopes. This is a soil region for developing

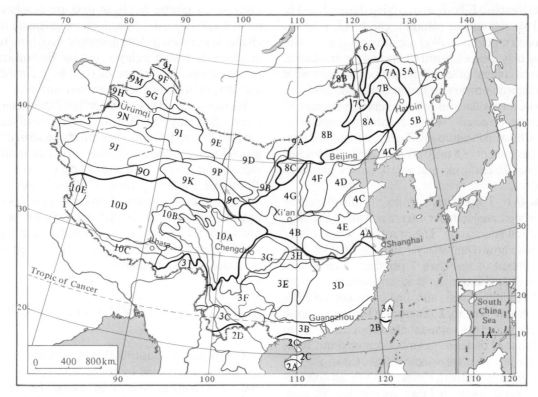

Figure 5-4 Soil geographical regions of China (Legends see table 5-3)

tropical crops; paddy rice may be triple-cropped annually, Hevea (rubber) and other tropical tree crops are widely planted. Soil conservation and overall planning are much needed in the mountainous and hilly areas. Four soil subregions can be identified.

3. Acrisols Region

This soil region is widely distributed in China and includes the extensive middle and southern subtropic zones. Major soil units are acrisols, located mostly on low mountains and hills. Gleysols are distributed in the valleys, intermontane basins, and lacustrine plains. In addition, rendzinas and ferric luvisols are extensively developed in the karst areas of Guangxi, Yunnan, and Guizhou provinces, and cambisols are found in the reddish sandstone and purple shale areas in the Sichuan Basin. In mountainous areas, cambisols, luvisols and lithosols are distributed, and along the coast there are small patches of solonchaks and regosols. In the plains, agriculture has been intensely developed since very ancient times, and generally there is triple-cropping (with two crops of rice). Measures must be taken for the better use of the land. On the mountains and hills forestry and economic tree crops are developed, although soil conservation and reforestation are two urgent

problems. Nine soil subregions can be identified.

4. Cambisols Region

This soil region is widely distributed in warmt emperate North China and northern subtropical Central China. Cambisols are the dominant soils and are extensively distributed in hilly lands. Fluvisols and gleysols are also widely distributed in valley plains and intermontane basins. Under various conditions of parent materials, vertisols are developed in the northern Huaihe Plain, planosols in the mesopotamia area between the Huaihe River and the Changjiang River, and solonchaks along the coast. Vertical distribution of soils is conspicuous on middle and high mountains. For example, on the northern slopes of the Qinling Mountains from the basic belt (calcic cambisols) upward, chromic cambisols, eutric cambisols, orthic luvisols, rankers, and lithosols appear in succession. This is one of the most important agricultural regions in China, generally with three crops in two years north of the Qinling Mountains and two crops each year south of the Qinling Mountains. The latter is now the chief grain-producing area in China, yet flood control is a big problem. The former is confronted with problems of soil conservation (especially the Loess Plateau) as well as drought, flood, and

Table 5-3 Soil Geographical Regions and Subregions in China

Soil geographical regions	Soil geographical subregions
1. Rendzinas region	1A South China Sea Islands — rendzinas subregion
2. Ferralsols region	2A Middle-southern Hainan Island — plinthic ferralsols, chromic luvisols subregion
	2B Southern Taiwan Island — rhodic ferrasols, eutric gleysols subregion
	2C Leizhou-Hainan — xanthic ferralsols, eutric gleysols subregion
	2D Southern Yunnan — rhodic ferralsols, dystric nitosols, eutric gleysols subregion
3. Acrisols region	3A Northern Taiwan Island — orthic acrisols, eutric gleysols subregion
	3B South Nanling Mountains — ferric acrisols, eutric gleysols subregion
	3C Middle-southern Yunnan — ferric acrisols, rendzinas subregion
	3D South Changjiang Hills and Basins — plinthic acrisols, eutric gleysols subregion
	3E Guizhou Plateau — humic acrisols, rendzinas, eutric gleysols subregion
	3F Yunnan Plateau — ferric acrisols, eutric gleysols subregion
	3I Sichuan Basin — calcic cambisols, eutric gleysols subregion
	3H Sichuan bordering mountains — humic acrisols subregion
	3G Qinghai-Xizang Plateau southeastern margin — orthic acrisols, humic cambisols subregion
4. Cambisols region	4A Lower Changjiang Plain — eutric cambisols, eutric planosols, eutric gleysols subregion
	4B Qinling-Dabie mountains — eutric cambisols, orthic luvisols subregion
	4C Liaodong-Shandong peninsulas — eutric cambisols subregion
	4D North China Plain — calcaric fluvisols, eutric gleysols, solonchaks subregion
	4E North Huaihe Plain — pellic vertisols, calcaric fluvisols subregion
	4F North China Mountains — chromic cambisols subregion
	4G Loess Plateau — calcic cambisols subregion
5. Luvisols region	5A Da and Xiao Hinggan Mountains — orthic luvisols, pheozems subregion
	5B Changbai Mountains — orthic luvisols subregion
	5C Sanjiang Plain — mollic planosols, eutric histosols subregion
6. Podzoluvisols region	6A Northern Da Hinggan Mountains — dystric podzoluvisols subregion
7. Pheozems-chernozems region	7A Eastern Song-Liao Plain — haplic pheozems, planosols subregion
	7B Middle Song-Liao Plain — calcaric pheozems, mollic gleysols subregion
	7C Southern Da Hinggan Mountains — luvic chernozems, greyzems subregion
8. Kastanozems region	8A West Liaohe Plain — calcic and luvic kastanozems subregion
	8B Eastern Nei Mongol Plateau — haplic and luvic kastanozems subregion
	8C Eastern Ordos Plateau — calcic kastanozems, sandy regosols subregion
9. Xerosols-yermosols region	9A Western Nei Mongol Plateau — haplic and luvic xerosols subregion
	9B Western Ordos Plateau — luvic xerosols, calcaric gleysols subregion
	9C Qinghai Lake — luvic xerosols, haplic kastanozems subregion
	9D Alashan and Hexi Corridor — haplic yermosols, sandy regosols subregion
	9E Nomin Desert — gypsic yermosols subregion
	9F Northern Junggar Basin — Haplic xerosols, haplic kastanozems subregion
	9G Junggar Basin — luvic yermosols, sandy regosols subregion
	9H Ili Valley — luvic xerosols, haplic kastanozems subregion
	9I Eastern Xinjiang — gypsic yermosols subregion
	9J Tarim Basin — gypsic Yermosols, solonchaks, sandy regosols subregion
	9K Qaidam Basin — Haplic yermosols, solonchaks subregion
	9L Altay Mountains — kastanozems, greyzems, podzoluvisols subregion
	9M Junggar western mountains — luvic xerosols, kastanozems subregion
	9N Tianshan Mountains — yermosols, kastanozems subregion
	9O Qilian Mountains — xerosols, cambisols subregion
10. Alpine cambisols-xerosols region	10A Eastern Tibetan Plateau — Albic luvisols, humic cambisols subregion
	10B Southern Qinghai — gelic cambisols, gelic gleysols subregion
	10C Southern Qinghai-Xizang Plateau — eutric and calcic cambisols subregion
	10D Qinghai-Xizang Plateau — haplic xerosols, regosols subregion
	10E Northwestern Qinghai-Xizang Plateau — haplic yermosols, haplic xerosols subregion

salinization. Seven soil subregions can be identified.

5. Luvisols Region

This soil region is restricted to humid, temperate Northeast China. Luvisols are most widely distributed. Other commonly seen soils are pheozems, planosols, gleysols, histosols, and lithosols. Vertical zonation is conspicuous. For example, in the Changbai Mountains area, gleysols and histosols dominate in flood plains, low terraces, and intermontane valleys below 800 m in elevation; luvisols on hills and mountains up to 1100 to 1200 m in elevation; podzoluvisols on slopes between 1200 to 1900 m, and gelic gleysools and lithosols above 1900 m to the mountain tops. Most of these soils are good both for agriculture and for forestry; they are important commercial grain producers and timber producers in China. Irrigation, drainage and soil improvement can be major problems in the plains and reforestation and soil conservation are chief requirements in hilly lands. Three soil subregions are demarcated.

6. Podzoluvisols Region

Restricted to the humid cool-temperate northern Da Hinggan Mountains and the upper Heilong River valley. Podzoluvisols are distributed in most parts of the region, and gleysols and histosols are restricted to low valleys. This is one of the most important timber producing areas in China, and it will likely remain so. It includes only one subregion.

7. Phaeozems-Chernozems Region

This soil region is mostly distributed in temperate subhumid Northeast China. Major soils are phaeozems and chernozems, including also gleysols, planosols, histosols, and solonchaks. These are the most fertile agricultural soils and commercial grain producers in China. Overall planning and better management are much needed here. Soil conservation is also an urgent problem in farmlands with slopes greater than 3°. Three soil subregions can be identified.

8. Kastanozems Region

This soil region is mainly distributed in the western Northeast China Plain and the eastern Nei Mongol Plateau. Dominant soils are kastanozems (or chestnut soil). Sandy regosols are also widely distributed in the Mu Us Sandy Lands and other sandy lands, whereas gleysols are found in low depressions. Small patches of solonchaks and solonetz are scattered here and there. These are the most important pastoral soils in China, although considerable parts have been cultivated. A balanced and better utiliza-

tion of farmland and pasture are the most pressing problems here. Three soil subregions can be identified.

9. Xerosols-Yermosols Region

The xerosols and yermosols are extensively distributed in the desert-steppe zone and the desert zone respectively. Sandy regosols are also widely distributed in the Taklimakan Desert and other sandy deserts, whereas solonchaks and gleysols are found in low depressions with impeded drainage. Surrounding high mountains are characterized by vertical zonation of soils; kastanozems, chernozems, cambisols, greyzems, podzoluvisols, regosols, and gelic gleysols appear successively with increasing elevation. It is a region of great contrast. Without irrigation it becomes a vast expanse of wild desert; with irrigation it turns into luxuriant, productive oases. Drought, salinization, aeolian action, and coarse parent materials are four major limiting factors for soil productivity. Combating these limiting factors is the key problem of this soil region. There are 15 soil subregions that can be demarcated.

10. Alpine Cambisols-Xerosols Region

This soil region is distributed mainly on the lofty Qinghai-Xizang Plateau and its neighboring high mountains. One feature is the prominent vertical zonation of soils. Taking the western slope of the Marla Mountains. (near Qamdo in eastern Xizang) as an example, cambisols dominate the terraces in the Lancang River valley bottom at elevations between 3500 to 3700 m, luvisols prevail on valley slopes between 3700 to 4200 m, and rankers and lithosols on water divides at about 4300 m. Gleysols are found on mountain tops at about 4500 m, and regosols predominate on extremely high peaks above 4500 m. Another feature is the horizontal zonation from southeast to northwest, which coincides with decreasing precipitation; luvisols, cambisols, xerosols, and yermosols appear successively. On the whole, owing to high, rugged topography and a frigid climate, the soils are mostly suitable only for pasture. Small patches of farmlands are concentrated on gleysols in valley bottoms between 3500 to 4100 m in elevation. Five soil subregions can be tentatively demarcated.

References

[1] Chinese Society of Soil Science, 1979, *Collective Papers on Soil Classication and Soil Geography,* Zhejiang People's Press, Hangzhou. (In Chinese)
[2] FAO-Unesco, 1977, *Soil Map of the World, 1:5000000,* Paris. (in English, French, Spanish, and Russian)
[3] *Physical Regionalization of China* Working Committee, Chinese

Academy of Sciences, 1959, *Physical Regionalization of China: Soil Regionalization of China,* Science Press, Beijing. (In Chinese)

[4] *Physical Geography of China* Compilation Committee, Chinese Academy of Sciences, 1981, *Physical Geography of China: Soil Geography,* Science Press, Beijing. (In Chinese)

[5] Nanjing Institute of Soil Science, Chinese Academy of Sciences, 1978, *Soils of China,* Science Press, Beijing. (In Chinese)

[6] Thorp, J., 1936, *Geography of the Soils of China,* National Geological Survey of China, Nanjing.

[7] Institute of Forestry and Soil Science, Chinese Academy of Sciences, 1980, *Soils of Northeast China,* Science Press, Beijing. (In Chinese)

[8] Institute of Soil and Soil Conservation, Chinese Academy of Sciences, 1961, *Soils of North China Plain,* Science Press, Bei-

jing. (In Chinese)

[9] Integrated Investigation Team of Xinjiang, Chinese Academy of Sciences, 1965, *Soil Geography of Sinjiang,* Science Press, Beijing. (In Chinese)

[10] Integrated Investigation Team of Nei Mongol and Ningxia, Chinese Academy of Sciences, 1980, *Soil Geography of Nei Mongol.* Science Press, Beijing. (In Chinese)

[11] Nanjing Institute of Soil Science, Chinese Academy of Sciences, 1982, *Soil Resources of Heilongjiang Province,* Science Press, Beijing. (In Chinese)

[12] Nanjing Institute of Soil Science, Chinese Academy of Sciences 1977, *Soils and Phosphate Mining in the Xisha Islands,* Science Press, Beijing. (In Chinese)

Chapter 6

Biogeography of China

Thanks to its vast territory, dominantly temperate and subtropic climate, mountainous topography, and complicated geological history, the biological resources of China are rich and varied. The human impact on the biogeography of China has been particularly strong since 7000 B.P., usually working great havoc on natural vegetation and the animal world. One instance is the phenomenal transformation of the ancient heavily forested country, with forests probably occupying more than one-half of its total land area, to the rather poorly wooded modern China, with forests accounting for only 12 per cent of its total land area. Scarcity of timber resources has been one of the acute economic problems in China, and reforestation ranks as one of the most urgent measures for environmental protection and reconstruction. Since 1949, much work has been done to improve cultural vegetation and domesticated animal husbandry, as has been briefly outlined in Chapter 1.

VEGETATION

Vegetation is the sum total of all plant formations in an area. It is a major component as well as a reliable mirror of the physical geographical environment.

Evolution of China's Vegetation

Modern vegetation is an endproduct of a long, complicated geological evolution, including the relatively recent human impact. The origin of China's modern vegetation can be traced back as far as the early Tertiary period although at that time the paleoclimate was much warmer and the modern monsoon climate was not yet formed. The Chinese mainland was then dominated with low, flat terrain and planetary wind systems, that is, mostly northeastern trade winds, and the climatic-vegetational zonation was rather weakly developed. From north to south, there appeared six successive broad climatic-vegetational zones.

1) Warm temperate mixed needle-and broad-leaved forest — distributed extensively in Northeast China and eastern Nei Mongol.

2) Warm-temperate mixed needle-and broad-leaved forest (forest-steppe and scrub-steppe west of the Helan Mountains), containing a considerable amount of subtropical plants — distributed widely in North China and Northwest China.

3) Northern subtropical mixed needle-and broad-leaved forest, predominantly subtropical plants — distributed in the southern part of North China.

4) Middle subtropical park savanna — distributed extensively from northern Taiwan and the Fujian coast, northwestwardly to the lower Changjiang valley and Central China, then up to Xinjiang in Northwest China.

5) Southern subtropical mixed evergreen and deciduous broad-leaved forest — distributed mostly in South China.

6) Tropical rainforest — distributed on the southern coasts of the Chinese mainland and the islands in the South China Sea.

Since the Pliocene epoch, owing to the large-scale uplift of the Qinghai-Xizang Plateau, the modern monsoon climate has gradually developed. In addition, the northern limits of the warm-temperate and subtropical climates have receded southward considerably, probably by more than 7 to 10° of latitude. Both Central China and South China have become humid, whereas Northwest China has become more desiccated.

Since the late Pleistocene or early Holocene epochs, the formation of China's modern climate and vegetation has taken place. Three vegetation realms have been identified: (1) the eastern humid forest realm, (2) the northwestern arid steppe and desert realm, and (3) the Qinghai-Xizang frigid plateau vegetation realm. Throughout history human activities have greatly modified these vegetation realms, so that practically no more "natural" vegetation types still exist.

Distribution of China's Vegetation

The distribution of China's modern vegetation is shaped by both zonal and azonal factors. Horizontal zonal distribution of vegetation stands out clearly in China; from southeast to northwest, as the distance from the sea increases and, consequently, the precipitation decreases, vegetation changes gradually from forests to steppes to deserts. For example, along 42°N from east to west, there first appears the broad-leaved deciduous oak (*Quercus mongolica*) forest on the eastern hills and mountains of Northeast China where the climate is humid with an annual rainfall of 600 to 700 mm. Then, the terrain becomes forb grass steppe (mostly *Stipa baicalensis, Festuca ovina,* etc.) on the eastern Nei Mongol Plateau, where the climate is subhumid with an annual rainfall of 400 to 550 mm. This changes again to needle-grass steppe (mostly *Stipa grandis, S. krylovii,* etc.) on the central Nei Mongol Plateau, where the climate is semiarid with an annual rainfall of 260 to 350 mm. Further westward on the Alashan Plateau (the western part of the Nei Mongol Plateau), where the climate is arid with an annual rainfall of only 100 to 200 mm, a shrubby desert appears sparsely dotted with *Artemisia sphaerocephala, A. ordosica, Calligonum mongolicum,* and other shrubs or semi-shrubs. In the extremely arid "black gobi" area of eastern Xinjiang and western Gansu where the annual rainfall is less than 50 mm, the ground is mainly bare or consists of very sparsely covered semi-shrub desert, containing mostly *Iljinia regelii, Ephedra przewalskii, Reumuria soongarica,* and so on.

In the eastern humid forest realm — chiefly owing to areal differentiation of solar incidence and consequently a diversity of temperature conditions — there is also conspicuous latitudinally horizontal zonal distribution of vegetation. From north to south, the sequence of cool-temperate needle-leaved taiga forest (composed chiefly of *Larix gmelinii* and *Pinus sylvestris*), temperate mixed needle-and broad-leaved forest, warm-temperate broad-leaved deciduous and evergreen forest, subtropical broad-leaved evergreen forest, tropical monsoon forest and tropical rain forest appears. In the northwestern arid steppe and desert realm, desert vegetation is also diversified into temperate and warm-temperate zones, although horizontal zonation is not so conspicuous as in the eastern humid forest realm.

Vertical zonation of vegetation — chiefly owing to areal differentiation in elevation and, consequently, the redistribution of temperature and moisture conditions — is most outstanding on the lofty Qinghai-Xizang Plateau and its surrounding high mountains. It is closely interrelated with horizontal zonation; literally, vertical vegetation zonation is "stamped" with horizontal zonal features and vice versa. Vertical vegetation zonation in China might be first divided into two structural types: maritime and continental. Figure 6-1 shows the maritime type of vertical vegetation zonation on 10 different mountains with different latitudes and altitudes; all of them are located in the eastern humid forest realm. They have different elevations for each vertical zone as well as different compositions in flora; yet, all of them are dominated with forests and with alpine meadow and dwarf scrub on their upper parts. For example, the upper limit of montane needle-leaved forest has an elevation of less than 1000 m in the cool-temperate zone; in the temperate zone, it is 1100 to 1800 m. The limit is extended to between 2000 to 2600 m in the northern part of the warm-temperate zone and 2600 to 3500 m in its southern part; it is between 3000 to 4000 m in the subtropical zone, and 2800 to 3800 (4000) m on tropical mountains. The structure of vertical vegetation zones becomes successively more complex from north to south. There are only two to three vertical vegetation zones on cool-temperate mountains. This increases to four to five on temperate and subtropical mountains and to six to seven on tropical mountains.

Figure 6-2 shows the continental type of vertical vegetation zonation that exists mainly in the northwestern arid steppe and desert realm as well as the northwestern Qinghai-Xizang frigid plateau vegetation realm. It is characterized by the dominance of montane steppe or montane desert, with alpine kobresia meadow and cushion vegetation or eternal snow on mountain tops. The upper limit of each vertical vegetation zone has a higher elevation as aridity increases rapidly from east to west. For example, the upper limit of montane desert reaches 3800 m on the northern slope of the Altun Mountains and directly above it appears alpine steppe or cushion vegetation. On the northern slope of eastern Kunlun Mountains the upper limit of montane desert reaches 4100 m, with alpine desert and eternal snow immediately above it.

Azonal vegetation types that occur in each broad vegetation zone are mainly determined by microlandforms and soil conditions. Major azonal vegetation types include saline vegetation marsh, aquatic and sandy vegetation, etc.

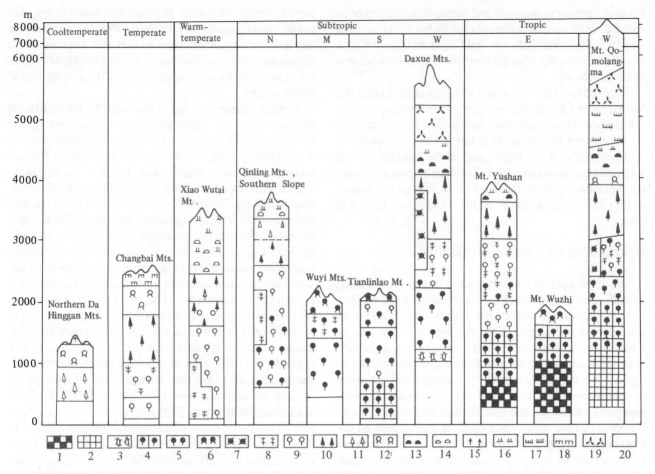

Figure 6-1 Vertical zonation of major vegetation types in China: martime type
(adapted from Prof. Hou Xueyu, et al)

1. Monsoon rain forest, rain forest; 2. Monsoonal rain forest; 3. Thorny scrub; 4. Monsoon evergreen broad-leaved forest;
5. Evergreen broad-leaved forest; 6. Evergreen broad-leaved dwarf forest; 7. Hard-leaved evergreen broad-leaved forest; 8.
Temperate needle-leaved forest; 9. Deciduous broad-leaved forest; 10. Cool-temperate evergreen needle-leaved forest; 11. Cool-
temperate deciduous needle-leaved forest; 12. Dwarf forest; 13. Sub-alpine evergreen scrub; 14. Subalpine deciduous broad-
leaved scrub; 15. Evergreen needle-leaved scrub; 16. Subalpine meadow; 17. Alpine Kobresia meadow; 18. Alpine tundra;
19. Sparse nival vegetation; 20. Eternal snow.

Major Vegetation Types of China

The vegetation types in China are quite varied and complicated, including nearly the whole array of vegetation types in the world, with the exception of tundra. Another feature is the extensively distributed alpine vegetation on the Qinghai-Xizang Plateau. According to a recent investigation, there are 29 major natural vegetation types, 52 subtypes and more than 600 formations in China. We will introduce briefly the major types and subtypes (see Table 6-1).

I. Needle-Leaved and Mixed Needle and Broad-Leaved Forests

These are the most important timber-producing areas in China, distributed widely in the cool-temperate, temperate, subtropical and tropical zones. They are mainly composed of *Abies, Picea, Pinus, Larix, Tsuga, Cupressus, Juniperus, Sabina, Cunninghamia,* and other genera. They may be classified into the following subtypes:

I(1) Cool-temperate needle-leaved forest This is the zonal vegetation of the cool-temperate zone, mainly

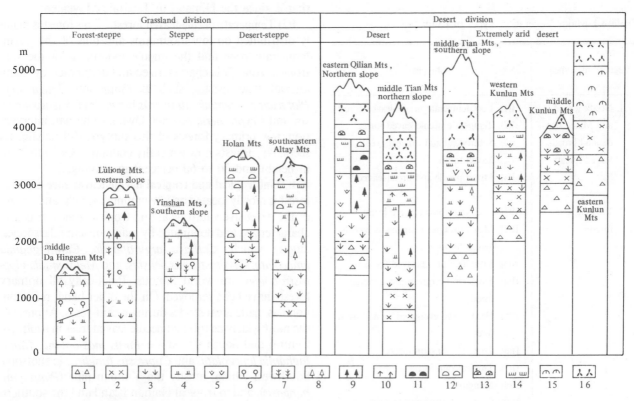

Figure 6-2 Vertical zonation of major vegetation types in China: Continental type
(adapted from Prof. Hou Xueyu, et al)

1. Semishrub desert; 2. Artemisia desert; 3. Montane grassland; 4. Alpine grassland; 5. Deciduous broad-leaved forest; 6. Deciduous broad-leaved forest; 7. Temperate needle-leaved forest; 8. Cool-temperate deciduous needle-leaved forest; 9. Cool-temperate evergreen needle-leaved forest; 10. Evergreen needle-leaved scrub; 11. Subalpine evergreen scrub; 12. Deciduous broad-leaved scrub; 13. Alpine cushion vegetation; 14. Alpine Kobresia meadow; 15. Alpine desert; 16. Sparse nival vegetation.

distributed on northern Da Hinggan Mountains (elevation: 400 – 1400 m) and on northeastern Altay Mountains (elevation: 1300 – 2100 m). It is also the southernmost extension of the taiga forest. In other parts of China, it serves as an important component of vertical vegetation zonation on high mountains, with elevations of its upper limit increasing from north to south and east to west. On the subalpine southeastern Qinghai-Xizang Plateau, it is particularly widely distributed. In extremely arid Northwest China, however, its distribution is usually intermittent and sometimes nonexistent.

The Chinese cool temperate and needle-leaved forest contains more numerous and varied species than the same forest type in other parts of the world. It is dominated by larch (*Larix gmelini*) on northern Da Hinggan Mountains and by fir (*Abies sibirica*), Picea (*Picea obovata*) and pine (*Pinus sibirica*) on the northwestern Altay Mountains. On the eastern hills and mountains of Northeast China, the dominant species are the *Abies nephrolepis, Picea jepoen-*

sis, and *P. koraiensis.* The mountains of North China and Nei Mongol are characterized by *Picea meyeri, P. wilsonii,* and *Larix principis-ruprechtii;* on the Qinling and Daba Mountains, there is mostly *Abies Chinesiensis, A. fargesii, A. sutchuenensis,* and *Larix chinensis.* The arid mountains in Northwest China contain the drought-tolerant *Picea asperata, P. neoveitchii, P. crassifolia, P. schrenkiana,* and *Sabina przewalskii.* On the southeast of the Qinghai-Xizang Plateau, there are many endemic species of fir, such as *Abies faxoniana, A. ernestii, A. squamata, A. fabri, A. georgei,* and also *Picea likiangensis, Pinus densata, Larix potaninii,* and *Sabina tibetica.* In the south of the Qinghai-Xizang Plateau and the Himalayas, there are many species that originated during the Quaternary or postglaciation period, such as *Abies spectabilis, Picea smithiana, Larix griffithiana,* and *Sabina wallichiana.* In the Taiwan Mountains, two endemic species (*Abies kawakamii* and *Picea morrisonicola*) are distributed on lower slopes between 2800 and 3300 m. Probably they have originated only recently,

Table 6-1 Major Vegetation Types of China

Vegetation types	Subtypes
I. Needle-leaved and Mixed nee-dle and broad-leaved forests	I(1) Cool-temperate needle-leaved forest
	I(2) Temperate needle-leaved forest
	I(3) Subtropical and tropical needle-leaved forest
	I(4) Mixed needle and broad-leaved forest
II. Broad-leaved forests	II(1) Deciduous broad-leaved forest
	II(2) Mixed evergreen deciduous broad-leaved forest
	II(3) Evergreen broad-leaved forest
	II(4) Monsoon evergreen broad-leaved forest
	II(5) Hard-leaved evergreen broad-leaved forest
	II(6) Tropical monsoon forest
	II(7) Tropical rain forest
	II(8) Mangrove
	II(9) Coral reef evergreen forest
III. Scrubs	III(1) Alpine evergreen and deciduous scrub
	III(2) Temperate deciduous scrub
	III(3) Subtropic deciduous scrub
	III(4) Evergreen broad-leaved scrub
	III(5) Evergreen thorny scrub
IV. Grassland and park savanna	IV(1) Steppe
	IV(2) Meadow-steppe
	IV(3) Desert-steppe
	IV(4) Alpine grassland
	IV(5) Park savanna
V. Deserts	V(1) Semi-arboreous desert
	V(2) Shrub desert
	V(3) Semishrub, small semishrub desert
	V(4) Small semishrub desert
VI. Alpine perma-frost and cushion vegetation	VI(1) Alpine permafrost vegetation
	VI(2) Alpine cushion vegetation
VII. Meadows and Swamps	VII(1) Typical meadow
	VII(2) Alpine meadow
	VII(3) Swampy meadow
	VII(4) Saline meadow
	VII(5) Scrub-forb swamp
	VII(6) Forb swamp

that is since the Himalayan Tectonic Movement.

I(2) Temperate needle-leaved forest   This forest is mainly distributed on low mountains and hills of the warm-temperate zone and the middle mountains of the subtropical zone. It is characterized and dominated by many ancient relic species, such as *Pinus* spp. *Tsuga* spp., *Platycladus orientalis, Juniperus rigida, Pseudotsuga sinensis,* and *Cryptomeria fortinei.* Owing to human intervention, few primary forests of this subtype still remain. Its secondary forests now generally contain trees 10 to 25 m in height and 30 to 60 per cent in coverage.

I(3)Subtropical and tropical needle-leaved forest   This forest is mostly located on low mountains, hills, and valleys of the extensive subtropical and tropical zones. It is also characterized and dominated by many ancient relic species, among which, *Cathaya argyrophylla, Cunninghamia lanceolata, Glyptostrobus pensilis,* and *Metasequoia glyptostroboides* are world famous. Practically all primary forests have been removed. On the other hand, reforestation has made great strides during recent years. At present, the most widely distributed needle-leaved trees in subtropic Central and South China are *Pinus massoniana, Cunninghamia lanceolata,* and *Cupressus funebris;* in subtropic Southwest China, mostly the Yunnan pine (*Pinus yunnanensis*), and in tropical Hainan Island and the southern Yunnan low valleys, mostly *Pinus latterii* and *P. khasya.* On the whole, this subtype has a tree-height of 15 to 30 m and a coverage of 50 to 70 per cent.

I(4) Mixed needle and broad-leaved forest   This is a transitional type between cool-temperate needle-leaved forest and temperate deciduous broad-leaved forest. It is distributed (1) as a horizontal zonal vegetation type on the eastern mountains and hills of Northeast China with elevations from 700 to 1100 m, and (2) as a vertical zonal vegetation type on elevations lower than the subtropic needle-leaved forest. The dominant species are two needle-leaved trees, the famous timber red pine (*Pinus koraiensis*) and *Abies holophylla,* as well as several important hardwoods — for example, *Tilia amurensis, Betula costata, Carpinus cordata, Fraxinus mandschurica, Phellodendron amurense,* and *Juglans mandschurica.*

II. Broad-Leaved Forests

These are the most extensively distributed forest types in China and also feature the most varied subtypes. They form zonal vegetation types, both horizontal and vertical, in warm-temperate, subtropical, and tropical zones of Eastern Monsoon China. In Northwestern Arid China, they are represented by the riparian forest.

II(1) Deciduous broad-leaved forest   This is the horizontal zonal vegetation type of the warm-temperate humid and subhumid region as well as the vertically zonal vegetation type of the subtropics, tropic middle mountains, and the lower parts of high mountains. Elevations of their upper limit increase from east to west. For example, these forests are mainly distributed on plains and hills below 400 m in the Liaodong Peninsula, on low mountains and hills below 1200 m in the Taihang Mountains, and below 1800 m on the Loess Plateau, whereas on the northern slope of the Qinling Mountains, they lie between 2000 to 2600 m. The dominant species in North China are oaks (*Quercus liaotungensis* and *Q. mongolica*). The riparian forest in Northwest China is dominated with species of *Betula*, *Populus*, and *Salix*, although in cultural vegetation the "sand date" (*Elaeagnus angustifolia*) is also important.

II(2) Mixed evergreen and deciduous broad-leaved forest This is the horizontal zonal vegetation of the northern subtropical zone, a transitional type between deciduous and evergreen broad leaved forest. The dominant evergreen species are *Cyclobalanopsis glauca*, *C. gracilis*, *C. multinervis*, *C. myrsinaefolia*, *Castanopsis sclerophylla*, *C. platycantha*, *Lithocarpus cleistocarpa*, *Schima superba*, and *Ilex chinensis*. The dominant deciduous species include *Fagus longipetiolata*, *F. lacida*, *Platycarya strobilacea*, *Corylopsis sinensis*, *Liquidambar formosana*, *Aphananthe aspera*, and *Davidia involucrata*. Some of these are species endemic to China. This subtype has usually a tree height of 10 to 30 m and a coverage of 80 to 90 per cent.

II(3) Evergreen broad-leaved forest This is the zonal vegetation type of the middle subtropic zone. Elevation of its upper limit increases from less than 200 to 1000 m in the east to about 1500 to 1800 m in the west. The composition species much more complex than the above-mentioned forest subtypes, and most of them are species endemic to China, such as *Castanopsis sclerophylla*, *C. fargesii*, *Cyclobalanopsis oxyodon*, *Lithocarpus chrysocoma*, *Lithocarpus chrysocomus*, *Schima superba*, *S. argentea*, *Machilus thunbergii*, *Monglietia fordiana*, and *Michelia* spp. This subtype has a tree height of 15 to 30 m and a coverage of 60 to 90 per cent.

II(4) Monsoon evergreen broad-leaved forest This is the transitional vegetation type between subtropical evergreen broad-leaved forest and tropical monsoon and rain forest. It is mainly distributed on low mountains of the southern subtropic zone and on middle mountains with elevations between 1000 to 1500 m in the tropical zone. The dominant species are abundant both in variety and quantity, of which *Castanopsis hystrix* is the most important and most widely distributed.

II(5) Hard-leaved, evergreen broad-leaved forest This is a peculiar vegetation type restricted to the western part of the subtropic zone, with elevations between 2000 to 3900 m in the south and between 2900 to 4300 m in the north. On montane slopes, the dominant species are *Quercus aquifolioides*, *Q. pannosa*, *Q. quyavaefolia*, and *Q. semecarpifolia*. In the hot, dry valleys the species are mostly dwarf-ed *Q. cocciferoides*, *Q. franchetii*, and *Q. senescens*.

II(6) Tropical monsoon forest As China's tropic area is mostly restricted to the northern margin of this extensive zone, the zonal vegetation type is more typically monsoon forest than rain forest. On hills and terraces lower than 500 m in China's eastern tropic zone, the evergreen monsoon forest is widely distributed. Its dominant species include several kinds of banyan trees (*Ficus microcarpa*, *F. locor*, *F. altissima*) and *Vitica astrotricla*, *Hopea chinensis*, *Chukrasia tabularis*, *Mesua ferrea*, *Gironniera cuspidata*, and *Sterculia*. In southern Yunnan and western Hainan Island, hot dry valleys with elevations between 700 to 1200 m are dominated by deciduous monsoon forest.

II(7) Tropical rain forest The upper limit of rain forest is below 500 m in China's eastern tropic zone, ascending to about 700 m in southwestern Yunnan and as high as 1000 m on the southern slopes of the Himalayas. The following species are most commonly seen: *Hopea hainanensis*, *H. mollissima*, *Dipterocarpus tonkinensis*, *D. pilosus*, *Myristica cagayanensis*, *M. simiarum*, *Pterospermum niveum*, *Reevesia thyrsoidea*, *Heritiera parvifolia*, *Artocarpus chaplasha*, and *Crypteronia paniculata*. In sheltered valleys with a conspicuous dry season, the "seasonal" rain forest is well developed and contains trees, such as the *Terminalia myriocarpa*, *Pometia tomentosa*, *Pterospermum lanceaefolium*, *Autiaris toxicaria*, *Ponteria grandifolia*, and *Parashorea chinensis*. On montane slopes between 500 to 1500 m, a "montane" rain forest has developed.

II(8) Mangrove The mangrove evergreen forest is mainly distributed along the clayey coast of Guangdong and Fujian provinces. The northernmost district along the Fujian coast, Fuding, (27°20'N) is its northern limit of distribution. The dominant species include *Bruguiera gymnorrhiza*, *B. sexangula*, *Ceriops tagal*, *Kandelia candel*, *Rhizophora apiculata*, *R. stylosa*, *Aegiceras corniculatum*, *Avicennia marina*, *Sonneratia caseolaris*, and *Nypa fruticans*. Owing to the marginal location of China's tropical zone and the heavy intervention of human activities, the mangrove evergreen forest is generally not well developed. It exists only as scrub forest with trees less than 5 m in height; nevertheless, in some favorable locations along the Hainan coast, mangrove trees may grow as high as 10 to 15 m.

II(9) Coral reef evergreen forest This peculiar vegetation type is mainly distributed on coral islands and islets of southern Taiwan Island and the South China Sea archipelago. It is rather poor in species, with *Pisonia grandis* and *Guettard speciosa* the most commonly seen trees, and *Scaevola servicea*, *Tournefortia argentea*, *Pemphis acidula*, and *Clerodendrum inerme* the dominant shrubs. These trees and shrubs adorn the coral islands with a beautiful greenish or silvery-green mantle.

III. Scrub

This vegetation type is distributed in China, especially in temperate forested areas, although most of the scrub formations do not represent a zonal vegetation type. The scrub vegetation might be divided into the following subtypes.

III(1) Alpine evergreen and deciduous scrub This is a vertical zonal vegetation subtype distributed widely on high mountains of temperate, subtropical, and tropical zones. Alpine evergreen needle-leaved scrub is mostly located on the Qinghai-Xizang Plateau and its surrounding high mountains, with elevations between 3500 m and 4800 m. It usually coexists with alpine meadow or alpine grassland. In vertical zonation, it is sandwiched between cool temperate needle-leaved forest and eternal snow. The dominant scrub species on the southern and southeastern Qinghai-Xizang Plateau include *Sabina squamata, S. pingii* var. *wilsonii,* and *S. wallichiana.* On the Tianshan Mountains and the Altay Mountains, the main species are *S. pseudosabina* and *Juniperus sibirica,* whereas at higher elevations of different high mountains, the predominant species are *Rhododendron* spp. and *Cassiope* spp.

Alpine deciduous scrub has an elevation ranging from 3000 to 5000 m. The dominant shrub species are *Betula rotundifolia, Caragana jubata, Desiphora fruticosa, D. globra,* and *Salix cupularis.* Both alpine evergreen and deciduous scrub have a coverage as high as 70 to 90 per cent.

III(2) Temperate deciduous scrub This is widely distributed on the Qinghai-Xizang Plateau and warm-temperate forested areas, mostly in the form of secondary vegetation after deforestation. Yet along river channels in the desert zone or in a saline soil habitat, it is the long-established primary vegetation. There are about 30 dominant species, including *Rosa* spp., *Spiraea* spp., *Prunus* spp., *Cotoneaster* spp., *Sorbus* spp., *Lespedeza* spp., *Campylotropis* spp., *Berberis* spp. and *Corylus heterophylla, Ostryopsis davidiana, Myriopnois dioica,* and *Forsythia suspensa.* The yellow rose (*Rosa hugonis*) is the characteristic species on the Loess Plateau, with scrub height of 1 to 3 m and coverage above 70 per cent. In drier habitats, especially in the intermontane valleys of Northwest China, drought-resistant thorny scrubs, such as *Caragana* spp., *Sophora vicifolia,* and *Spiraea hypericifolia* are more commonly seen.

III(3) Subtropic deciduous scrub This is a secondary vegetation type that has arisen after continuous deforesting in the subtropic forested areas. The dominant scrub species in the karst region, include *Coriaria sinica, Vitex negundo, Rosa cymosa,* and *Pyracantha fortuneana.* In dry hot valleys, the main species are *Phyllanthus emblica, Sageretia pycnophylla,* and *Bauhinia faberi* var. *microphylla.* If the deforesting process stops, this vegetation type will soon be restored to an evergreen broad-leaved forest.

III(4) Evergreen broad-leaved scrub This is also a secondary vegetation type, distributed widely on hills and low mountains of tropical and subtropical zones. The dominant species in the middle subtropic zone are *Leropetalum chinense, Vaccinium bracteatum,* and *Rhododendron simsii;* in the southern subtropic zone and northern part of the tropic zone are *Rhodomyrtus tomentosa, Baeckea frutescens, Helicteres angustifolia,* and *Macaranga denticulata.* In karst areas of Guangxi and Guizhou, the dominant species are particularly rich and varied and include *Viburnum cinnamomifolium, Fanthoxylum planispinum, Pleomele cambodiana, Myrsine africana,* and *Pistacia weinmannifolia.*

III(5) Evergreen thorny scrub On the sandy beaches of China's tropic coast, evergreen thorny scrub vegetation is distributed. The dominant species are *Pandanus tectorius, Atalantia buxifolia* and *Opuntia dillenii.*

IV. Grassland and Park Savanna

Extensively distributed in China, grassland and park savanna occupy about 30 per cent of the total land area. The immense Nei Mongol Plateau has been the home of many nomad peoples ever since Neolithic times, and its landscape is aptly described in Mongolian folklore, "The sky is boundless, the earth is endless; when greenish grass bows before the whistling wind, innumerable cattle and sheep are seen." This area was, during the thirteenth century, the homeland of Genghis Khan.

IV(1) Steppe Temperate steppe is widely distributed in the western part of Northeast China, most parts of Nei Mongol, the northern part of the Loess Plateau, the central part of the Qinghai-Xizang Plateau, and the montane areas of North China. It is the eastern extension of the immense Eurasian grassland. It is also the most important pastureland in China.

Steppe is the zonal vegetation type of the temperate semiarid region. It is characterized by bunch grass. The dominant species are *Stipa grandis, S. krylovii* (both being typical of the eastern part), *S. capillata, Festuca sulcata* (both being typical of the western part), *Cleistogenes squarrosa, Koeleria cristata,* and *Agropyron cristatum.* As a whole, it has a coverage of 30 to 60 per cent, and an annual grass production of 2000 to 4500 kg/ha.

IV(2) Meadow-steppe This is a transitional vegetation type between grassland and forest and is mainly distributed in western Northeast China and eastern Nei Mongol with *Stipa baicalensis, Aneurolepidium chinense,* and *Filifolium sibiricum* as the dominant species. In the arid Junggar Basin, it is located high up montane slopes of 1400 to 1600 m in elevation as a vertical zonal vegetation type, and with *Stipa kirghisorum* and *Aneurolepidium angustum* as the dominant species. This is probably the best natural pastureland in China, with a coverage of 60 to 80 per cent and an annual grass production of 4000 to 6000 kg/ha.

IV(3) Desert-steppe This is a transitional vegetation type between grassland and desert, with mixed bunch grass and small semishrubs as the dominant species. These species include *Stipa gobica, S. breviflora, S. glareosa, S. orientalis, S. caucasica, Hippolytia trifida, Ajania achilleoides,* and *A. fruticulosa.* It has generally a moderate coverage of 20 to 40 per cent and an annual grass production of 1000 to 2000 kg/ha. However, owing to the high protein content of its dominant species, it is still a good pastureland and is especially fit for sheep and goats.

IV(4) Alpine grassland This is mainly distributed on the lofty Qinghai-Xizang Plateau and high mountains above 2700 to 3500 m in elevation. It is characterized by cold and drought-resistant small bunch grass, *Carex* spp. and small semishrubs. The most commonly seen species are *Stipa purpurea, S. subsessiliflora, Festuca kryloviana, F. pseudovina,* and *Carex moorcroftii.* On the northern Tibetan Plateau and the northern slopes of the Himalayas, there are also *Artemisia salsoloides* var. *wellbyi, A. minor,* and *A. younghusbandii.* This vegetation subtype has generally a coverage of 20 to 80 per cent and an annual grass production of 400 to 500 kg/ha. It serves as good late summer and autumn pastureland.

IV(5) Park savanna This is a vegetation subtype of the semiarid tropical and subtropical zones and is restricted to small patches in some dry, hot valleys or gorges in Southwest China and South China. It is characterized by tall grass, with scattered drought-resistant shrubs, and isolated trees. It is mostly a secondary growth after deforestation of a tropical monsoon forest and other primary vegetation types. The dominant grass species are *Heteropogon contortus, Aristida chinensis,* and *Eragrostis elongata;* these are generally 30 to 80 cm in height, with a coverage of 80 to 90 per cent. The shrubs and trees are mostly composed of *Woodfordia fruticosa, Wendlandia scabra, Acacia farnesiana,* and "wood cotton" (*Bombax malabarica*). Seasonal variation is conspicuous, with meager yellow vegetation in the dry season and luxuriant green growth in the rainy season. Once harmful human intervention is stopped or reduced, these areas will likely be restored to evergreen scrub or tropical monsoon forests.

V. Deserts

Widely distributed in Northwest China, desert vegetation types occupy about 22 per cent of China's total land area. They are characterized by low coverage (usually less than 20 to 30 per cent or even less than 1 to 5 per cent), and by poverty of species and simplicity in structure, with drought-resistant shrubs, semishrub and small semishrubs predominating. They may be subdivided into the following sub-types.

V(1) Semi-arboreous desert This vegetation subtype, sometimes called "desert forest", is mainly composed of *Haloxylon ammodendron* and *H. persicum,* generally with a tree height of 3 to 5 m and a coverage of 30 to 50 per cent. The distribution of *H. persicum* is restricted to the Junggar Basin. Together with *H. ammodendron,* it fixes shifting sand dunes effectively. Therefore, shifting sands occupy less than 3 per cent of Xinjiang's Gurbantünggüt Desert; this is quite different from the Taklimakan Desert, where shifting sands account for more than 85 per cent of the area.

V(2) Shrub desert This is the most widely distributed zonal desert vegetation in China, and endures severe natural conditions. The dominant species include *Ephedra przewalskii, Zygophyllum xanthoxylum, Nitraria sphaerocarpa, N. roborowskii, Calligonum mongolicum,* and *C. leucoeladum.* Generally, shrubs are less than 1 m in height and less than 5 to 20 per cent in coverage.

V(3) Semishrub, small semishrub desert The deciduous semishrub desert is extensively distributed on low mountains and gravel gobi, with *Reumuria soongarica, sympegma regelii, Iljinia regelii,* and *Ceratoides latens* as the dominant species. The halophytic semishrub desert, sometimes called "salt desert", is distributed chiefly on saline depressions, with *Halostachys belangeriana, Halocnemum strobilaceum, Kalidium cuspidatum, K. foliatum, Atriplex cana, Suaeda physophora,* and *S. microphylla* predominating. The Artemisia desert is located on sandy land and alluvial-diluvial fans and rises to 2600 to 3200 m on the southern slopes of the Kunlun Mountains.

V(4) Small semishrub desert This is a desert located on the lofty Qinghai-Xizang Plateau and its surrounding high mountains, with elevations between 4000 to 5500 m. It is characterized by cushion small semishrubs, such as *Ceratoides campacta* and *Ajania tibetica.* They are usually less than 5 to 15 cm in height and have less than 10 to 30 per cent in coverage.

VI. Alpine Permafrost and Cushion Vegetation

VI(1) Alpine permafrost vegetation On the upper parts of the Da Hinggan, Changbai, and Altay mountians, alpine permafrost vegetation has developed. It is usually less than 10 to 20 cm in height and has coverage of less than 20 per cent. The dominant species are *Vaccinium vitis-idaea, Rhododendron xanthostephanum, Phyllodoce caerulea, Dryas octopetala,* and *Salix tschanbaischanica,* which are found on the Changbai Mountains at elevations above 3000 m. *Drepanocladus* spp., *Hygrohypnum luridum, Bryum* spp., and Anomobryun filiforme are found on the Altay Mountains at elevations above 2100 m.

VI(2) Alpine cushion vegetation This is extensively distributed on the Qinghai-Xizang Plateau, Tianshan

Mountains, and other high mountain slopes immediately below the eternal snow line at elevations of between 3200 to 5500 m. The dominant species are many kinds of cushion perennial grass or small shrubs, such as *Artemisia musciformis, Thylacospermum caespitosum, Androsace tapete, A. squarrosula, Sibbaldianthe tetrardra,* and *Acantholimon hedinii.* Most of them are about 10 to 20 cm in height and 20 to 30 cm in diameter.

VII. Meadows and Swamp

These are azonal vegetation types, whose habitat is determined chiefly by high ground water tables or flooded environments. Yet they also show conspicuous areal differentiation and zonal "stamp".

VII(1) Typical meadow   This is located either on cool-temperate plateaus and high mountains or on lowlands and seacoasts with a high ground water table. Typical meadow is mainly distributed on lowlands of temperate forest and grassland zones as well as on montane slopes of desert and subtropic forest zones. The dominant species in the forest-meadow of Northeast China are Sanguisorba officinalis, Potentilla fragarioides, Artemisia laciniata, and Vicia pseudo-orobus. There are usually more than 20 to 30 plants in each square meter, sometimes even more than 50 plants. The specimens are generally 30 to 50 cm in height with a coverage of 60 to 100 per cent. During summer, many flowers bloom abundantly, which has given rise to the local name "multiflower grass meadow". This meadow has a high capability for both agricultural and pastoral development.

VII(2) Alpine meadow   This is mainly distributed above the alpine grassland belt at an elevation of between 3200 to 5200 m. It is the zonal vegetation type of the eastern Qinghai-Xizang Plateau, characterized by different species of *Kobresia,* such as *K. pygmaea, K. humilis,* and *K. capillifolia.* On the Tianshan Mountains Altay Mountains, and other high mountain systems, the dominant species are *Carex atrata, C. melanantha, Polygonum viviparum, P. sphaerostachyum,* and *Saxifraga* spp. The alpine meadow usually has a height of only 3 to 5 cm but a coverage of 30 to 80 per cent.

VII(3) Swampy meadow   This is an intermediate vegetation type between swamp and meadow and is located mainly in wet, low depressions. The dominant species are *Kobresia tibetica, K. littledalei, K. pamiroalaica, Carex atrofusca,* and *C. caespitosa.* This vegetation type has usually a height of 10 to 30 cm and a coverage of 60 to 90 per cent.

VII(4) Saline meadow   This is mainly distributed on saline soil areas, with a ground water table of less than 1 to 3 m in depth. The dominant species are salt-tolerant graminoids and forbs, such as *Achnatherum splendens, Aneurolepidium dasystachys, Aeluropus littoralis* var. *sinensis, Phragmites communis, Trachomitum lancifolium,* and *Iris lactea chinensis.* This vegetation type has usually a height of 40 to 60 cm and a coverage of 20 to 70 per cent.

VII(5) Scrub-forb swamp   On wet or flooded lowlands in the Da and Xiao Hinggan Mountains and Changbai Mountains scrub-forb swamp has developed. The predominant shrubs are *Ledum palustre* var. *angustum, Vaccinium uliginosum,* and *Betula fruticosa;* the predominant forbs are *Carex* spp., *Sphagnum* spp., and *Eriophorum vaginatum.*

VII(6) Forb swamp   This is widely distributed on heavily overwet or flooded lowlands, particularly on the Sanjiang Plain in Northeast China and the Zoigê area in the eastern Qinghai-Xizang Plateau. Here, even shrubs cannot survive; the ground is densely covered with *Carex* spp. and other forbs, usually with a height of 30 to 50 cm and a coverage of 60 to 80 per cent. If adequately drained, this area can be turned into good cropland.

ZOOGEOGRAPHY

As stated above, the faunal resources of China are quite rich and their areal differentiation conspicuous. Owing to limited space in this textbook, we will discuss terrestrial vertebrates only.

According to a preliminary survey, there are more than 2000 species of terrestrial vertebrates in China, accounting

Table 6-2 Number of Species, Genera, and Families of Terrestrial Vertebrates in China and in the World

Terrestrial vertebrates	China			World		
	families	genera	species	families	genera	species
Amphibians	10	34	196	18	>300	2800
Reptiles	21	105	315	42	>758	5700
Birds	81	392	1166	156	?	8590
Mammals	44	183	414	122	1017	4237
Total	**156**	**714**	**2091**	**338**	**?**	**21327**

for about 10 per cent of the total terrestrial vertebrate species in the world (Table 6-2). Many of them are endemic or principally distributed in China, such as the red-crowned crane (*Grus japonensis*); long-tailed pheasant (*Syrmaticus* spp.), mandarin duck (*Aix galericulata*), and the eared pheasant (*Crossoptilon*). The endemic mammals include the golden monkey (*Rhinopithecus* spp.), plum-flower deer (Cervus nippon), tufted deer (*Elaphodus cephalophus*), and takin (*Budorcas taxicolor*). Some of them are post-Quaternary glaciation relic species, such as the world-famous giant panda (*Ailuropoda melanoleuca*) in the northern *Hengduan Mountains* of Southwest China and the wild horse (*Equus przewalskii*) and wild camel (Camelus bactrianus) in the desert zone of Xinjiang. Two peculiar reptiles, the Chinese alligator (*Alligator sinensis*) and the crocodile lizard (*Shinisaurus crocodilurus*) in the middle and lower Changjiang River are other examples. These rare terrestrial vertebrates have now been carefully protected by the Chinese government according to the following categories: (1) hunting absolutely prohibited, (2) hunting strictly prohibited, (3) hunting restricted (see the map — Distribution of Rare Terrestrial Vertebrates in China and Table 6-3).

Based on areal distribution of major terrestrial vertebrates, six zoogeographical realms have been identified world wide. They are: Paleoarctic, Neoarctic, Paleotropic, Oriental, Neotropic, and Australian. In China, two realms are represented: north of the Qinling Mountains-Huaihe River line is the Paleoarctic realm; south of that line is the Oriental realm.

Origin and Evolution of China's Terrestrial Vertebrates

The origin of modern terrestrial vertebrates in China can be traced back to the Pliocene's three-toed horse (*Hipparion*), which was at that time distributed widely in most parts of Eurasia and Africa. Within its extensive distribution area, faunal areal differentiation was not conspicuous. By the time of the Pliocene epoch, most currently existing animals in China had already appeared, including primarily grassland-inhabitating species in the north and arboreal species in the south.

During the late Tertiary and early Quaternary periods, the large-scale uplifting of the Qinghai-Xizang Plateau and its surrounding high mountains played an important role in the environmental change and areal differentiation of fauna in China. Two great faunal groups were thus formed: the Gigantopithecus fauna in the south and the Nihowan fauna in the north, comprising features of the Oriental realm and Paleoarctic realm respectively. In the middle Quaternary period, these faunal differentiations were further enhanced. The former developed into the Ailuropoda-Stegodon fauna that were widely distributed

Table 6-3 A List of Rare Terrestrial Vertebrates in China (See the map — Distribution of rare terrestial vertebrates in China)

1. *Rangifer tarandus* (reindeer) 驯鹿
2. *Alces alces* (moose) 驼鹿
3. *Moschus moschiferus* (musk deer) 麝
4. *Gulo gulo* (wolverine) 狼獾
5. *Martes zibellina* (sable) 紫貂
6. *Panthera tigris* (tiger of Northeast China) 东北虎
7. *Cygnus* spp. (swan) 天鹅
8. *Grus japonensis* (red-crowned crane) 丹顶鹤
9. *Aix galericulata* (mandarin duck) 鸳鸯
10. *Cervus nippon* (Sika deer, plum-flower deer) 梅花鹿
11. *Naemorhedus goral* (green goat; goral) 青羊
12. *Otis tarda* (great bustard) 大鸨
13. *Procaprra gutturosa* (mongolian gazelle) 黄羊
14. *Gazella subgutturosa* (goitered gazelle) 鹅喉羚
15. *Mustela erminea* (ermine) 扫雪
16. *Equus przewalskii* (wild horse) 野马
17. *Equus hemionus* (wild ass) 野驴
18. *Camelus bactrianus* (wild camel, Bactrian camel) 野骆驼
19. *Saiga tatarica* (saiga antelope) 赛加羚羊
20. *Ovis ammon* (Big horn sheep) 盘羊
21. *Martes foina* (stone marten) 石貂
22. *Tetraogallus tibetanus* (Tibetan snowcock) 高原雪鸡
23. *Crossoptilon* spp. (eared pheasant) 褐（白）马鸡
24. *Castor fiber* (beaver) 河狸
25. *Megalobatrachus davidianus* (giant salamander) 大鲵
26. *Macaca mulatta* (Rhesus macaque) 猕猴
27. *Rhinopithecus roxellanae* (golden-hair monkey) 金丝猴
28. *Cervus albirostris* (white-lip deer) 白唇鹿
29. *Pseudois nayaur* (blue sheep) 岩羊
30. *Pantholops hodgsoni* (Tibetan antelope) 藏羚羊
31. *Bos grunniens* (wild yak) 野牦牛
32. *Grus nigricollis* (black-necked crane) 黑颈鹤
33. *Tragopan* spp. (tragopan pheasant) 角雉
34. *Ailuropoda melanoleuca* (giant panda) 大熊猫
35. *Ailurus fulgens* (lesser panda) 小熊猫
36. *Budorcas taxicolor* (takin) 羚羊
37. *Bos gaurus* (wild cattle; gaur) 野牛
38. *Nycticebus coucang* (slow loris) 懒猴
39. *Presbytis* spp. (white-headed langur) 叶猴
40. *Hyiobates concolor* (crested gibbon) 黑长臂猿
41. *Elephas maximus* (wild elephant; Indian elephant) 野象
42. *Buceros bicornis* (great pied hornbill) 双角犀鸟
43. *Pavo muticus* (green peacock) 绿孔雀
44. *Hydropotes inermis* (river deer) 獐
45. *Syrmaticus* spp. (long-tailed pheasant) 长尾雉
46. *Elaphodus cephalophus* (tufted deer) 毛冠鹿

47. *Muntiacus crinifrons* (black Muntjac deer) 黑麂子
48. *Lipotes vexillifer* (Chinese river dolphin) 白鳍豚
49. *Alligator sinensis* (Yangtze alligator) 扬子鳄
50. *Shinisaurus crocodilurus* (crocodile lizard) 鳄蜥
51. *Panthera tigris* (tiger of South China) 华南虎
52. *Cervus eldi* (Eld's deer) 海南坡鹿
53. *Macaca cyclopis* (Taiwan macaque) 台湾猴

in the south as well as partly in the north. Some of its genera have become extinct now, such as Pongo, Hyaena, and Tapirus, whereas the distribution of others, such as Elephas, Hylobates, and Ailuropoda, have become restricted. The Nihowan fauna developed into the Peking Man faunal group, then further diversified into *Mammuthus-Coeledonta* fauna in Northeast China and eastern Nei Mongol, and *Shantindong* Fauna in North China. A much warmer and more humid climate existed in North China then rather than now, and the region's fauna included many current genera, such as *Macaca, Moschus, Bos, Marmota,* and *Myospalax.*

Since the early Holocene, China's modern terrestrial vertebrate fauna has been fully developed. Their habitat is essentially divided into two realms: the Oriental realm in the south and the Paleoarctic realm in the north. Their boundary runs generally along the Qinling Mountains-Huaihe River line, which represents the northern limit of the evergreen broad-leaved forests as well as the northern limit of many tropical and subtropical terrestrial vertebrates. In the Paleoarctic realm, the adjustment to the diversification of climatic-vegetational environments between the more humid east and drier west has brought about further areal differentiation. In the Oriental realm, the transitional features shown by the tropical animals might illustrate the successive retreat of subtropical and tropical climatic-vegetational types from the north to the south since the early Tertiary period.

Distribution of China's Terrestrial Vertebrates

Distribution and areal differentiation of terrestrial vertebrates in China corresponds closely to the three great natural realms of China.

Eastern Monsoon China

Chief features of faunal distribution are the dominance and dispersion of hydrocols (terrestrial vertebrates adjusted to a humid environment) all over Eastern Monsoon China. The tropical and subtropical species, such as *Macaca mulatta* and *Paguma larvata,* extend their distribution area northward to North China, whereas tiger (*Panthera tigris*) and black-naped oriok, (*Oriolus chinensis*) are to be found even

farther into the northernmost part of China — Heilongjiang Province. On the other hand, species originally distributed in the north, such as the toad (*Bufo bufo*), European nuthatch (*Sitta europaea*), and flying squirrel (*Pteromys volans*) are now also dispersed all over Central and South China.

Owing to the different temperature characteristics from north to south, there exists areal differentiation between the north and the south, and some distribution limits are quite conspicuous. The southern boundary of the cool-temperate zone is the southern limit of typical cool-temperate terrestrial vertebrates, such as reindeer (*Rangifer tarandus*), as well as the northern limit of some temperate or subtropical families and orders, such as Microhylidae (in amphibians), Testudindae and Scincidae (in reptiles), Coraciiformes, Caprimulgiformes, Timaliinae (in birds), and Rhinolophidae (in manmals). The northern boundary of the warm-temperate zone is the northern limit of some typical tropical species, such as *Macaca mulatta.* The northern boundary of the northern subtropic zone, which corresponds to the Qinling Mountains-Huaiho River line, is roughly the boundary between the Paleoarctic and the Oriental realms. The northern boundary of the tropic zone is the northern limit of the tropical rain forest and some peculiar tropical vertebrate families, such as *Eurylaimidae, Psittacidae, Bucerotidae* (in birds), *Lorisidae, Hylobatidae, Tragulidae,* and *Elephantidae* (in mammals).

Northwestern Arid China (Nei Mongol-Xinjiang Region)

Terrestrial vertebrates in Northwestern arid China are adjusted to an arid habitat and are, thus, called Xerocols. They include most mammal species of the Dipodidae and Gerbillinae families; the bird species, *Otis* spp. and *Syrrhaptes paradoxus* (Pallas' sandgrouse); and the reptile species, *Phrynocephalus* spp. They are mostly endemic species. Some hydrocols, however, penetrate from Eastern Monsoon China into the eastern part of this region along river valleys.

Qinghai-Xizang Frigid Plateau

Natural conditions are severe on the Qinghai-Xizang Plateau and hence, only those species adjusted to a frigid habitat (the so-called Cryocols) can survive here. The birds include the Tibetan rosefinch (*Kozlowia roborowskii*) and Koslow's Bunting (*Emberiza koslowi*); the animals include the wild yak (*Bos grunniens*) and the Tibetan antelope (*Pantholops hodgsoni*). The barrier action of the lofty Himalayas is outstanding; some southern families such as Campephagidae and Nectariniidae (in birds) and primates and pholidota (in mammals) penetrate further northward to the Qinling Mountains-Huaihe River line in Eastern

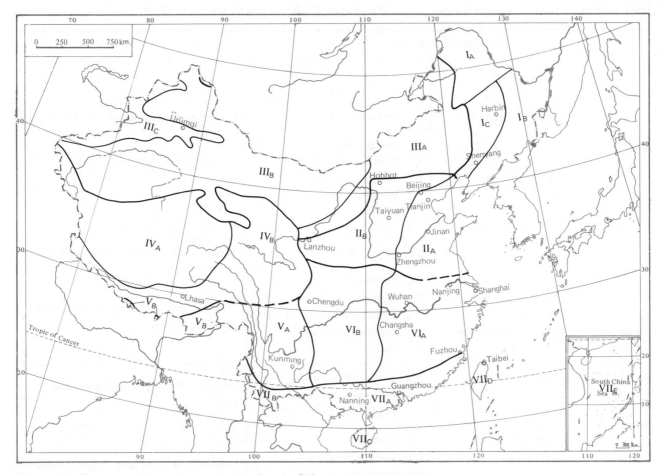

Figure 6-3 Zoogeographical regions and subregions in China (Legends see table 6-4)

Monsoon China, but they fail to cross the Himalayas and enter the Qinghai-Xizang Plateau. Neither Ochotona spp. nor some other Qinghai-Xizang species can move southward across the Himalayas. Yet, between this region and Northwestern Arid China, many components of the terrestrial vertebrate fauna are similar, particularly in the steppe zone. Such a close relationship and the recency of the uplifting of the Qinghai-Xizang Plateau might explain the phenomena of the inconspicuous differentiations (only on the level of species or lower) between these two regions.

From a global point of view, terrestrial vertebrates of China can be divided between the Paleoarctic and Oriental realms. The former includes the northern part of Eastern Monsoon China (North China and Northeast China) as well as Northwestern Arid China and a greater part of the Qinghai-Xizang Frigid Plateau; the latter is mainly restricted to the southern part of Eastern Monsoon China (Central China, South China, and Southwest China). These

zoogeographical regions can again be subdivided according to specific habitats. Thus, seven zoogeographical regions and 21 subregions have been identified (Figure 6-3, Table 6-4) in China.

Ecological Faunal Groups of China's Terrestrial Vertebrates

The adjustment of terrestrial vertebrates to their habitats is also clear from the distribution patterns of ecological faunal groups. Similar ecological faunal groups are usually distributed in similar habitats. Consequently, China's three great natural realms are roughly identical with the three great ecological faunal groups. Again, as each great natural region is composed of several different climatic-vegetational zones and each is dominated by a specific ecological faunal group; in all, there are seven ecological faunal groups that have been identified (Figure 6-4). Their close interrelationship with zoogeographical regionalization

Table 6-4 Interrelationship Between Zoogeographical Regions and Ecological Faunal Groups in China

Zoogeographical regions			Ecological Faunal groups
Realm (zero level)	Regions (first level)	Subregions (second level)	
A. Paleoarctic realm	I. Northeast China	I$_A$ Da Hinggan Mts.	Cool-temperate needle-leaved forest faunal group
		I$_B$ Changbai Mts. I$_C$ Northeast China Plain II$_A$ North China Plain II$_B$ Loess Plateau	Temperate forest, forest-steppe, and farmland faunal group
	II. North China		
	III. Nei Mongol-Xinjiang	III$_A$ Eastern steppe	Temperate steppe faunal group
		III$_B$ Western desert	Temperate desert and semidesert faunal group
B. Oriental realm	IV. Qinghai-Xizang Plateau	III$_C$ Tianshan Mts. IV$_A$ Qinghai-Xizang Plateau IV$_B$ Southern Qinghai-Xizang Plateau V$_A$ Southwest Mountains V$_B$ Himalayas	Highland forest-steppe, meadow-steppe, alpine desert faunal group
	V. Southwest China		
	VI. Central China	VI$_A$ Eastern hills and plains VI$_B$ Western mountains and plateaus	Subtropical forest, scrub, grassland faunal group
	VII. South China	VII$_A$ Southern coasts VII$_B$ Southern Yunnan VII$_C$ Hainan Island VII$_D$ Taiwan Island VII$_E$ South China Sea Islands	Tropical forest, scrub, savanna, and farmland faunal group

is shown in Table 6-4.

I. Cool-temperate needle-leaved forest faunal group   This group is characterized by great quantity of mammals that are adjusted to cold climate and taiga vegetation. They are usually winter dormant and have heavy fur. The faunal composition is rather simple. The most commonly seen mammals are deer (*Alces alces, Cervus elaphus, Moschus moschiferus, Capreolus cepreoslus*) and wild boar (*Sus scrofa*). Other commonly seen species are squirrel (*Sciurus vulgaris*), sable (*Martes zibellina*), ermine (*Mustela erminea, M. altaica*), brown bear (*Ursus arctos*), and badger (*Meles meles*). Several famous minority peoples live in the Da Hinggan Mountains and specialize in hunting. They have also successfully domesticated reindeer and some species of deer. The dominant birds are *Tetrastes bonasia, Tetrao wogalloides,* and *Garrulus glandarius.* The most typical species of reptiles are *Lacerta vivipara* and *Elaphe schrenckii.* Amphibians are rather few, with *Bombina orientalis, Hyla japonica, Kaloula borealis* being the most commonly seen.

II. Temperate forest, forest-steppe, and farmland faunal group   On the eastern hills and mountains of Northeast China, forest and grassland-inhabiting animals are still commonly seen, such as deer, wild boar, wild rabbit (*Lepus mandshuricus*), *Capreolus capreolus, Cervus elaphus, Moschus moschiferus, Erinaceus europaeus, Apodemus sylvaticus,* and *Sarex araneus.* Hunting is still an important occupation here. In North China, most forests have been cut down and farmlands are widely distributed. Hence, few large wild animals survive. The most commonly seen species are different kinds of rats (*Apodemus agrarius, Citellus daurica, Cricetulus barabensis, Cricetalus triton, Myospalax* spp., and *Mus musculus*). The dominant birds are *Parus major, P. palustris,* and *Emberiza cioides.* The commonly seen reptiles are *Natrix tigrina, Coluber spinalis, Dinodon rufozonatus,* and *Eremias argur.*

III. Temperate steppe faunal group   The dominant mammal species are herbivorous rodents, including *Microtus* spp., *Citellus* spp., *Ochotona* spp., and *Marmota* spp. Their populations vary with great amplitude accor-

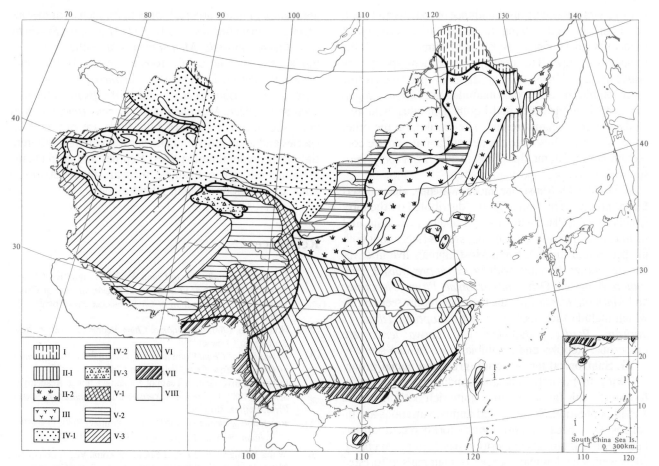

Figure 6-4 Distribution of ecological fauna groups of terrestrial vertebrates in China

   I. Cool-temperate needle-leaved forest faunal group
  II. Temperate forest, forest-steppe, and farmland faunal group
      II (1) Temperate forest faunal group; II (2) Temperate forest-steppe faunal group.
 III. Temperate steppe faunal group
 IV. Temperate desert and semidesert faunal group
      IV (1) Desert faunal group;   IV (2) Semidesert faunal group; IV (3) Plateau desert faunal group.
  V. Highland forest-steppe, meadow-steppe, alpine desert faunal group
      V (1) Highland forest-steppe faunal group; V (2) Highland meadow-steppe faunal group; V (3) Alpine desert
         faunal group.
 VI. Subtropical forest, scrub, grassland faunal group
VII. Tropical forest, scrub, savanna, and farmland faunal group
VIII. Farmland (oases) fauna in different faunal groups

ding to the supply of foodstuffs, which, in turn, depends closely on the amount of annual precipitation. In dry years, they migrate, like nomads, for long distances in search of vegetation. Their population density often reaches 3000 to 6000 burrows/ha, causing great havoc amid the pastureland. The control and elimination of these rats has become an urgent problem in the Nei Mongol Plateau. Herbivorous hoofed mammals also exist in great numbers; the galloping mongolian gazelle (*Procapra gutturosa*) is most commonly seen. Their western limit of distribution coincides roughly with that of the temperate steppe. Birds are few in species, with *Alauda arvensis, Melanocorypha mongolica, Eremophila alpestris,* and *Oenanthe oenanthe,* the most commonly seen. Reptiles and amphibians are rather rare.

IV. Temperate desert and semidesert faunal group This

is also characterized by an abundance of rodents and hoofed mammals, yet, owing to the relative scarcity of foodstuffs, their population is not so numerous as in the temperate steppe. The predominant rodent species are *Allactaga* spp., Dipus spp. and *Meriones* spp., whereas the most commonly seen hoofed mammals are *Gazella subgutturosa* (goitered gazelle), and Equus hemionus (wild ass). Birds and amphibians are few. The reptiles, on the other hand, are well adjusted to the desert habitat and are consequently quite numerous. They include *Phrynocephalus* spp., *Eremias* spp., and many others.

V. Highland forest-steppe, meadow-steppe, alpine desert faunal group This faunal group, distributed on the lofty Qinghai-xizang Plateau and its surrounding mountains (not including the Hengduan Mountain area) is rather poor both in species and in population. Most animals live in caves and are winter dormant, with *Procapra prezwalskii, Pantholops hodgsoni, Ochotona curzoniae, Pitymys leucurus,* and *Marmota himalayana* the most commonly seen. Birds, which include such specimens as *Crossoptilon Tibita, C. auritum, Perdix hodgsoniae,* and *Lerwa lerwa* are numerous. Reptiles and amphibians are seldom seen.

VI. Subtropical forest, scrub, grassland faunal group
Owing to favorable climatic conditions and plentiful foodstuffs, this faunal group is quite rich both in species and in population. Seasonal and annual variations are inconspicuous. Because most primary forest vegetation has been depleted, faunal composition has been much changed as well. The dominant species of mammals in forested areas are monkeys (*Macaca mulatta, M. speciosa*), squirrels (*Callosciurus* spp., *Dremomys* spp., and *Tamiops* spp.) In deforested hills and low mountains there are deer, *Muntiacus reevesi, Elaphodus cephalophus, Hydropotes inermis,* and so on. Rodents, such as *Mus musculus, Apodemus agrarius,* and *Rattus norvegicus* live mostly in cultivated areas as do birds, such as the sparrow, crow (*Corvus macrorhynchos, C. frugilegus*), *Hirundo daurica,* and *Motacilla alba.* Reptiles are represented by many kinds of snakes (*Natrix natrix, Agkistrodon halys, Trimeresurus* spp.), lizards (Lacerta, *Takydromus septentrionalis, Eumeces chinensis*), and turtles (*Chinemys reevesii*). Amphibians such as frogs (*Rana limnocharis, R. nigwmaculata, R. plancyi,* and *Bufo* spp.) are also commonly seen.

VII. Tropical forest, scrub, savanna, and farmland faunal group This is characterized by very complicated faunal composition and many endemic species. Practically no species can be said to predominate. Another feature is the abundance of many arboreal vertebrates, such as numerous species of primates (especially *Hylobates* spp., *Presbytis* spp., *Macaca mulatla, M. assamensis* and other species of monkeys), bats, and carnivorous mammals, Hylidae and Rhacophoridae (in amphibians), and *Draco* spp. (in reptiles). After deforestation, species living on the ground increase, such as the deer, *Muntiacus* spp., *Cervus unicolor,* wild boar and different kinds of rats. Birds are rich both in species and in quantity and also show great areal differentiation in predominant species. Reptiles and amphibians are numerous too; for example, there are about 120 to 130 kinds of snakes on Hainan Island alone.

References

[1] *Physical Regionalization of China* Working Committee, Chinese Academy of Sciences, 1959, *Physical Regionalization of China: Vegetation Regionalization of China,* Science Press, Beijing. (In Chinese)

[2] Hou Xueyu, 1960, *Vegetation of China,* People's Education Press, Beijing. (In Chinese)

[3] *Vegetation of China* Compilation Committee, 1980, *Vegetation of China,* Science Press, Beijing. (In Chinese)

[4] Integrated Investigation Team of Xinjiang, Chinese Academy of Sciences, 1980, *Vegetation of Xinjiang and Its Utilization,* Science Press, Beijing. (In Chinese)

[5] *Vegetation of Sichuan* Compilation Committee, 1980, *Vegetation of Sichuan,* Sichuan People's Press. Chengtu. (In Chinese)

[6] Wang Yifeng, et al., 1979, "Features of Zonal Vegetation of Nei Mongol", *Journal of Botany,* Vol. 21, No. 3. (In Chinese)

[7] Wang Hesen, 1979, "Basic Features of China's Flora." *Acta Geographica Sinica,* Vol. 34, No. 3. (In Chinese, with English abstract)

[8] Guangdong Institute of Botany, 1976, *Vegetation of Guangdong,* Science Press, Beijing. (In Chinese)

[9] Zheng Du et al., 1979, "On Natural Zonation of Qinghai-Xizang Plateau," *Acta Geographica Sinica,* Vol. 34, No. 1. (In Chinese, with English abstract)

[10[Zhang Houqiao, 1980, "Distribution Features of Yunnan's Vegetation," *Yunnan Botanical Research,* Vol. 2, No. 1. (In Chinese)

[11] *Physical Regionalization of China* Working Committee, Chinese Academy of Sciences, 1959, *Physical Regionalization of China: Zoogeographical Regionalization of China,* Science Press, Beijing. (In Chinese)

[12] *Physical Geography of China* Compilation Committee, Chinese Academy of Sciences, 1979, *Physical Geography of China: zoogeography,* Science Press, Beijing. (In Chinese)

[13] Zhang Yongzu, 1978, "On the Zoogeographical Characteristics of China", *Acta Geographica Sinica,* Vol. 33, No. 2. (In Chinese, with English abstract)

Chapter 7

Comprehensive Physical Regionalization and Land Classification in China

So far, we have studied and discussed different physical elements separately — climate, landform, surface and ground water, soil and biological geography one by one, and these topics might be termed topical physical geography. Now we are going to focus on the integration of these physical elements. This combined approach might be called comprehensive physical geography.

There are three major approaches to the study of integrated physical geography. The first one is based on comprehensive physical regionalization and regional study in which all physical elements are integrated in terms of natural regions. The second approach combines all physical elements according to land types. The third consists of integrating all physical elements within modern geographical processes, such as (1) the distribution and balance of heat and moisture conditions on the earth's surface and their impact on the physical geographical environment, (2) migration and distribution of chemical elements, (3) biosystems, with special emphasis on the interchange of matter and energy between biota and their habitats, and (4) human impacts and feedbacks on the geographic environment. We have not yet, however, gone far enough to make an accurate systematic presentation of this last category. For the time being, we will concentrate our attention on the first and second approaches.

COMPREHENSIVE PHYSICAL REGIONALIZATION AND REGIONAL STUDY

Comprehensive physical regionalization is an effort to identify differentiation among different areas on the earth's surface and document the similarities within the same area. It reflects the sumtotal of all attributes of all the physical elements and, hence, might serve as a guideline for land-use planning within these areas.

Comprehensive physical regionalization, as pointed out at the beginning of this book, is one of the oldest traditions of Chinese geographical study. Since 1949, in China physical regionalization has been considered not only to be a major scientific area, but also as one of the chief aspects of regional planning, especially for agricultural development. A series of physical regionalization works on both national and local scales has been conducted. So far, the most comprehensive study was completed in 1958 by the Working Committee of Physical Regionalization of China, Chinese Academy of Sciences, headed by Professor Zhu Kezhen (Chu Ko-chin) and Professor Huang Bingwei. According to this Committee, China can be divided into 3 natural realms, namely, the Eastern Monsoon Realm, the Nei Mongol-Xinjiang Arid Realm and the Qinghai-Xizang Alpine Realm. Subsequently, 18 natural regions, 28 natural zones, and 90 natural provinces are identified.

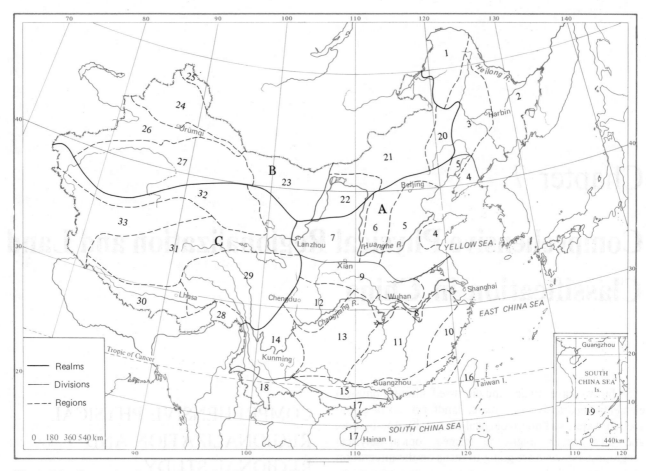

Figure 7-1   Comprehensive physical regionalization in China (Legends see Table 7-2)

This is the scheme, with some important modifications, that we adopt for comprehensive physical regionalization. In this textbook, 3 natural realms, 7 natural divisions and 33 natural regions are demarcated in China (Figure 7-1).

Three Natural Realms

Using the most significant areal differential features in China as delimiting criteria — that is geographical location, topography and neotectonic movements, chief climatic characteristics, areal differentiation in geological history, and impact of human activities on physical geographical environment — the country is first of all divided into three natural realms: Eastern Monsoon China, Northwest Arid China, and the Qinghai-Xizang Frigid Plateau.This is the starting point of comprehensive physical regionalization in China.

Eastern Monsoon China

Eastern Monsoon China occupies about 45 per cent of

the total land area of China and has a population accounting for 95 per cent of the country's total population.It has the following chief characteristics:

1) It is located near neighboring seas and the Pacific Ocean and, hence, the influence of summer maritime monsoons is strong, with sharp seasonal variations both in wind direction and in precipitation.

2) The climate is humid or subhumid, with forest as the dominant vegetation.

3) The chief parameter for areal differentiation is temperature, decreasing from south to north; although in North China and Northeast China, precipitation also decreases conspicuously from east to west.

4) The neotectonic uplift is not great; few mountains rise more than 2000 m above sea level. The mountains are entirely without modern glaciation. East to the Qinzhou-Zhengzhou-Beijing-Huma line is a broad belt of down-warping with alluvial plains extensively distributed (mostly below 500 m in

elevation).

5) Modern geomorphological processes are predominautly normal fluvial action, and surface water is mainly fed by precipitation.

6) Owing to the absence of Quaternary continental glaciation, both fauna and flora are rich in species and a reddish weathering crust is distributed extensively.

7) With the exception of a few localities, the impact of human activities is great. Practically all arable lands have been cultivated and all natural vegetation has been modified.

Northwest Arid China

Northwest Arid China is the eastern part of the immense Eurasian desert and grassland; it occupies about 30 per cent of the total land area of China, but its population accounts for only 4 per cent of the country's total population. It is characterized as follows:

1) In recent geological times, there has been conspicuous differential uplifting, with the widespread formation of plateaus and inland basins of approximately 500 to 1500 m in elevation. In addition, a considerable portion of the area has been violently uplifted to form lofty mountains that surround or traverse neighboring plateaus and inland basins (e.g. Tianshan Mountains at more than 3500 m and the Altay Mountains at more than 3000 m).

2) Because the area is located deep in the Eurasian continent and is surrounded by a series of lofty mountains, the influence of summer maritime monsoons is rather weak, and, hence, the climate is arid or semiarid.

3) Modern geomorphological processes consist mostly of denudation and deflation, with shamo (sandy desert) and gobi (gravel and stony desert) extensively distributed. However, along larger rivers and on piedmont plains, linear or sheet water erosion can be quite severe after rainstorms.

4) There is mostly inland drainage, with few perennial rivers, which are fed by runoff and melting snow in the surrounding mountainous areas. Lakes are numerous and are mostly saline.

5) There has been gradual desiccation (with some fluctuation) ever since the late Mesozoic era. Both fauna and flora are rather poor in species as well as in number. In comparison with Eastern Monsoon China, human impact is not so conspicuous. Yet, many fertile oases have been developed along the middle and lower reaches of the perennial rivers and large tracts of grassland have been used for pasture since ancient times.

The Qinghai-Xizang Frigid Plateau

The Qinghai-Xizang Plateau is the highest and largest plateau in the world, it occupies about 25 per cent of the total land area of China, for less than 1 per cent of the country's total population. It is characterized as follows:

1) Recent large-scale uplift has given this plateau a mean elevation of above 4000 m. It is interspersed with a series of high snow-capped mountains above 7000 and even 8000 m in elevation.

2) As a result of the high elevation, the atmosphere is generally thin, the temperature low, and permafrost is widely distributed. Solar radiation is intense and winds are strong.

3) Most parts have inland drainage, with numerous lakes distributed in the inland basins. The modern geomorphological process is chiefly characterized by nivation, although fluvial action is also strong along the southeastern border, and the process of desiccation dominates the arid northwestern part. Glaciers, both Quaternary and modern, are widely distributed.

4) Both fauna and flora are rich in species. Vertical zonation is conspicuous; in the northwestern Qinghai-Xizang Plateau, desert, montane grassland, montane coniferous forest, alpine meadow, periglacial dwarf shrub, and continual snow appear in succession from footslope to mountain top.

5) Owing to the influence of strong glaciation and weak chemical weathering, soil parent materials are usually coarse and thin, Soils are generally young, modern soil-forming processes having started only after the last glaciation. Consequently, soil profiles are poorly developed and fertility is rather low.

6) Natural conditions as a whole are unfavorable to human activities. The density of population is even lower than in the Northwest Arid China.

Seven Natural Divisions

Based on the above three geographical realms, seven natural divisions are identified. A natural division is characterized by having within its borders similar temperature and moisture conditions as well as similar broad types of soil and vegetation. Eastern Monsoon China can be subdivided, according to areal differentiation of temperature conditions, into four natural divisions: temperate humid and subhumid, warm-temperate humid and subhumid, subtropical humid, and tropical humid. Northwest Arid China can be subdivided, mainly on the basis of moisture and vegetation, into temperate grassland and warm-temperate and temperate desert. The Qinghai-Xizang Frigid Plateau forms a single natural division because the

Table 7-1 Seven Natural Divisions of China and Their Chief Climatic Indices

Realms	Natural divisions	Accumulated temperature during ≥ the 10°C period*	Aridity index**	Frost-free days
A. Eastern Monsoon China	I. Temperate humid and subhumid	1400 – 3200°C	0.5 – 1.2	< 145
	II. Warm-temperate humid and subhumid	3200 – 4800°C	0.5 – 1.5	150 – 220
	III. Subtropical humid	4500 – 7500°C	0.5 – 1.0	230 – 330
	IV. Tropical humid	>7500°C	0.5 – 1.0	no frost
B. Northwestern Arid China	V. Temperate grassland	2000 – 3200°C	1.2 – 4.0	< 180
	VI. Warm-temperate and temperate desert	3200 – 4500°C	>4.0	200 ±
C. Qinghai-Xizang Frigid Plateau	VII. Qinghai-Xizang Plateau	<2000°C vertical distribution	0.5 – 0.4 vertical distribution	< 130

* The accumulated temperature is the sumtotal of temperatures in the period when the temperature is ≥ 10°C. The term is also known as active temperature because most plants begin to grow only when the temperature is above 10°C. It may be pointed out that the accumulated temperature for spring wheat to mature is more than 1400°C, whereas that for corn is more than 2000°C.

** The aridity index is computed according to the following empirical formula: $I = 0.16(\Sigma t \geq 10°C)/(\Sigma t \geq 10°C)$, where $\Sigma t \geq 10°C$ denotes accumulated temperature during the ≥ 10°C period, $\Sigma r \geq 10°C$ is the total precipitation in mm during the same period, whereas 0.16 is a constant.
When $I < 1.0$, the region is humid, with forests as the dominant natural vegetation;
$I = 1.0 - 1.5$, the region is subhumid, with forest-steppe predominating;
$I = 1.5 - 2.0$, the region is semiarid, with steppe predominating;
$I = 2.0 - 4.0$, the region is arid, desert-steppe predominating;
$I > 4.0$, the region is extremely arid, desert predominating.

whole realm is subjected to frigid conditions and vertical zonation. These seven natural divisions and their chief climatic indices are shown in table 7-1. In the second section of this chapter, we will give a brief account of each natural division and in the final seven chapters of this book, each natural division will be discussed individually in greater detail.

Natural Zones and Natural Regions

The Physical Regionalization of China Working Committee, Chinese Academy of Sciences (1958–1959) further divided China into 28 natural zones on the basis of either vertical zonation or horizontal variation of climatic-biological-soil features. Natural zones have relatively uniform temperature and mosisture conditions as well as similar zonal soil and vegetation, consequently, they have similar suitabilities and capabilities for land use. Such a demarcation into natural zones is of great significance in scientific analysis both for agricultural production and for economic planning.

To simplify our classification system, however, we omit the natural zones and subdivide China into 33 natural regions directly on the basis of the 7 natural divisions. A natural region has not only uniform zonal features (climatic, biological, and soil), but also fairly uniform azonal

characteristics (geological and geomorphological). Hence, a natural region reflects more fully the total physical geographical environment and can be used much better than either natural divisions or natural zones as a geographical unit for regional planning. In Table 7-2 and Figure 7-1, 33 natural regions are shown. They will be discussed in more detail in Chapters 8 through 14.

Lower Level Regional Units

Lower level regional units are mainly based on azonal features. A natural region can be again subdivided into natural subregions. A natural subregion has rather homogeneous natural features, both zonal and azonal; hence, it is most useful in regional planning. However, owing to the unbalanced condition of scientific data, it is sometimes impractical to further divide natural regions into natural subregions, for example, on the Qinghai-Xizang Plateau.

The lowest level regional unit in comprehensive physical regionalization might be considered identical to land or landschaft (landscape) in land classification. This will be discussed in the second section of this chapter. We may simply point out here that comprehensive physical regionalization and land classification are closely interrelated; natural divisions serve as the starting point for land classification, whereas first-level land types or their com-

Table 7-2 Thirty-three Natural Regions of China

Natural divisions	Natural regions
I. Temperate humid and subhumid division (Northeast China)	1. Da Hinggan Mts. — needle-leaved forest region 2. Northeast China mountains — mixed needle-and broad-leaved forest region 3. Northeast China Plain — forest-steppe region
II. Warm-temperate humid and subhumid division (North China)	4. Liaodong-Shandong peninsulas — deciduous broad-leaved forest region 5. North China Plain — deciduous broad-leaved forest region 6. Shanxi-Hebei mountains — deciduous broad-leaved forest and foreststeppe region 7. Loess Plateau — forest-steppe and steppe region
III. Subtropical humid division (Central and South China)	8. Middle and lower Changjiang plain — mixed forest region 9. Qinling-Dabie mts. —mixed forest region 10. Southeast coast — evergreen broad-leaved forest region 11. South Changjiang hills and basins — evergreen broad-leaved forest region 12. Sichuan Basin — evergreen broad-leaved forest region 13. Guizhou Plateau — evergreen broad-leaved forest region 14. Yunnan Plateau —evergreen broad-leaved forest region 15. Lingnan hills — evergreen broad-leaved forest region 16. Taiwan Is. — evergreen broad-leaved forest and monsoon forest region
IV. Tropical humid division (South China)	17. Leizhou-Hainan — tropical monsoon forest region 18. Southern Yunnan —tropical monsoon forest region 19. South China Sea islands — tropical rain forest region
V. Temperate grassland division (Nei Mongol)	20. Xi Liahe Basin - steppe region 21. Nei Mongol Plateau — steppe and desert-steppe region 22. Ordos Plateau — steppe and desert-steppe region
VI. Temperate and warm-temperate desert division (Northwestern China)	23. Alashan Plateau -temperate desert region 24. Junggar Basin — temperate desert region 25. Altay Mts. — montane grassland and needle-leaved forest region 26. Tianshan Mts. — montane grassland and needle-leaved forest region 27. Tarim Basin — warm-temperate desert region
VII. Qinghai-Xizang Plateau division	28. Southern Himalayas slope — tropic and subtropic montane forest region 29. Southeastern Qinghai-Xizang Plateau — montane needle-leaved forest and alpine meadow region 30. Southern Qinghai-Xizang Plateau — shrubby grassland region 31. Central Qinghai-Xizang Plateau — montane and alpine grassland region 32. Qaidam Basin and Northern Kunlun Mts. slopes — desert region 33. Ngari-Kunlun Mts. — desert-steppe and alpine desert region

bination correspond to the lowest level regional units of physical regionalization.

LAND CLASSIFICATION AND MAPPING

The term land or terrain is broadly construed as the total physical environment of a certain section on the earth's surface, including climate, landform, surface water, ground water, soil, vegetation and other physical elements. It also includes past and present human activities and their impacts on the physical environment.

Land classification is another old tradition in China's geographical studies, as has been pointed out in the introduction to this book. Since 1978 a serious and comprehensive land-classification and mapping program has

Table 7-3 A Series of Land Types From the Nenjiang Plain Northwestward to the Da Hinggan Mountains

First-level land types	Second-level land types
1. Swampy depression	1(1) *Phragmites* heavy swamp
	1(2) *Carex, Salix* swamp
	1(3) Shrubby, high swamp
	1(4) Gramineous, *Carex,* swampy meadow
2. Meadow — low flatland	2(1) Meadow, lacustrine and alluvial plain
	2(2) Gramineous meadow, gully and small valley
	2(3) Needle-and broad-leaved mixed forest, gully and small valley
	2(4) Gramineous deep black earth piedmont
	2(5) *Salix, Artemisia* half-fixed sand dune
	2(6) Shifting sand dune.
3. Forest-meadow — terrace and slopeland	3(1) Lush meadow, leached chernozem terrace and slopeland
	3(2) *Salix, Carex* meadow, leached chernozem flat ridge and gentle slope
	3(3) Gramineous meadow chernozem terrace and slopeland
	3(4) Shrubby grassland, chernozem terrace and slopeland
	3(5) *Pinus sylvestris* fixed sand dune
	3(6) *Quercus, Betula* lava terrace
	3(7) Sparse grassland, lava terrace
4. Needle-and broad-leaved mixed forest — hill and low mountain	4(1) *Quercus,* dark brown forest soil, hill and low mountain
	4(2) *Populus, Betula,* grey forest soil, hill and low mountain
	4(3) Mixed forest, dark brown forest soil, low mountain
	4(4) Sparse woodland, brown forest soil, low mountain
	4(5) Needle-leaved forest, podzolized brown forest soil low mountain
5. Taiga forest-middle mountain	5(1) *Larix* taiga forest, middle mountain
	5(2) Dwarf taiga forest, middle mountain

Table 7-4 Land Classification System in the Xifeng Area, Loess Plateau

First-level land types	Second-level land types
1. Loessic low flat land (*Chuan*)	1(1) Loessic valley bottom plain
	1(2) Moist loessic soil, low terrace
	1(3) Black loessic soil terrace
	1(4) Loessic cropland
2. Loessic high plain (*Liang* and *Yuan*)	2(1) Loessic high plain
	2(2) Loessic gentle slope
	2(3) Loessic flat ridge
	2(4) Loessic gentle ridge slope
	2(5) Loessic gully
	2(6) Loessic small valley
3. Deciduous broad-leaved forest — hill and low mountain	3(1) Shrubby, drab soil hill
	3(2) Deciduous broad-leaved forest, drab soil low mountain
4. Needle-and broad-leaved mixed forest — middle mountain	4(1) Needle- and broad-leaved mixed forest, leached drab soil middle mountain
	4(2) Shrubby, leached drab soil middle mountain

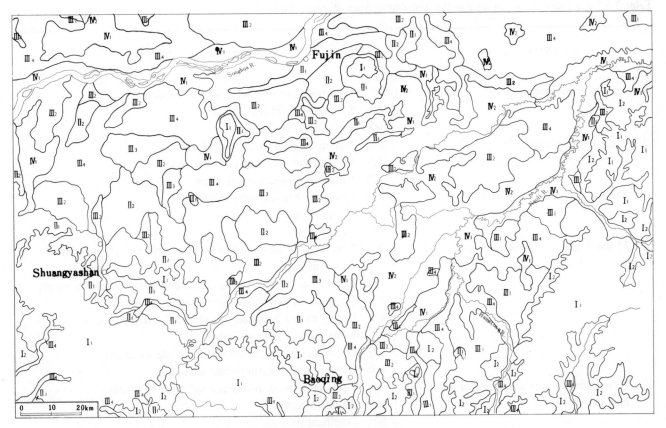

Figure 7-2 Land types in western part of the Sanjiang Plain, Northeast China

I. Broad-leaved forest — hill and low mountain
 I (1) *Quercus,* dark brown forest soil — hill and low mpuntain; I (2) *Populus, Betula* brown forest soil — hilly slope.
II. Forest meadow — terrace and gentle slope
 II (1) Lush meadow, leached chernozem — terrace and gentle slope; II (2) Gramineous meadow, chernozem — terrace and gentle slope.
III. Meadow — low flat land
 III (1) Gramineous meadow Bai-jiang (planosol) soil — low flat land; III(2) Gramineous meadow soil — low flat land;
III (3) Saline meadow soil — low flat land; III (4) Glei Bai-jiang soil — low depression.
IV. Swampy depressions
 IV (1) Gramineous *Salix* swamp; IV (2) *Carex* heavy swamp.

been in progress. This national program has two basic purposes: to provide solid and detailed data on China's geographical environment and to evaluate the nation's land resources as well as agricultural capabilities. A series of maps on land classification will be produced. The nation will be mapped at the scale of 1:1000000, the major provinces and regions at the scale of 1:200000, and the sample study areas at 1:50000. The project is scheduled for completion before 1990.

A Scheme for Land Type Classification and Mapping

As mentioned above, the natural divisions are the foundation, or zero level, of land classification. Within each natural division, we adopt a two-level scheme of land-type classification.

The first-level land types within a natural division approximate the land system of the Australian CSIRO (Commonwealth Scientific and Industrial Organization). They

Table 7-5 Land Classification System in the Tengchong Area, Western Yunnan

First-level land types	Second-level land types
1. Low intermontane basin	1(1) Lacustrine and riverine swampy depression
	1(2) Meadow diluvial-alluvial plain
	1(3) Paddy flatland
2. High intermontane basin	2(1) Shrubby, high intermontane basin
	2(2) Dry farming, high intermontane basin
	2(3) Paddy, high intermontane basin
3. Evergreen broad-leaved forest — hill	3(1) Grassland, red and yellow earth, hill
	3(2) Evergreen broad-leaved forest, red and yellow earth, hill
	3(3) Oil crops, hill
4. Mixed forests — middle mountain	4(1) Needle-and broad-leaved mixed forest, red and yellow earth middle mountain
	4(2) Pine forest, slightly podzolized red and yellow, middle mountain
	4(3) Shrubby, red and yellow earth, middle mountain
	4(4) Evergreen and deciduous broad-leaved mixed forest, yellowish brown earth, middle mountain
	4(5) Grassland, red and yellow earth, middle mountain
	4(6) Shrubby, brown earth middle mountain
5. Alpine meadow and needle leaved forest — high mountain	5(1) Picea-Abies forest, podzolized soil, high mountain
	5(2) Alpine meadow
	5(3) Continual snow

Table 7-6 Land Classification System in Southwestern Hainan Island

First-level land types	Second-level land types
1. Swampy depression	1(1) Lacustrine and riparian swampy depression
	1(2) Clay coastal beach
	1(3) Sandy coastal beach
	1(4) Saline coastal depression
	1(5) Mangrove coastal swamp
2. Meadow — low flat land	2(1) Meadow marine plain
	2(2) Meadow marine low terrace
	2(3) Meadow alluvial and lacustrine plain
	2(4) Meadow diluvial-alluvial plain
	2(5) Paddy low, flatland
3. Tropical monsoon forest — terrace and slopeland	3(1) Savanna, laterite terrace and slopeland
	3(2) Shrubby, laterite terrace and slopeland
	3(3) Tropical monsoon forest, laterite terrace and slopeland
	3(4) Dry farming gentle slopeland
	3(5) Paddy-terraced land
4. Tropical monsoon forest — hill	4(1) Tropical monsoon forest, laterite hill
	4(2) Tropical evergreen forest and monsoon forest, red earth, hill
5. Tropical montane rainforest — low mountain	5(1) Tropical evergreen forest, red earth low mountain
	5(2) Tropical montane rainforest, yellow earth low mountain
6. Needle-and broad-leaved mixed forest — mountain	6(1) Needle-and broad-leaved mixed forest, podzolized yellow earth, middle mountain
	6(2) Montane dwarf forest, yellow earth, middle mountain

Table 7-7 Land Classification System in Eastern Nei Mongol Plateau

First-level land types	Second-level land types	
1. Swampy depression	1(1)	*Phragmite* swamp
	1(2)	*Carex* swamp
	1(3)	Shrubby swampy depression
2. Meadow — low flat land	2(1)	Saline low flatland
	2(2)	Gramineous meadow gully and small vally
	2(3)	Carex meadow, lacustrine and alluvial plain
	2(4)	Gramineous meadow lacustrine and alluvial plain
	2(5)	Gramineous deep dark chestnut soil, piedmont
3. Steppe — high plain	3(1)	Gramineous dark chestnut soil, high plain
	3(2)	Gramineous and shrubby chestnut soil, high plain
	3(3)	*Pinus sylvestris* fixed sand dune
	3(4)	Sparse psammophile half-fixed sandy land
	3(5)	Shifting sand dune
4. Shrubby steppe — hill	4(1)	Shrubby, stony hill
	4(2)	Gramineous dark chestnut soil, hill
	4(3)	Gramineous and shrubby chestnut soil, hill
	4(4)	Basalt and lava terrace
	4(5)	Volcanic cone
5. Deciduous broad-leaved forest—low mountain	5(1)	Deciduous broad-leaved forest, dark brown forest soil, low mountain
6. Needle-and broad-leaved mixed forest — middle mountain	6(1)	Needle-and broad-leaved mixed forest, dark brown forest soil middle mountain

are identified mainly on the basis of macrolandforms or mesolandforms and are listed according to the elevation of each type, from low depression to mountain top. The first-level land types are the major mapping units for the nation as a whole at a scale of 1:1000000.

The second-level land types approximate the land units of the Australian CSIRO. They are generally identified and delimited by relatively homogeneous mesolandforms or microlandforms and surface materials in the hilly regions and by similar soil and vegetation subtypes in the level plains. These suggest a similar land use suitability and capability and are the major mapping units for China's provinces and large regions. Several adjacent and genetically similar second-level land types may be grouped into a first-level land type.

In the development of our land type classification system, we have carefully selected for detailed sample studies some small areas located in different natural divisions. Such small areas have been mapped at scales larger than 1:50000. They have been studied mainly on the basis of microlandforms, soil species, and vegetation formation and may be used for the identification of third-level land types should the need arises.

Our approach to land classification is based on the processes of landscape formation and their interrelationship. We feel that all the physical factors of the environment must be viewed collectively in the determination of land types at all levels. It is the interplay of all environmental factors that gives rise to the land types or landscapes on the earth's surface.

Land Types in the Temperate Humid and Subhumid Division

The temperate humid division includes all of Northeast China except southern Liaoning, it occupies 8.1 per cent of China's total land area. This natural division has flat fertile plains and extensive virgin land awaiting reclamation. The rugged forested mountains are China's important timber-producing areas. The landform features stand out as the major factors in land classification, although horizontal zonation from south to north and from east to west is also an important factor. From the Nenjiang River Plain northwestward to the northern Da Hinggan Mountains, a series of land types appear in succession (Table 7-3). The land types of the Sanjiang Plain, China's largest land

Table 7-8 Land Classification System in the Hexi Corridor

First-level land types	Second-level land types
1. Clay and silt level land	1(1) Nonsaline clay and silt level land
	1(2) Saline clay and silt level land
	1(3) Takyr
	1(4) Salty marsh
	1(5) Yardang
2. Sandy level land (Shamo)	2(1) Shifting sandy desert
	2(2) Half-fixed sandy desert
	2(3) Fixed sandy desert
3. Gravel and stony level land (Gobi)	3(1) Alluvial-diluvial sandy gravel gobi
	3(2) Diluvial-alluvial gravel gobi
	3(3) Colluvial-diluvial gravel gobi
	3(4) Denudational stony gobi
4. Denudational mountain and hill	4(1) Montane desert
	4(2) Montane grassland
5. Erosional high mountain	5(1) Montane coniferous forest
	5(2) Alpine meadow
6. Nival extremely high mountain	6(1) Dwarf cushion vegetation
	6(2) Continual snow
7. Oases	7(1) Irrigated oases
	7(2) Non-irrigated oases

Table 7-9 Land Classification System in the Semiarid Qinghai-Xizang Plateau

First-level land types	Second-level land types
1. Swampy depression	1(1) Kobresia swampy meadow
2. Meadow — level land	2(1) Meadow floodplain
	2(2) Dry farming floodplain
	2(3) Dry farming low terrace
	2(4) Fixed and half-fixed sand dune
	2(5) Shifting sand dune
3. Shrubby steppe — high plain	3(1) Shrubby steppe, colluvial-diluvial fan
	3(2) Shrubby steppe high terrace
	3(3) Dry farming high plain
4. Alpine meadow and shrubby meadow-steppe — hill and high mountain	4(1) Shrubby meadow-steppe hill
	4(2) Shrubby meadow-steppe high mountain
	4(3) Alpine steppe high mountain
	4(4) Dwarf-cushion vegetation collovial-diluvial fan
	4(5) Dwarf-cushion vegetation and alpine meadow high mountain
5. Periglacial alpine meadow — extremely high mountain	5(1) Sparse cushion vegetation, extremely high mountain
	5(2) Continual snow

reclamation area at the present time, have been studied in detail (see Figure 7-2, Landsat image 2).

Land Types in the Warm-temperate Humid and Subhumid Division

The warm-temperate humid and subhumid division consists of a large part of North China and the southern portion of Liaoning Province. This natural division occupies 10.2 per cent of the total area of China. It is essentially a transitional belt between the subtropical south and the temperate north and between the humid east and the subhumid or semiarid western divisions. Table 7-4 shows the patterns of land types in the sample area of Xifeng on the Loess Plateau.

Land Types in the Subtropical Humid Division

The subtropical humid division is extensively distributed in central China and South China, occupying 26.1 per cent of the nation's total land area. The southern and middle subtropical subdivisions are characterized by evergreen broad-leaved forests and red earths and yellow earths; the northern subtropical subdivision has deciduous broad-leaved forests and yellow brown earths. As a whole, the entire division is hilly, a feature basic to the classification of land types in this division. Table 7-5 displays a classification system for Southwest China based on the Tengchong area, western Yunnan Province.

Land Types in the Tropical Humid Division

The area of the tropical division in China is not extensive, occupying only 1.6 percent of the nation's total area. Two subdivisions can be recognized: (1) the equatorial rain forest — laterite subdivision restricted to the southernmost islands in the South China Sea and (2) the tropical monsoon forest — laterite subdivision characterized by its coastal location and hilly landscape. Table 7-6 lists the major land types in southwestern Hainan Island.

Land Types in the Temperate Grassland Division

The temperate grassland division located mostly on the Nei Mongol Plateau eastward from the Helan Mountains. This natural division occupies 7.4 per cent of China's total land area. The major factors in land classification are landform features and their spatial variations. A land-classification system in the eastern Nei Mongol plateau is shown in Table 7-7.

Land Types in the Temperate and Warm-Temperate Desert Division

The temperate and warm-temperate desert division is an extensive natural division and located northwest of the Helan-Qilian mountains line, it occupies about 22 per cent of China's total land area. It can be divided into two subdivisions: the temperate and the warm temperate. The region is composed mainly of arid, level plateaux and inland basins, with sandy deserts and gobi extensively distributed. Consequently, the major factors for land classification are surface materials, although vertical zonal differentiation is also conspicuous in the surrounding high mountains. The Hexi Corridor's land-classification system is shown as an example in Table 7-8.

Land Types in the Qinghai-Xizang Plateau Division

The Qinghai-Xizang Plateau has an area of about 2.5 million sq km and an average elevation of more than 4000 m. Except for the southern Himalayan slopes, which belong to the northern margin of the tropical humid division, the plateau proper can be divided into three subdivisions: (1) the subhumid Qinghai-Xizang Plateau, which occupies about 4 per cent of China's total land area; (2) the widely distributed semiarid Qinghai-Xizang Plateau, which occupies about 13.3 per cent of the nation's total land area; and (3) the arid Qinghai-Xizang Plateau, which occupies about 9.4 per cent of the total land area of China. In each subdivision, the major factors of land classification are invariably the landform features and their impact on the other physical factors. Taking the most widely distributed semiarid subdivision as an example, a land-classification system is tentatively presented in Table 7-9.

References

[1]   *Physical Regionalization of China* Working Committee, Chinese Academy of Sciences, 1959, *Comprehensive Physical Regionalization of China*, Science Press, Beijing. (In Chinese)

[2]   Zhu Kezhen (Chu Ko-chin), 1958, "Subtropical zone of China", *Science Bulletin*, 1958, No. 17. (In Chinese)

[3]   Huang Bingwei (Huang Pin-wei), 1959, "A scheme for comprehensive physical regionalization of China", *Science Bulletin*, 1959, No. 18. (In Chinese)

[4]   Hou Xueyu et al, 1963, "On agricultural development for different natural regions of China", *Science Bulletin*, 1963, No. 9. (In Chinese)

[5]   Ren Mei'e (Jen Mei-o) et al., 1979 *An Outline of Physical Geography of China*, Commercial Press, Beijing. (In Chinese)

[6]   Zhao Songqiao et al., 1979, "Thirty years in integrated physical geography in People's Republic of China", *Acta Geographica Sinica*, Vol. 34, No. 3. (In Chinese, with English abstract)

[7]   Zhao Songqiao, 1981, "Thirty years in Physical geography in the

People's Republic of China". *Chili* (Japanese Geographical Journal), 1981, No. 3. (In Japanese)

[8] Zhao Songqiao, 1983, "A new scheme for comprehensive physical regionalization in China", *Acta Geographica Sinica,* Vol. 38, No. 1. (In Chinese, with English abstract)

[9] Zhao Songqiao, 1983, "Land Classification and Mapping in China", *Land Resources of the People's Republic of China* , The United Nations University, Tokyo.

Chapter 8

Humid and Subhumid Temperate Northeast China

This natural division is located in northeastern China, delimited by Sino-Russian and Sino-Korean international boundaries on the north and east. Its western boundary is the aridity isopleth of 1.2 and merges into the vast temperate grassland of Nei Mongol. Its southern boundary is the 3200°C isotherm of accumulated temperature during $\geqslant 10$°C period that demarcates the temperate and warm-temperate thermal zones and, thus, differs somewhat from the traditional southern boundary of Northeast China which roughly corresponds with the Great Wall. Administratively, this natural division includes the whole of Heilongjiang and Jilin provinces, the northern part of Liaoning province and a small part of the Nei Mongol Autonomous Region. It has an area of about 1.1 million sq km and a population of 40 million. It is one of the most important bases of commercial grains and economic crops (soybeans, sugar beets etc.) as well as the largest timber- and petroleum-producing area in China (Figure 8-1).

CHIEF PHYSICAL FEATURES

This natural division is characterized by the following physical features.

Humid Temperate Climate

Since the humid temperate division is within the northernmost part of the Eastern Monsoon Realm, its climate is entirely dominated by the alternating continental and maritime monsoons. Owing to its higher latitudinal location (from about 41 to 53°N), it is of course the coolest area in China, with accumulated temperature during the

$\geqslant 10$°C period about 3200°C in the southern part, decreasing to less than 1400°C in the northwestern part. On the whole, the winter is long and severe, with a frost-free season of less than 150 days, resulting in a one crop-one year system. In the most northwestern part, the frost-free season is less than 90 days, making this area generally unfit for field crops. According to recent statistics, low temperature and frost rank as the first natural hazards for agricultural development in this area; frequency of the low-temperature hazard in the spring occurs once every three years, in the summer and autumn, it occurs once every five years, and for the whole growing season, it occurs once every 20 years.

Precipitation decreases from southeast (Tonghua, 900 mm) to northwest (about 400 mm) as the distance from the sea increases (Figure 8-2). In many low plains and during the rainy season. There is usually too much rather than too little annual precipitation, Thus, drainage becomes a greater problem than irrigation. Relief bears importantly on both precipitation and temperature, with much heavier precipitation and lower temperatures in the bordering mountains — Da Hinggan Mountains, Xiao Hinggan Mountains, Changbai Mountains, and so on. Therefore, the last two montane areas have a temperate humid climate, with annual precipitation of more than 500 mm and an aridity index of less than 1.0. The Northern Da Hinggan Mountains have a cool-temperate humid climate, with annual precipitation of more than 400 mm and an aridity index of less than 1.0, whereas the extensive Northeast China Plain, sandwiched between these mountains, has a temperate subhumid climate, with annual precipitation around 450 mm and an aridity index 1.0 to 1.2. Precipitation is generally concentrated in summer and autumn, coinciding with the growing season, which is a

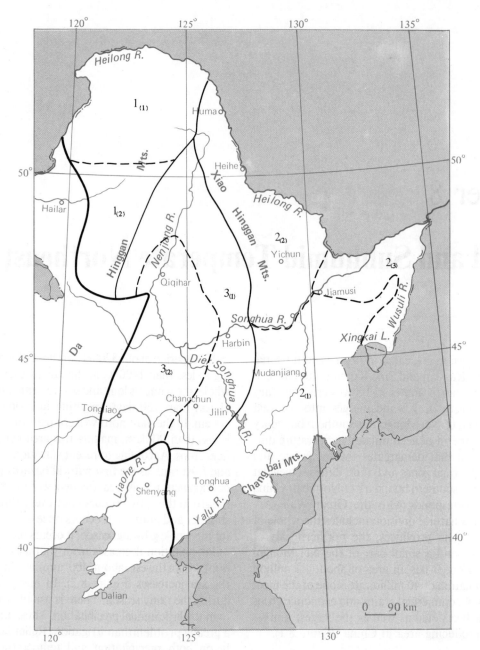

Figure 8-1 A location map of Humid and subhumid Temperate Northeast China

favorable condition for agricultural development. Yet, because of the flood hazard in low plains and depressions, harvesting is sometimes at risk. Spring drought is often a problem, and supplementary irrigation is necessary.

Extensive Low Plain Surrounded by Middle and Low Mountains

This natural division is essentially composed of the Songhua-Nenjiang Plain or the northern part of the Nor-

theast China Plain, which is surrounded, like a huge horseshoe, by a series of NE-SW trending middle and low mountains — the Da Hinggan Mountains on the west (Photo III-1), the Yilehuli Mountain on the north, and the Xiao Hinggan-Changbai mountain system on the east (Photos III-2, III-5). Only on southern side does the horseshoe open and merge into the southern part of the Northeast China Plain. A NW-SE topographic profile of this natural division reveals it as essentially of two humps-one saddle type, with the Da Hinggan Mountains and the

Xiao Hinggan-Changbai mountain system as the two humps and the Songhua-Nenjiang Plain as the saddle (Figure 8-2).

Geologically, the Songhua-Nenjiang Plain is a huge synclinorium that has been subjected to periodical sinking since the Mesozoic era. Geomorphologically, it is an undulating low plain with an elevation decreasing from the peripheral piedmonts of 250 to 300 m to below 200 m around the confluence of the Songhua River and its largest tributary, the Nenjiang River. The surrounding middle and low mountains are generally strongly folded belts. The Da Hinggan Mountains are geologically ancient, with generally smooth rounded profiles. Its crest line has an elevation of around 1000 m, with steep eastern slopes overlooking the Northeast China Plain and relatively gentle western slopes descending to the Nei Mongol Plateau. The Xiao Hinggan Mountains, uplifted as late as the middle Quaternary, are composed of low mountains and hills below 1000 m in elevation. The Changbai Mountains, however, are characterized by a series of NE-SW trending middle and low mountains, alternating with broad intermontane basins and valleys. Its highest peak, Mount Baitou towers to 2744 m. The scenic volcanic crater lake, Tianchi Lake, is located on its top. Amid the Xiao Hinggan-Changbai mountain system, there lies also the graben-structured Sanjiang Plain between Heilong River and its two tributaries, the Songhua River and Wusuli River, with an elevation generally below 80 m. At the foot of all these surrounding mountains, the undulating and fertile piedmont plains are well developed. They can be considered as the transitional belt between the Northeast China Plain and its surrounding mountains.

Bountiful Water Resources

Comparatively rich precipitation and low evaporation as well as densely forested mountains bring bountiful water resources to this natural division. The Heilong River and its largest tributary the Songhua River along with numerous other tributaries form one of the greatest river networks in China, second only to the Changjiang River in annual discharge. There is no deficiency in irrigation water when there is such a need. The swampy Sanjiang Plain is especially representative of the bountiful surface water resources. The annual runoff depth reaches 500 to 600 mm in the southeastern part of this natural division where annual precipitation is heaviest. It decreases northwestwardly to less than 200 mm in the central part of the Northeast China Plain where many small inland drainage basins are formed, becoming somewhat higher again on the Da Hinggan Mountains (more than 200 mm). On the whole, the variation coefficient (Cv) is small, about 0.3 to 0.4 in surrounding montane areas, increasing to 0.6 to 0.8 on the northern Songhua-Nenjiang Plain and Sanjiang Plain. Thanks to the surrounding rugged topography, hydroelectric power potential capacity is high in the upper and middle reaches of many swift rivers of this division. There are many excellent sites for the construction of reservoirs. The Second Songhua River is especially famous for its hydroelectric power station; the Fengman Hydroelectric Station, together with its huge water reservoir, the Songhua Lake.

The ground water resources are also quite rich, both in quantity and in quality. They have been outlined already

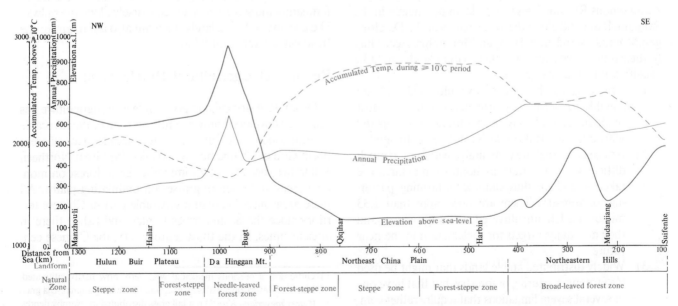

Figure 8-2 A comprehensive physical profile along the Manzhouli-Harbin-Suifenhe Railway, Northeast China

in Chapter 4.

Bountiful Land Resources

The above-mentioned climatic and geomorphological conditions lead also to bountiful land resources. Both the Songhua-Nenjiang Plain and Sanjiang Plain (Photo III-3) comprise the largest modern pioneer settlement belt in China, and the surrounding mountains form the largest timber-producing area.

There are now about 8.7 million hectares (ha) of cropland in Heilongjiang Province alone, of which more than two thirds are concentrated in the Songhua-Nenjiang Plain. Here, deep and fertile black earth and chernozem are widely distributed; after decades of large-scale reclamation, croplands occupy more than 50 to 60 per cent of the total land area in some well-developed districts. As these croplands with loessic parent materials are mainly located on the gentle rolling slopes, soil erosion has been rather severe. According to one estimate, about one half of the total cropland in the province is subjected to erosion, about one fourth of which is subjected to severe erosion. A blackish soil horizon, originally 60 to 70 cm in thickness, on gently sloping cropland — with a slope of about 3 to 5° might be reduced to a thickness of only 20 to 30 cm or less, after 40 to 50 years of cultivation. Therefore, a recent trend in agricultural development has been the downward movement of croplands from gentle sloping lands to the low-lying meadows (the so-called *dintzee*) and the upward migration of woodlands along both gentle and steep slopes.

There are still about 5 million ha of arable virgin lands in Heilongjiang Province. The province ranks first in quality and second in area only to the Xinjiang Uigur Autonomous Region. These virgin lands are mainly in the Sanjiang Plain and along the lower flanks of the Da Hinggan Mountains and Xiao Hinggan Mountains. According to their productive capability, these virgin lands might be classified into four categories:

1) Arable lands that are easily cultivated and are suitable for farming, pasturing, or forestry. Most of them have already been turned into cropland with only about 333,000 ha not yet cultivated.

2) Arable lands that are cultivatable without too much difficulty — if certain meliorating measures are taken — and are, thus, suitable for farming, pasturing, or forestry. There are now more than 1.33 million ha left uncultivated in the province, and they are major targets for reclamation in the near future.

3) Widely distributed arable lands that might be used for farming, pasturing or forestry, but that have one or several severe limitations that require rather complicated meliorating measures. After the adoption of these measures, however, such lands would become good cropland. For example, marshes and meadows that are presently unworkable because of flood hazards have been classified as Category 3 lands; yet, if adequate hydrologic and other agricultural improvement measures were taken, they could be transformed into excellent cropland with fertile soils and good irrigation facilities.

4) Arable lands with severe limitations for agricultural use. Unless urgently required, they remain better uncultivated.

Bountiful Biological Resources

As stated in Chapter 6, biological resources, both fauna and flora, are quite rich in this natural division. It has been famous for timber products and "three marvels" — ginseng, marten fur and ula grass[1]. Before a large fire which broke out at the turn of the century, even the streets of old Jilin city were paved with large timbers.

Taiga forest is distributed in the Da Hinggan Mountains, with larch (*Larix dahurica*) as the chief timber resource. It is also the most important hunting ground in China. In the Xiao Hinggan-Changbai mountain system, mixed needle-and broad-leaved forests dominate, with red pine (*Pinus koraiensis*) and many hardwoods as the major timber resources. Hunting is also an important occupation here. In the extensive low plains luxuriant, productive forest-steppe and meadow-steppe are distributed and provided excellent pasturing and hunting grounds before the nineteenth century (Photo III-4). Even as late as 1953, when the author traveled by train from Harbin to Halar, a vast greenish sea of tall grasses still dominated the landscape of the Songhua-Nenjiang Plain, with small patches of farmlands and settlements dotted sparsely along the railway like a string of tiny islands. Fish are also abundant in the freshwater rivers and lakes.

Recency of Agricultural Development

Owing to historical and political factors, agriculture has been well developed only recently, in spite of the favorable natural conditions. Until only a few decades ago, all areas northward of Changchun were called the Great Northern Wilderness because of an immense sea of forest on montane areas and a sea of grasses on undulating plains. All these areas have been an inseparable part of Chinese territory since the Soshin tribes hunted and fished there in ancient times. In the third century A.D., the Chinese cen-

---

1) Ginseng is a precious medicinal wild herb, now partly cultivated. Marten (Martes zibellina) is an excellent fur-bearing animal. Ula grass (*Carex meyeriana Kunth*) is a tall grass distributed in swampy depressions, excellent raw material for making shoes.

tral government introduced farming into the region. And in 1858 A.D., the Qing imperial government was finally forced to give up its no-allowance policy for the Chinese farmers moving into the region. Yet, up to 1897, there were only about 5300 ha of croplands and 25000 inhabitants in the whole Heilongjiang Province. Between 1897 and 1949, pioneer settlement by small-holding farmers developed rapidly; in 1930 (the eve of the Japanese invasion of Northeast China), Heilongjiang Province had already a total cropland area of 3.85 million ha and a population of 3.7 million. In 1949, the cropland had reached a new level of 5.7 million ha. In about 30 years after 1949, the croplands had increased again by more than 8.7 million ha, with total croplands occupying about 12 per cent of the total land area in the province.

Such a recency of agricultural development implies that human impact on the physical geographical environment has been relatively slight here and that there are still abundant natural resources waiting for profitable use.

NATURAL REGIONS

Based on the integration of the above-mentioned physical features and their areal differentiation, this natural division can be subdivided into three natural regions (Figure 8-1).

1. The Da Hinggan Mountains — Needle-Leaved Forest Region

This needle-leaved forest region is located in the northern Da Hinggan Mountains — the northern and northwestern borders are delimited by the Sino-Russian international boundary, the southwestern and southern borders coincide with the 1.2 isopleth of the aridity index, and the eastern border follows the mountain footline. Its northeastern boundary with the Northeast China Mountains mixed needle-and broad-leaved region is chiefly demarcated by vegetation features: to the east mixed forest (characterized by red pine) dominates, whereas on the western side, needle-leaved forest (characterized by larch) predominates.

The region was first uplifted during the Hercynian Tectonic Movement with a NNE-trend and accompanying large-scale granite intrusion. During the Mesozoic era, there again occurred faulting and magmatic activities as well as volcanic eruption along old structural lines. During the Cenozoic era, faulting and basalt flow dominated. And during the late Tertiary and early Quaternary periods, neotectonic movements uplifted all post-Yanshan and post-Himalayan erosion surfaces to their present elevation, about 1000 m and 500 to 600 m respectively. As a result of its geological history, the Da Hinggan Mountains are generally around 1000 m in elevation, with a few peaks

between 1000 to 1500 m. One topographic feature is its asymmetrical slopes; its eastern slopes are much steeper — overlooking the Songhua-Nenjiang Plain like a huge wall — with conspicuous steplike terraces and erosion surfaces as well as numerous deep gorges. Its western slopes, on the other hand, are much gentler, merging with the Nei Mongol Plateau at about 700 m above sea level. Another feature is the conspicuous longitudinal areal differentiation: northward to 51°N, granites predominate, with rounded peaks, gentle slopes, and well-preserved erosion surfaces; whereas southward to 51°N, bedrocks are more varied with greater relative relief and narrower width. Permafrost and glaciation are also less commonly seen.

Climatically, it is unique for being the only cool-temperate zone in China. Annual solar radiation totals less than 110 kcal/cm². Winter is long and severe, with a mean July temperature below 19°C and a mean January temperature below -28°C, and the winter season lasts more than eight months. Except in some intermontane valleys below 700 m in elevation, there is practically no summer season. Accumulated temperature during the ≥ 10°C period totals generally less than 1600°C, and the frost-free season is less than 90 days. Seasonal frozen soil during wintertime usually has a thickness of more than 2.5 m, and permafrost is distributed continuously from 51°N northward and in isolated islands from 51°N southward. As a whole, it is a humid region with annual precipitation of about 450 mm on the eastern slopes and 350 mm on the western slopes, of which about 80 per cent of the precipitation is concentrated in the rainy season from May to September. Owing to low temperature and, consequently, low evaporation, the aridity index is generally lower than 1.0.

Surface water resources are quite rich. Rivers on western and northern slopes flow into the Heilong River and its headwaters, the Ergun River whereas on the eastern slopes, they flow into the Nenjiang River. The annual runoff depth increases from 150 mm in the south to 250 mm in the north, the Cv value decreases from 0.6 to 0.3. About 80 per cent of the annual runoff is concentrated in a period from June to September.

Distribution of vegetation is determined not only by geographical location and the accompanying climatic features, but also by the uplift of the Da Hinggan Mountains. As a result it is here that the taiga forest has its southernmost extension in the world with the excellent softwood larch (*Larix dahurica*) predominant. Other needle-leaved trees include pine (*Pinus sylvestris,* mostly distributed in sandy areas), spruce (*Picea obovata, P. microsperma),* and at higher elevations, dwarf pine (*Pinus pumila*) and juniper (*Juniperus dahurica*). Broad-leaved trees including birch (*Betula platyphylla*), poplar (*Populus davidiana, P. suaveolens),* willow (*Salix rorida*), and oak (*Quercus mongolica*), grow on lower slopes. In swampy areas there are often scattered shrubby clumps of dwarf birch (*Betula fruticosa)* and some species of willow (*Salix brachypoda, S. sibirica*). On montane slopes, the most commonly seen shrubs are Rhododendron (*R. macromulata, R. dahurica*) — whose name in Chinese literally means make-mountain-red because of their brilliant red flowers — cover nearly all the montane slopes in autumn. Other shrubs include *Vaccinium vitis-idaea* and *Ledum palustre*. Vertical zonation and differentiation between eastern and western slopes is conspicuous. For example, at about 50°N, broad-leaved forests predominate below 700 m; from 700 to 1000 m, the forests become *Larix-Rhododendron,* whereas above 1000 m, the dwarfed pine becomes an important component in the larch forests.

This natural region can be subdivided into two subregions, the northern and the central Da Hinggan Mountains.

1(1) Northern Da Hinggan Mountains

The southern boundary of the northern Da Hinggan Mountains with the next subregion is roughly demarcated at 51°N. Areal differentiation between these two subregions is chiefly characterized by the following: (1) permafrost is distributed continuously in the Northern Da Hinggan Mountains, but it is found only in isolated islands in the central Da Hinggan Mountains; (2) larch forests are distributed extensively in the northern subregion, whereas they are found only in patches in the central subregion.

Vertical zonation and areal differentiation between different slopes is quite conspicuous. First, precipitation and temperature decrease gradually from southeast to northwest. Second, land types and vegetation types vary according to different altitudinal zones and different slopes. For example, shrubby *Pinus pumila* forest is found on stony montane slopes. *Larix dahurica-Pinus pumila* forest is on steep slopes between 900 and 1000 m, *Larix-Rhododendron* forest is on lower steep slopes, and Larix grass forest is found on gentle slopes.

The dominant landscape is the larch-covered middle mountain type. Larch forests are, and will be, the chief flora here; lumbering will become the chief economic activity in this subregion. However, great care must be taken to protect and to renew these forest resources so that a permanent lumbering ground can be secured. Hunting is also an important occupation, especially for minority groups in the area. All precious and rare vertebrates should be strictly protected; hunting should be reorganized into a joint hunting-domesticating enterprise. Small patches of farmlands can be cultivated in the valley plain below 700 m. They should be restricted to producing vegetables and some foods for local consumption only, and effective methods of frost control and soil conservation must be undertaken.

1(2) Central Da Hinggan Mountains[1]

In the central Da Hinggan Mountains subregion, precipitation and temperature also decrease from southeast to northwest. The distribution of vegetation in terms of land types is generally shrubby *Pinus pumila* forest on stony montane tops, *Larix-Rhododendron* forest on steep slopes above 1200 m, mixed needle-leaved (*Larix dahurica*) and broad-leaved (*Quercus mongolica*) forest on lower slopes, broad-leaved forest (*Quercus, Populus, Betula*) on low mountains and hills, and shrubby meadow and meadow-steppe on lower gentle slopes and intermontane valleys.

The land-use pattern is essentially the same as in the northern Da Hinggan Mountains subregion, with forestry being paramount. Yet owing to a warmer climate and better pasture, agriculture and animal husbandry could occupy a somewhat larger proportion of the subregion, whereas forestry and hunting should be more restricted.

2. The Northeast China Mountains — Mixed Needle and Broad-Leaved Forest Region

The Northeast China Mountains region is located in eastern Heilongjiang and Jilin provinces and northeastern Liaoning province. It includes three distinct physical units: the Xiao Hinggan Mountains, the Sanjiang Plain and the Changbai Mountains. Its eastern and northern limits are international boundaries with Korea and Russia. Its southern boundary corresponds to the 3200°C isotherm of accumulated temperature during $\geq 10°C$ period. This humid montane region is separated on its western border from the subhumid Northeast China Plain by the 1.0 isopleth of the aridity index.

Geological and geomorphological features play an important role in areal differentiation. The Changbai Mountains and their neighboring folded mountains were first faulted and uplifted during the Hercynian Tectonic Movement, then folded and faulted again during the Yanshan Tectonic Movement. They were peneplaned in the early Tertiary period, then covered with erupted basalt. Now they are characterized by a series of NE-SW trending middle and low mountains alternating with broad intermontane basins and valleys. The highest peak, Mount Baitou, is surrounded by a basalt plateau at an elevation of more than 2600 m. The Xiao Hinggan Mountains were first uplifted along a NE-SW direction during the Hercynian Tectonic Movement. In the early Tertiary period, a broad graben cut it into two parts, the southern and the northwestern. It was then peneplaned and uplifted again dur-

---

1) The southern Da Hinggan Mountains, just like the southern Northeast China Plain, lie beyond this natural division.

ing the middle Quaternary period along a NW-SE direction. It is now composed of a series of low mountains and hills. The low, flat, and swampy Sanjiang Plain is a graben amid these folded mountains, with ground-surface materials composed chiefly of loessic clay.

Climatically, this is a typical humid temperate region, except for some subhumid rainshadow areas (such as the western half of the Sanjiang Plain). Owing to the region's extending through 12 degrees of latitude, there exists considerable thermal variation from south to north. Annual solar radiation totals 130 kcal/cm^2 in the south, decreasing to 110 kcal/cm^2 in the north. The growing season, thus, decreases from 150 to 110 days, whereas accumulated temperature during the $\geq$ 10°C period falls from 2500 to 1800°C. Rainfall, on the other hand, is determined mainly by distance from the sea as well as by topographic features. Annual precipitation totals more than 800 mm in the southeast (Tonghua at 894 m in elevation recorded a high of 1217 mm in 1954), about 500 mm in the central part (Yanji, 515 mm), and only 450 to 550 mm in the Xiao Hinggan Mountains and the Sanjiang Plain. About 60 to 80 per cent of the annual precipitation occurs in growing season.

The combined effects of climatic and topographic conditions have resulted in a well-developed river system in the region. The lofty and dome-shaped Mount Baitou, with its heavy precipitation, is characterized by spectacular annular drainage pattern. Besides the mighty Heilong River and several independent rivers, there are three large river systems in the region — the Songhua, the Wusuli and the Yalu rivers — of which, two (the Wusuli River and Yalu River) serve as internationally boundaries, and two (the Songhua River and Yalu River) both originate in Mount Baitou. Annual runoff depth corresponds with annual precipitation, as high as 500 to 600 mm on the southern slopes of Mount Baitou, decreasing to less than 200 mm in the northwest and to less than 150 mm in the Sanjiang Plain. The Cv value generally ranges from 0.3 to 0.4, increasing to 0.6 to 0.7 in the Sanjiang Plain.

Mixed needle and broad-leaved forests dominate the region, and vertical distribution of vegetation is prominent. In the basic vegetation belt (500 to 1000 m in elevation), the dominant needle-leaved tree species is red pine mingled with other needle-leaved trees (*Abies holophylla,* and *Picea obovata*), and many broad-leaved trees *(Tilia amurensis, Fraxinus mandshurica,* and *Quercus mongolica).* Above this basic vegetation belt lies the needle-leaved forest belt dominated by red pine and spruce (*Picea obovata, P. koraiensis*). Its lower limit has an elevation of 1000 to 1200 m on the western slope of Mount Baitou, lowering to 800 to 900 m on the Zhangguangcai Mountain and 650 to 750 m on the Xiao Hinggan Mountains. Further up is the spruce-fir forest belt, dominated by *Picea jeppensis*, with its lower limit of 1700 to 1800 m on the western slope of

Mount Baitou and descending to 1350 to 1450 m on the Zhangguangcai Mountain and 900 to 1000 m on the Xiao Hinggan Mountains. Again higher up there appears the subalpine dwarf birch (*Betula ermanii)* belt, restricted on Mount Baitou to elevations between 1800 to 2100 m. Finally, alpine tundra mainly composed of cushion shrubby vegetation, such as *Dryas octopetala, Vaccinium vitisidaea,* and *Phyllodoce caerulea,* occurs on the top of Mount Baitou.

By the integration of all the above-mentioned physical conditions, this natural region can be subdivided into the following three subregions:

2(1) Changbai Mountains

In the Changbai Mountains, a humid temperate subregion, the characteristic landscape is one of beautiful tree-clad middle mountains. These mountains consist of a series of NE-SW trending ranges, mostly 500 to 1500 m in elevation, where water and biological resources are quite rich. There are also many broad basins and valleys interspersed among these mountains, such as the Mudan River valley where the famous ancient Bohai Kingdom was founded in the eighth century. Since the late nineteenth century this has been one of the most prosperous pioneer settlement belts in China.

Vertical zonation is conspicuous, and a better land-use scheme should be based on the specific physical conditions of different altitudinal belts. Three groupings might be determined on the basis of these altitudinal belts:

1) The top altitudinal belt is the alpine tundra belt,which is limited in land area, yet unique and scenic. With an elevation of 1800 to 2100 m and a location immediately below the eternal snow line, it is the loftiest area in Northeast China and enjoys special and precious biological resources. It should be reserved mostly as a national park; one famous natural protected area has already been established.

2) All altitudinal forested belts, including the mixed forest belt, the needle-leaved forest belt, the spruce-fir forest belt, and the subalpine dwarf birch forest belt should be devoted mainly to forestry with scientific management. Restricted hunting and pasturing in the forests might also be conducted. The most serious problem at the moment is deforestation, especially in the basic vegetation belt (mixed forest belt). It should be strictly prohibited as soon as possible and reforestation should be encouraged.

3) In the low basins and valleys where luxuriant meadow, swamp, and other azonal vegetation dominates as well as where water and land resources are rich, farming and animal husbandry have the potential for being intensively developed. Paddy rice has long been one of the important crops here. On

either side of the larger rivers, there usually stretch broad floodplains and two or three steps of terraces with deep, fertile meadow soils. They are excellent sites for intensive farming.

2(2) Xiao Hinggan Mountains

Just like the above-mentioned Changbai Mountains subregion, the Xiao Hinggan Mountains comprise a humid-temperate, and tree-clad montane subregion. Owing to its higher latitude, however, the temperature is much lower and seasonal frozen soils are extensively developed. Furthermore, owing to a much smaller amplitude of uplift and much better preserved ancient erosion surfaces, the land surface is usually less dissected and less well drained. Consequently, swamps are widely distributed.

Mainly on the basis of geological and topographical features, the subregion can be subdivided again into three parts:

1) The southeastern part has a much more ancient geological history and a higher relief. It is mainly composed of low mountains and hills, with an elevation of around 500 m and an annual precipitation of about 600 mm. There are also several peaks rising above 1000 m, with much heavier precipitation. The dominant vegetation is mixed needle-and-broad-leaved forest, with a large red pine component. It is essentially a land for lumbering, with some broad valleys, such as the Tangwang valley, suitable for farming.

2) The middle part, located along the Heihe-Bei'an line, is a Tertiary graben basin. It is mainly composed of undulating plain, with an elevation of about 400 m and an annual precipitation of about 500 mm. The dominant vegetation is also mixed forest, with both larch and red pine the major components. On the whole, it is cold and swampy and is more suitable for forestry than for farming.

3) The northwestern part of the subregion is geologically associated with the Da Hinggan Mountains. It is composed mainly of hills, with an elevation of about 400 m and an annual precipitation of about 450 mm. It is the northernmost part of the region, with a larger larch component and also some *Pinus sylvestris* trees although mixed needle and broad-leaved forest still predominates. It should be confined to forestry except for some cultivable river valleys such as the narrow, dissected plain along the main Heilong River.

2(3) The Sanjiang Plain

With an area of 42,500 sq km, the Sanjiang Plain is geologically a Mesozoic era graben. Since the Quaternary period, it has undergone periodic sinking, with a total thickness of accumulated sediments of more than 200 m. The general topography dips from southwest to northeast. The southwestern part has an elevation of 60 to 80 m, with ground-surface materials composed mainly of silt. Drainage is good, and the distribution of marsh is quite limited. The northeastern part, however, has an elevation of 40 to 60 m and contains several sluggish, meandering rivers with extensive swampy floodplains. There is also a clay layer with a thickness of 3 to 17 m on the ground surface that makes drainage difficult and results in the extensive distribution of swamps. Surrounding the plain on the western and southern sides are a series of low mountains and hills, generally between 500 to 1000 m in elevation, with a number of their offshoots stretching into the margins of the plain as isolated small "mountains without roots", under such a geomorphological background, landform types are quite varied and are sequenced from floodplains and terraces to hills and low mountains. Each landform type has its specific temperature, moisture, vegetation, soil, and other physical features, and hence, provides an excellent index for classifying comprehensive land types (Figure 7-2, Landsat image 2).

Climatically, the subregion is entirely controlled by alternating monsoons. It has a mean January temperature of -18 to -22°C, a mean July temperature of 21 to 22°C, a frost-free season of 120 to 145 days, an accumulated temperature during the ≥ 10°C period of 2200 to 2600°C, and an annual precipitation of 500 to 650 mm. Owing to the rain shadow effects of the surrounding mountains, it is reduced to a subhumid zone in its southwestern part, with forest-steppe and fertile black earth as zonal vegetation and zonal soil respectively. In the northeastern part, however, it is still a temperate humid zone with deciduous broad-leaved forest (chiefly oak) and dark brown earth as zonal vegetation and zonal soil respectively. Such an areal differentiation reflects clearly on land use. In the southwestern part, most lands have been cultivated, resulting in one of the most flourishing agricultural areas in China. In the northeastern part, however, swamps are still extensively distributed, but many state farms are now working efficiently to transform this wilderness into a granary.

From the national viewpoint, this subregion certainly should be devoted extensively and exclusively to agriculture and reclamation. On arable plains, forestry should be restricted to shelter belts. Again, overall planning for the subregion should be designed and lands should be allocated for certain development according to their special attributes. For example, on mountain and hill slopes, forests should be dominant with no allowance for cultivation. Besides, reclamation should go hand in hand with the establishment of a permanent agricultural base. Draining away surplus surface water and clearing up luxuriant wild grass cover are certainly two necessary measures in transforming swamps into farmlands. But care must also be taken to use adequate fertilizer and to meliorate some infertile soils. Irrigation engineering works should be introduced at the beginning of reclamation, otherwise the surface water resources will be depleted and soil moisture will deteriorate rapidly.

3. The Northeast China Plain — Forest-Steppe Region

The Northeast China Plain is a natural region sandwiched between two humps — the Da Hinggan Mountains and the Northeast China Mountains. Its southern boundary is a climatic line (the 3200°C isotherm of accumulated temperature during the ≥ 10°C period) that merges into the southern part of the Northeast China Plain without distinct topographic features.

Geologically, it corresponds to the so-called Songhua-Nenjiang synclinorium that has been periodically sinking since the Mesozoic era, with a greater amplitude of sinking on the western side. Quaternary deposits of about 30 to 50 m in thickness have accumulated. Such a sinking and accumulating process is still going on in its central part — the Harbin-Qiqihar-Baicheng triangle. Geomorphologically, this triangle is the lowest part of the Songhua-Nenjiang Plain, with an elevation of only 130 to 140 m. Numerous lakes and swamps dot the area. On the periphery of this depressed alluvial-lacustrine central part, is a broad belt of diluvial-alluvial piedmont plains with an elevation of 250 to 300 m. The piedmont plain was slightly uplifted during the late Quaternary period, then dissected into a mosaic of gentle ridges, undulating terraces, and low hills, with a relative relief of less than 50 to 100 m. The broad low NWW-trending water divide between the Songhua River and the Liaohe River represents the southern part of this piedmont belt. It was uplifted in the late Quaternary period and is accompanied by numerous small volcanic cones.

Owing to its relatively low topographic location, precipitation is less abundant than in the surrounding mountains. The annual precipitation is about 600 mm in the southeast, decreasing to about 400 mm in the northwest. The aridity index ranges from 1.0 to 1.2; hence, in physical regionalization, it is delimited as subhumid and is different from the surrounding humid mountains. The annual runoff depth also decreases from about 200 mm in the southeast to about 25 mm in the northwest.

In terms of temperature this region differs from the other two montane regions by being a little bit warmer and having no conspicuous altitudinal zonation. Winter is still long and cold, with a mean January temperature of -17 to -24°C and a pronounced decrease of temperature from south to north. Summer is short and rather warm, with mean July temperatures of 21 to 25°C, the thermal difference between south and north being insignificant. Sometimes, when the Pacific Subtropical High extends far westward and shifts somewhat northward, the weather may be quite hot, with maximum high temperatures above 35°C. Compared with similar latitudes, this is an area with the greatest seasonal variation of temperature in the world, showing the rather large factor of continentality in China's east coast climate. A warm summer coupled with bountiful precipitation in that season (accounting for about 60 to 70 per cent of the total annual precipitation) creates a favorable climatic environment for agricultural production. Spring wheat, corn, millet, soy beans, sugar beets, and other crops grow with a high annual yield.

Vegetation is luxuriant and the soil fertile. In peripheral piedmonts, the zonal vegetation is forest-steppe, with a varied flora. The dominant grass species are "sheep's grass" (*Aneurolepidium chinense*), *Stipa baicalensis* and *Fillifolium sibiricum*. The most commonly seen trees are oak *(Quercus monglica)* and birch *(Betula dahurica)*, and the common shrubs are *Lespedeza bicolor* and *Corylur hetephyua*. Owing to heavy human intervention, few needle-leaved trees still exist. The corresponding zonal soil type is black earth, usually with a humus layer more than 60 to 80 cm deep, or even more than 1 m deep. In the central part of the Songhua-Nenjiang Plain, owing to a higher water table, meadow-steppe and chernozem soils dominate. There are also many azonal vegetation and azonal soil types, such as sandy, saline, and swampy vegetation and soil types. Near the western border, the plain gradually becomes characterized by steppe vegetation and a chestnut soil belt.

In terms of areal differentiation, two subregions can be clearly defined.

3(1) Piedmont Forest-Steppe and Black Earth Subregion

The Piedmont forest-steppe and black earth subregion is distributed in a great arc (Figure 8-1); it is a transitional belt between the Northeast China Mountains (mixed needle and broad-leaved forest region) and the Northeast China Plain (forest-steppe region). Geomorphologically, it is essentially a gently sloping plain, composed mostly of diluvial-alluvial deposits, with a thin veneer of Quaternary loessic materials. In the widest part, such as the area between the Wuyur River and the Hulan River, it stretches more than 100 km from northeast to southwest, with an elevation around 300 m and with an extensive level land surface. Yet, in most other areas it has been dissected into undulating or hilly land, with severe soil creep and soil erosion.

The subregion has an accumulated temperature during the ≥ 10°C period of 2400 to 3100°C and an annual precipitation of 500 to 600 mm. The dominant vegetation type is forest-steppe, which has been largely turned into farmland, leaving only small patches of mixed forest. The zonal soil type is the fertile, deep black earth with a top layer humus content of 3 to 6 per cent, decreasing usually from north to south.

As far as agricultural development is concerned, this subregion is nearly ideal. Here, land resources are rich; precipitation is sufficient for crop growing and is not excessive. Temperature conditions are generally adequate for a single annual crop system, although sometimes low temperature and early frost are harmful. Thanks to good drainage, salinization and flood hazards are nonexistent. During the last 100 years, this subregion has been developed rapidly into one of the most prosperous pioneer settlement areas in China. It is now one of the most important commercial grains and soy bean producing areas in China. Future trends for agricultural development lie chiefly in the domain of higher yield per unit area, which again depends mainly on control and melioration of limiting factors, such as low temperature and soil erosion. An overall plan is very much needed.

3(2) Songhua-Nenjiang Plain Meadow-Steppe and Chernozem Subregion

The Songhua-Nenjiang Plain meadow-steppe and cher-

nozem subregion is the central part of Northeast China Plain, a forest-steppe region with comparatively low elevation and poor drainage; consequently there are widely spread swamps and saline soils. It is composed of Quaternary deposits 30 to 50 m in thickness, mostly alluvial-lacustrine and partly diluvial and aeolian. It also has great reserves of petroleum and natural gases.

Compared to the above-mentioned piedmont subregion (see Figure 8-3), it is a little bit warmer and more arid. It has an accumulated temperature during the ≥ 10°C period of 2500 to 3100°C and an annual precipitation of 400 to 500 mm with an aridity index of 1.0 to 1.2. The dominant vegetation type is meadow-steppe, characterized by its drought-tolerant and salinity-tolerant grass species. On its western margin, this subregion merges into the semiarid climate and steppe vegetation of Nei Mongol. Apart from growth along riversides and cultural vegetation, few trees exist, hence, such place names as One Tree, Two Trees, and Three Trees, are often used. The important terminal station of the Beijing-Harbin Railway is Sankeshu (literally, Three Trees), a northeastern suburb of Harbin. The zonal soil type is chernozem, which is quite fertile but not as good as black earth. In depressed areas, it is liable to carbonation and salinization.

As a whole, this is an area for comprehensive develop-

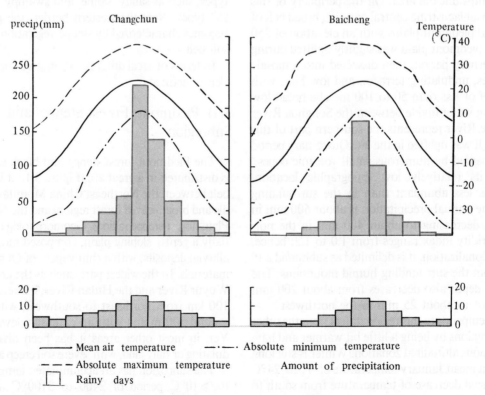

Figure 8-3 A comparison of climate between the Piedmont forest-steppe subregion (Changchun) and the meadow-steppe subregion (Baicheng)

ment of farming, pasturing, and forestry. First of all, there should be overall planning to coordinate land-use patterns in direct association with specific land types. For example, the meadow-terrace, meadow-alluvial plain and other good arable lands should be devoted to farming; the swampy and saline low depressions are better kept as pasture. River margins and sandy dunes, on the other hand, are the best places for forestry. Reeds (*Phragmites communis*), widely distributed in the swamps and around numerous lakes are also an important source of pulp. It is also worth pointing out a special land-use combination pattern in the Daqing area, the location of one of the largest oil fields in China. It was originally an immense wild meadow inhabited by a few herdsmen. Now a gigantic oil complex has been built as well as numerous gardens of vegetables and other crops that are cultivated chiefly by the family members of petroleum workers.

References

[1]   *Physical Regionalization of China* Working Committee, Chinese Academy of Sciences, 1958 – 1959, *Physical Regionalization of China,* 8 Vols, Science Press, Beijing. (In Chinese)

[2]   Integrated Investigation Team of Heilong River Basin, 1961, *Natural Conditions of the Heilong River Basin and its Neighboring Areas,* Science Press, Beijing. (In Chinese)

[3]   Institute of Forestry and Soil, Chinese Academy of Sciences, 1980, *Soils of Northeast China,* Science Press, Beijing. (In Chinese)

[4]   Shen Yuancun, 1980, "Land Types and Their Transformation Measures in the Sanjiang Plain", *Acta Geographica Sinica,* Vol. 35, No. 2. (In Chinese, with English abstract)

[5]   Changchun Institute of Geography, Chinese Academy of Sciences, 1981, "A Preliminary Study on Change of Natural Environment After Current Large-scale Reclamation in the Sanjiang Plain", *Acta Geographica Sinica,* Vol. 36, No. 1. (In Chinese, with English abstract)

[6]   Chao Sung-chiao (Zhao Songqiao), 1981, "Transforming Wilderness into Farmland: An Evaluation of Natural Conditions for Agricultural Development in Heilongjiang Province", *China Geographer* (Westview Press, Boulder, Colorado), No. 11.

[7]   Zhao Songqiao et al., 1983, *Natural Zones and Land Types in the Heilongjiang Province,* Science Press, Beijing. (In Chinese)

Chapter 9

Humid and Subhumid Warm-temperate North China

Humid and subhumid warm temperate North China lies between the north latitudes 32 to 43° and the east longitudes 104 to 115°. It has an area about 1 million sq km and an population of nearly 250 million. It includes the Liaodong and Shandong peninsulas, the Liaohe Plain, the North China Plain, the Shanxi-Hebei Mountains and the Loess Plateau (Figure 9-1). Administratively, it consists of the whole of Beijing and Tianjin municipalities, the provinces of Shandong and Shanxi, the major parts of Hebei, Henan, and Shaanxi provinces, and a part of Liaoning, Jiangsu, Anhui, Gansu and Ningxia. Historically, it is the site of Lantian man and Peking man (Table 1-4) and the cradle of Chinese civilization, and it has been the political economic, and cultural center of China for thousands of years. Of the six ancient capitals of China (Beijing, Xi'an, Luoyang, Kaifeng, Nanjing and Hangzhou), four are located in this natural division. Yet, it is also the scene of many natural hazards, such as floods, droughts, salinization, soil erosion, and earthquakes. For the 1000 years since the Song Dynasty, foods have been imported from Central China and South China into North China.

This natural division is bounded by the Bohai Sea and the Yellow Sea on the east and by the marginal mountains of the Qinghai-Xizang Plateau on the west. Its southern boundary runs along the Qinling Mountains-Huaihe River line, which will be discussed in more detail in the next chapter. The northern boundary generally follows the 3200°C isotherm of accumulative temperature during the ≥ 10°C period. This sets the northern limit of winter

wheat, cotton, and warm temperate fruit trees.

CHIEF PHYSICAL FEATURES*

Structure and Relief

The geological history of the North China Platform shows that it has a common foundation and has undergone similar major tectonic movements. It is characterized by long periods of continental building and relatively short periods of transgression. Its basement is formed by metamorphic rocks, such as gneiss, crystalline schist, slate, and phyllite. From the Cambrian to the middle Ordovician periods there was a time of transgression when marine sediments several-thousand meters thick were deposited. Then followed a period of land denudation during the upper Ordovician period through the middle Carboniferous period. From then to the end of the Permian period, North China was brought under a shallow and fluctuating sea; shales, sandstones, and thin beds of coal and limestone were deposited. Since the end of the Permian period, North China as a whole has emerged as a continent.

The Mesozoic era was distinguished by igneous intrusions, volcanic outbursts and tectonic movements. Intensive folding and faulting accompanied by intrusion of enormous masses of granite and outpouring of andesite and rhyolite occurred in the Yanshan Mountains and the

---

* The first draft of this section was prepared by Professor Lin Chao of Peking University.

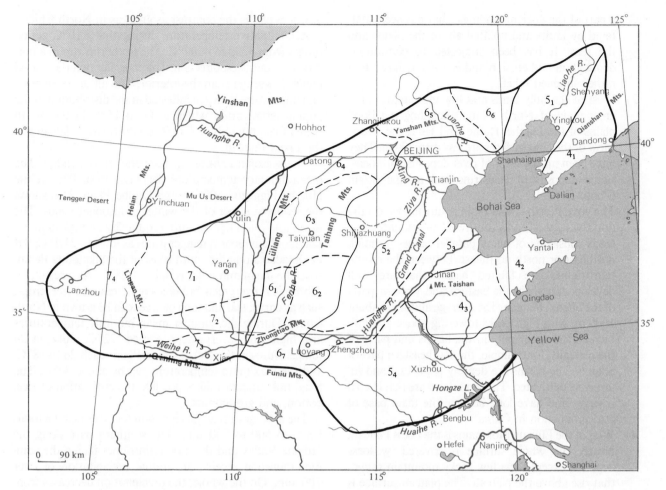

Figure 9-1 A location map of humid and subhumid warm-temperate North China

Shanxi Plateau as well as the Liaodong and Shandong peninsulas and the Loess Plateau.

The Cenozoic era was marked by uplifting and subsidence. The mountains of the Liaodong and Shandong peninsulas and the Yanshan and Liupan mountains as well as the Shanxi Plateau were regions of uplifting, whereas, fluvial and lacustrine deposits were developed in the North China Plain and the intermontane basins.

An event of paramount importance to the landscape of North China was the accumulation of loess deposits during the Quaternary period (see Figure 3-5). In northern Shaanxi and eastern Gansu, the loess forms a nearly continuous mantle that covers the underlying topography. This is the Loess Plateau par excellence. East of the Loess Plateau, loess also covers the basins and lower mountain slopes of the Shanxi Plateau. Through water erosion, loess was carried by the streams and redeposited in the surrounding plains and the seas. In fact, this silty material is found nearly everywhere in North China. The significance of the

loess in the development of the landscapes of North China can never be overestimated.

Tectonic activities did not cease during the Quaternary period. This is shown by volcanic eruption and the outpouring of lava, especially in and around the Tertiary basins in northern Shanxi. Earthquakes are a frequently occurring phenomenon in North China.

The land surface of North China consists of four fundamental types of landforms, that is, plains, hills, plateaus, and mountains. These types are grouped into units that conform to geological structures. On the whole, the topography descends from the plateaus and mountains in the west to low plains and hills in the east and finally submerges under the seas. It trends generally NE to SW or NNE to SSW. From east to west, four major units may be distinguished.

1) The first unit consists of hilly lands extending from the Liaodong Peninsula to Shandong Peninsula and the central Shandong Mountains and Hills. This is

a part of the ancient landmass that has been shattered by faults and uplifted above the plains and the seas. It has been subjected to continuous denudation and erosion and is, thus, reduced to a rather subdued relief.

2) West of the hilly lands and low mountains is the great plain of North China, including the flood plains of the Liaohe, Haihe, and Huanghe rivers. Most of the land surfaces are less than 100 m. They have been mostly built of river deposits with occasional deposits of lacustrine and marine origin (Photo III-7).

3) The great North China Plain is surrounded on the northwest, west, and southwest by mountainous lands consisting of many parallel ranges trending northeast or north northeast. Between the mountain ranges are faulted basins that are well developed, both in the eastern and western parts of the Shanxi Plateau. The average altitude is about 1000 m, with high peaks towering above 2000 m. Level lands occur only in the basins and parts of the plateaus, for example, the southeastern part of Shanxi. The basins are densely populated and intensively cultivated. The mountains are rich in coal reserves and have long become the main base of coal production in China.

4) West of the Lüliang Mountains is the Loess Plateau, Nearly the whole surface is covered by loess deposits, except for a few rocky mountain ranges that rise above the plateau. The plateau surface is level with an elevation of over 1000 m. Owing to the nature of loess deposits and the deforestation during historical times, the Loess Plateau has suffered from severe erosion and has been dissected into a maze of gullies and hills. The chief elements of the landforms are broadly defined as yuan (level land, Photo III-9), liang (hilly ridge, Photo III-10), and mao (gentle hillslope).

Climate and Hydrology

North China belongs to the continental monsoon-type of subhumid warm-temperate climate. This features great seasonal contrast, great daily and annual range in temperature, as well as moderate but highly concentrated precipitation.

North China lies in the middle latitudes of Eastern Monsoon China. In winter, it is dominated by the Mongolian-Siberian high pressure system. The cold and dry prevailing northerly winds sweep over North China. In summer, the conditions are completely reversed. The continent is dominated by a low pressure system, and warm moisture-laden southerly winds prevail. Although such reversals in monsoons are noticed all over Eastern Monsoon China,

nowhere else is the contrast as great as in North China.

Accumulative temperature during the $\geqslant 10°C$ period ranges from 3200 to 4500°C. The isotherms run in a northeast to southwest direction. This indicates that the coastal region is warmer than the interior. The influence of relief on temperature is also manifested in the distribution of the annual temperature which is 11 to 15°C in the North China Plain but only 8 to 11°C on the Shanxi Plateau and the Loess Plateau.

The seasonal contrast in temperature is striking. The mean January temperature decreases from 0°C in the Huaihe Plain to –10°C in the Liaohe Plain. This shows the reducing effect of the winter monsoonal winds. The mean minimum temperature is quite low. Yingkou in the Liaohe. Plain records temperatures as low as –31.1°C, but in Qingdao along the south coast of the Shandong Peninsula, the mean minimum is only –16.5°C. As a result, no subtropical evergreen broad-leaved trees and tree crops, such as citrus, tea, and tung oil trees can be grown in North China. During the summer mean monthly temperatures are high and everywhere exceed 22°C. July mean temperatures in the North China Plain reach 26 to 28°C, and the maximum temperature can be above 40°C. The long, hot summer makes possible the cultivation of rice, cotton, and groundnuts.

The average annual precipitation for North China totals between 500 to 700 mm. The southern part along the Huaihe Valley and the mountain slopes facing the summer monsoon receive an annual precipitation of over 700 mm. On the whole, the precipitation decreases from the coast to the interior, which shows the reducing effect of the summer monsoon. The aridity index is mostly 1.0 to 1.5; it is less than 1.0 along the coast and over 1.5 in the western part of the Loess Plateau. This shows that the major part of North China belongs to the subhumid climate, with a small humid area in the east and a semiarid climate in the west.

There are great differences in the seasonal distribution of precipitation. Of the total amount, 60 to 70 per cent precipitates in summer (June to August); only 5 to 10 per cent in winter (December to February). Spring and autumn receive 15 per cent and 30 per cent of the precipitation respectively. Summer precipitation often comes in the form of heavy showers, sometimes as much as several hundred millimeters in one day. The result is disastrous, causing soil erosion and mudflows in the mountains and floods on the plains.

The annual variability of rainfall in North China is great. The annual amount of rainfall in a rainy year may reach 1400 mm; a dry year may yield only a little more than 100 mm.

The major rivers of North China are the Huanghe (Photo II-7, II-8), Huaihe, Haihe, and Liaohe rivers. Most of these rivers flow from the west to the east; only the

Liaohe River flows from the north to the south. All of them have their sources in the mountains and highlands, then flow to the lowlands and empty into the seas. The discharge of the rivers of North China is not large because of the moderate amount of regional precipitation and the high evapo-transpiration and percolation. Total discharge amounts to 150 billion cubic meters per year (m3/yr).The runoff depth is generally less than 200 mm and, in some parts, even less than 50 mm. Considering the increasing demand of population, agriculture, and industry on water, the potential shortage of the water supply is evident. In recent years, underground water has been exploited and even somewhat overexploited. A scheme for diverting water from the Changjiang River northward to the North China Plain has been seriously considered.

A special problem of the hydrology of North China's rivers is the enormous sediment load that originates in the loessic lands of the plateaus and mountains. The average loessic silt load of the Huanghe River is 37.7 kilograms per cubic meters (kg/m3), and 60.8 kg/m3 for the Yongding River (a tributary of the Haihe River). The total sediment load per year amounts to 1.6 billion tons for the Huanghe River and 80 million tons for the Yongding River. The Huanghe River (Yellow River) is named after its yellowish water as is the Yellow Sea; the Yongding River was once called the Small Yellow River.

Vegetation and Soils

The basic vegetation type in North China is the summer green deciduous broad-leaved forest. It represents a response to the continental monsoon climate with a cold and dry winter and a warm and wet summer. The trees shed their leaves during the winter and turn to a dense and green canopy during the summer. The most commonly seen trees are various species of oak (*Quercus*), maple (*Acer*), poplar (*Populus*), birch (*Betula*), elm (*Ulmus*) and willow (*Salix*). Some coniferous trees, such as pine (*Pinus*), juniper (*Juniperus*), fir (*Abies*), spruce (*Picea*), and larch (*Larix*), are also important constituents.

Following the distribution of precipitation and humidity, the vegetation types change from broad-leaved trees in the east to forest-steppe and steppe in the Loess Plateau. There are also differences in composition of the vegetation from east to west. More varieties of *Quercus* are found in the peninsulas of Liaodong and Shandong than in the mountains west of them. Species of pine are also different: *Pinus deniflora* is found in the east, *P. tabuliformis* flourishes in the west.

Landform is another important factor that influences the distribution of vegetation. For example, in the mountainous region around Beijing, the lower slopes are covered by *Quercus* and *Pinus tabuliformis*, and there are undergrowths of mesophytic to semixerophytic types.

Above 1200 to 1800 m, the deciduous forest is dominated by poplar, birch, and oak. Above 1800 m, some fir, spruce, and larch are found. The tops of the mountains are covered by alpine shrubby meadow.

Natural vegetation in the North China Plain was cleared for many thousands of years for agricultural purposes, so that few examples of it can be found. Common deciduous trees that are sparsely grown around the settlements and along the roads and riverbanks are willow (*Salix babylonica, S. matsudana*), poplar (*Populus canadensis, P. tomentosa*), maple (*Acer negundo*), *Robinia pseudoacacia*, *Sophora japonica*, and *Paulownia*. The last one is widely planted in the plain south of the Huanghe River along the sandy tracts seasonally flooded by the rivers. It is interplanted with the crops as windbreaks. Because its canopy is sparse and its roots deep, it does not interfere with the crops. The common shrubs are *Vitex chinensis, Zizyphus jujuba* var. *spinosus, Lespedeza*, and *Caragana sinica*. All of them are native plants of North China. *Amorpha fruticosa* is widely planted on saline soils as fodder and fertilizer; its branches are also used for weaving handicrafts.

On the Loess Plateau, most of the level lands on the plateau surface and along the river valleys have been cultivated. The slopes of gullies are covered by shrubs and grasses. The common shrubs are lespedeza (*Lespedeza bicolor*), *Spiraea pobescens, Vitex negundo* var. *heterophylla, Zizyphus jujuba*, and *Caragana microphylla*. The predominant species of grasses are *Brothriochloa ischaeman* and *Themede triandra* var. japonica. Toward the western margin of the Loess Plateau, the grass species are replaced by stipa (*Stipa grandis, S. krylovii*). The spread of these steppe elements is attributed to the semiarid climate.

Owing to the influence of climatic-botanical factors, the distribution of soil types coincides in a general way with the distribution of vegetation. Zonal soils are represented by brown forest soils (luvisols), drab soils (cambisols) and heilu tu (dark loessial soil). Brown forest soils are distributed in the humid coastal regions of the Liaodong and Shandong peninsulas. They also appear in vertical zonation on the montane regions. The altitude of the upper limit of brown forest soils increases from several hundred meters above sea-level in the east to more than 1500 m in the west. Drab soil are extensively distributed west of the brown forest soil belt. They are developed under a subhumid climate and are mostly located on the piedmont plains, foothills, and mountain slopes to over 1000 m. In the North China Plain, under the influence of a high ground water table, wet soils (fluvisols) are developed. Tracts of saline soils (solonchaks) are found along the coastal belt as well as in the central parts of the great plains.

On the Loess Plateau the zonal soils are dark loessial soils (calcic cambisols). They are characterized by a thick upper layer of dark soil in profile and are found on the level land on the top of the plateau. These soils have been sub-jected to continual cultivation. The dark color is the result of humus accumulation. On the slopes of the rolling hills, mainly in the northern or western parts of the Loess Plateau, cultivated loessial soils are developed under con-tinuous soil erosion. They are characterized as being little differentiated from the parent loess material. Along the Weihe and the Fenhe valleys, the fertile alluvial soils under intensive cultivation are stratified old loessial soils. The soil profile is characterized by two distinctive layers, an upper layer developed under continuous manuring and a lower layer developed under natural conditions.

NATURAL REGIONS

The above analysis of the physical features of North China shows that areal differentiation occurs in every aspect. Within this natural division, 4 natural regions and 19 subregions may be subdivided (see Figure 9-1).

4. The Liaodong-Shandong Peninsulas—Deciduous Broad-Leaved Forest Region

This region includes the Liaodong Peninsula, the Shan-dong Peninsula and the central Shandong mountains and hills. It is the door to North China, with long seacoasts and numerous seaports, of which Qingdao, Dalian and Yantai are the most important. Its climate is much more mild and humid than interior areas with similar latitudes. On the whole, its physical conditions are favorable to agricultural development as well as to general economic development.

Geologically and geomorphologically, the region is a part of the so-called North China Platform, with pre-Cambrian metamorphic rocks serving as its base. During the Yan-shan Tectonic Movement, there occurred intense NE-SW trending faulting and warping as well as widespread intru-sion of granite and eruption of andesite. A great fault, star-ting from the mouth of the Liaohe River southward along the western coast of the Liaodong Peninsula and the Miaodao Islands and up to Laizhou Bay, resulted in a dif-ferential neotectonic movement, the eastern side uplifting, the western side subsiding. A series of lakes formed along the fault line during the Tertiary period and later further subsiding formed the present Baohai Sea—a territorial sea as well as one of the major oil-bearing areas in China (Photo II-1). The Himalayan Tectonic Movement was characteriz-ed by faulting and basalt eruption. Since the Tertiary period, the region has experienced a long period of denuda-tion resulting in the dominance of undulating mountainous and hilly lands. There are a few peaks towering above 1000 m. as well as some narrow strips of coastal and in-termontane plains. Three geomorphological units can be identified in the region:

1) The Liaodong Peninsula has the NE-SW trending Qianshan Mountains as its backbone, with a total length about 340 km and an average elevation of 500 m.
2) The Shandong Peninsula is composed mainly of un-dulating hills, mostly under 300 m in elevation, with a few higher rocky mountains (the highest be-ing Mount Laoshan: 1133 m). Its western margin is the peneplaned and alluvial Jiaolai Plain, which is intensively cultivated and densely populated.
3) The central Shandong mountains and hills are com-posed mainly of horsts and grabens with many peaks above 1000 m in elevation, of which the sacred Mount Taishan (1524 m) is the highest. Numerous rivers radiate away from Mount Taishan to form an impressive fluvial network. Low moun-tains and hills bordering Mount. Taishan usually have rounded tops and gentle slopes and are quite rich in coal reserves.

The climate is mild and humid. The mean January temperature ranges from 14 to 0°C, with an absolute minimum temperature of –32°C in the northern part of the Liaodong Peninsula and –20°C in the southern part of the Shandong Penisula. The mean July air temperature shows little areal differentiation from south to north and is generally around 25°C. The frost-free season lasts from 5.5 to 8.5 months, with an accumulative temperature of 3200 to 4500°C during the $\geq 10°C$ period. Double crop-ping is generally practiced. Annual precipitation totals 600 to 900 mm, decreasing from the coast inland. About 60 per cent of the annual precipitation is concentrated in the summer. Ther aridity index is generally around 1.0, which represents the marginal humid condition—yet, owing to rather high annual variability, about two-thirds of the year deficient in moisture during the $\geq 10°C$ period.

Rivers are mostly short and rapidly flowing. The annual runoff depth ranges from 200 to 400 mm, with the eastern Liaodong Peninsula and central Shandong mountains at-taining 500 mm. Cv values range from 0.4 to 0.5. About 80 to 90 per cent of the annual runoff is concentrated in summer and autumn. Floods are not infrequent in the lowlands.

The original vegetation has been long since removed. Zonal vegetation is deciduous broad-leaved forest, with oak (*Quercus dentata, Q. acutissima, Q. liaodongensis, Q. mongolica*) dominating. The most commonly seen con-iferous tree is *Pinus densiflora*, differing both from that of the Changbai Mountains (with *Pinus koraiensis* dominating) and from the North China Plain (*P. tabulifor-mis* dominating). The common shrubs are *Rhododendron dauricum* and *Deutzia amurensis*. In the northern or moun-tainous area, mixed broad and needle-leaved forests are also seen. The region ranks as one of the important tussah silk producing areas in China, with *Q. acutissima* widely planted to feed the tussah silk worms. Fruit trees, such as apple, pear, peach, chestnut, and walnut are also widely and productively grown.

The zonal soil is brown forest soil developed under warm-temperate deciduous broad-leaved forest. Argillation and leaching are the chief soil-forming processes. The soil profile is characterized by a top litter layer and a humus layer (with humus content 2 to 9 per cent) overlying a brownish transitional B-horizon, which has generally a clay content of about 50 per cent. Under mixed broad-and needle-leaved forest, podzolization is an important soil-forming process; in the karst area, calcification is also important. Owing to the extensive distribution of granite and gneiss, which are composed largely of silica, as well as the large tracts of sandy areas along the rivers and coasts, sandy soils are widely developed in the region, which is planted largely in groundnuts.

The region enjoys quite rich land resources and well-managed agricultural development. The most widespread land type is low mountains of 500 to 1000 m, covered mostly with deciduous broad-leaved trees. High hills have elevations between 200 to 500 m, with slopes of generally more than 10°. They are covered with oak trees, shrubs and orchards; these are also some dry farming lands on the more gentle slopes. Low hills have an elevation between 50 to 200 m, with a slope of less than 10°, and they are largely planted with fruit trees. All mountains and hills are liable to severe soil erosion when vegetative cover has been removed. Therefore, soil conservation and reforestation are of paramount importance. Plains lie below 50 m, including the alluvial plain, undulating peneplaned plain, the marine-eroded terrace, and the marine deposited plain. They are characterized by level lands and thick soils as well as better temperature and moisture conditions. Hence, farmlands and population are concentrated here. Beaches include both marine beach and fluvial beach. The marine beach is flat and marshy and is suitable for salt production, reed planting, or fishery. Part of the marine beach might be also reclaimed as cropland. The fluvial beach is mostly composed of sandy land and is fit for planting forest-shelter belts.

In conjunction with the three geomorphological units, three natural subregions can be demarcated within the region.

4(1) Liaodong Peninsula

Located between the Bohai Sea and the Yellow Sea and separated from the Shandong Peninsula by the 57 nautical mile width of the Bohai Sea Strait, the subregion consists mainly of hills below 500 m in elevation and has a width of about 150 km, narrowing down to about 10 km at its southern tip (only 5 km near Jinxian County) and

submerging into the sea south of Dalian. The topography also descends from north to south. The coastline extends for 900 km, with numerous rocky islands as well as extensive marine beaches. Annual precipitation totals 600 to 1000 mm (as high as 1200 mm in the northeastern mountainous area). Accumulative temperature during the $\geqslant 10°C$ period ranges from 3000 to 3700°C, and the frost-free season is from 160 to 200 days. Winter wheat and fruit trees can survive without special protection in the winter. The zonal vegetation is broad-leaved forest, with brown forest soils as zonal soils; at higher elevations, mixed broad-and needle-leaved forests and dark brown forest soils are developed. In terms of land-use, this is one of the most important fruit, fish, and tussah silk-producing areas in China.

4(2) Shandong Peninsula

Encircled by the Bohai Sea and the Yellow Sea to the north, east, and south, this subregion is connected with the central Shandong mountains and hills through the broad Jiaolai Plain in the west. About two thirds of the subregion consists of undulating hills 200 to 300 m in elevation. Shorelines are bold and rocky; wave-cut cliffs rise abruptly from the water's edge and isolated hilltops project out of the sea as rocky islands. There are many excellent harbors, yet they are handicapped by the limited hinterland. Annual precipitation totals 700 to 900 mm; the accumulative temperature during the $\geq 10°C$ period is 4000 to 4600°C, and the frost-free season lasts 180 to 210 days. Qingdao has a mean January air temperature of –1.1°C, and an absolute minimum temperature above –16°C; it is the warmest part of North China during winter time. Zonal vegetation is deciduous broad-leaved forest, and the zonal soil is brown forest soil. Practically all level lands and a part of the montane slopes have been cultivated. This is one of the major fruit, fish, and groundnut producing areas in China.

4(3) Central Shandong Mountains and Hills

Located west of the Shandong Peninsula and roughly separated from the North China Plain by the 200 m contour line, the central Shandong mountains and hills subregion is a vast area of block mountains with Mount Taishan and other high peaks over 1000 m in elevation rising majestically above the surrounding plains. Annual precipitation totals 600 to 900 mm, and accumulative temperature during the $\geq 10°C$ period ranges from 4000 to 4500°C. Zonal vegetation is deciduous broad-leaved forest, with some coniferous trees at higher elevations. Zonal soils are brown forest soils and leached brown forest soil. This is mainly a land of forestry, with farming restricted to intermontane valleys. Much work in soil con-

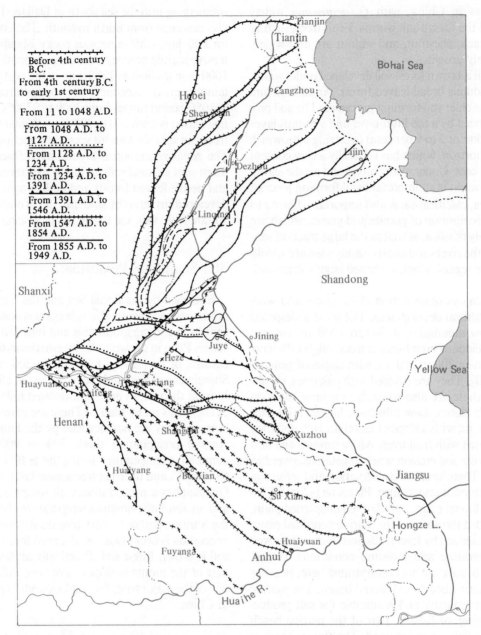

Figure 9-2 Changes of channels in the lower Huanghe River since 602 B.C.

servation and reforestation needs to be done here to stabilize the land use and increase productivity.

5. The North China Plain—Deciduous Broad-Leaved Forest Region

The North China Plain—deciduous broad-leaved forest region stretches from the Taihang Mountains and the Funiu Mountains eastward up to the Bohai Sea and the Yellow Sea, and from the Liaohe valley and the Yanshan Mountains southward up to the Qinling Mountains-Huaihe

River line for an area of about 400,000 sq km. The region is composed essentially of flat low alluvial plains under 50 mm in elevation, and it is characterized by rich land and water resources as well as a subhumid warm temperate climate. For thousands of years, it has been the political, economic, and cultural center of China. The famous German geographer and geologist, von Richthofen, once aptly compared it with the great drawing room of an apartment to which all other rooms open and through which they unite.

Geologically the North China Plain is a part of the

Figure 9-3 A sketch of diluvial fan in southern Taihang Mountains piedmont plain (drawn by Lu Renwei)

North China Platform that has been submerged since the late Cretaceous period and overlain with thick Cenozoic sediments. The thickness of these sediments varies from several hundred to more than 5000 m. The lower Liaohe Valley has also been a submerged zone since the Yanshan Tectonic Movement, with Cenozoic sediments of about 2000 m thick. It is still submerging, yet the Liaohe River silts it even faster. The mouth of the Liaohe River was located at Niuzhuang 200 years ago, but now it is about 40 km to the south.

The river systems are well developed, and they are the agents that nourish the immense North China Plain. The Huanghe River is, of course, the most important river in the region. It flows through the loose, easily eroded Loess Plateau, carrying a silt load of about 1.6 billion tons annually. It then suddenly drops into the North China Plain after passing through the Sanmen Gorge and starts the building of a gigantic delta of about 250,000 sq km. Hence, the Huanghe River is notorious for shifting, flooding, and silting in its lower reaches. According to historical documents, the Huanghe River has broken its banks 1573 times since 602 B.C. 26 of the breaks have been of a major significance (Figure 9-2). Generally speaking, the Huanghe River broke its banks northward before A.D. 1194, captured the Haihe River system and flowed into the Bohai Sea; from A.D.1194-1280, and again since 1946, the river shifted southward to near its present channel; during A.D. 1280—1854 and again from 1938—1946, the river moved to the south of the Shandong Peninsula, incorporated the Huaihe River basin and emptied into the Yellow Sea. The present channel has been silted up as a river above ground, generally 5 to 6 m above the surrounding plain (maximum, 13 m), and it serves as a water divide between the Haihe River and the Huaihe River. According to an estimate, about a quarter of its silt load is deposited on the North China Plain; the remaining three quarter transported into the sea. The current delta of 5450 sq km has been created since A.D. 1855 when the Huanghe River assumed its present channel (except for the

period 1938–1946). In addition, the Haihe River and its tributaries (the Yongding, Ziya, and Weihe rivers), the Huaihe River and its tributaries (the Yinghe and Guohe rivers), as well as the Liaohe River contribute a part to the plain building.

The region is characterized by monotonous flatness. Few localities have an absolute relief above 50 m and a relative relief of above 20 m. Nevertheless , from the piedmont plain eastward to the sea, three geomorphological and hydrologic belts can be demarcated.

1) Diluvial-alluvial inclining plains (Figure 9-3) are Located at the piedmonts of the Taihang Mountains and other mountains. They are composed mainly of numerous alluvial fans, with a width of 10 to 55 km, an elevation of 50 to 100 m, and a gradient of 1/200 to 1/2000. Ground water resources are rich in this belt, with the ground water table varying from 2 to 10 m. Drainage is good, mineralization of ground water is less than 1 gram/liter (g/l). But on the margin of the alluvial fans, the ground water table is shallow and sometimes emerges as springs. Numerous marshy lands and fresh water lakes (such as the Baiyang Lake) exist on the interalluvial fan depressions.

2) Alluvial low plains are located to the east of the above-mentioned belt, these are major components of the North China Plain, with elevations below 50 m and sometimes as low as 3 m. The gradient is negligible (1/5000 to 1/10,000) and drainage is poor. Rivers often flow sluggishly "above" the ground, resulting in frequent dam breaks and flood hazard. Marshy depressions and shallow lakes are widely dotted on the alluvial plains. Aeolian sandy dunes are also formed along the river channels. The ground water table is shallow (generally 2 to 3 m from the surface); the water quality is fairly good (mineralization 1 to 2 g/l), although the quantity is not as plentiful as in the above-mentioned belt.

3) Coastal plains are low plains bordering the Bohai

Sea, with an elevation generally lower than 4 m and a gradient smaller than 1/10000. They are formed both by fluvial and marine deposits. Clay dominates the land. The ground water table ranges from 1 to 1.5 m deep, with high mineralization (10–30‰ or more, composed mainly of NaCl). The landforms along the coast are quite varied, including old and new deltas, sandy beaches, and dunes.

The climate is essentially subhumid warm-temperate characterized by a dry, windy spring, a warm rainy summer, a fine, calm autumn and a cold, dry winter. The hottest month (July) is quite subtropical, with a mean temperature of 24 to 29°C and an absolute maximum temperature above 40°C (Beijing, 42.6°C; Luoyang, 44.2°C). The coldest month (January) is, however boreal, with a mean temperature of 0 to -14°C, and an absolute minimum temperature below –20°C (Beijing –22.8°C; Shenyang, –30.5°C). Accumulative temperature during the ≥ 10°C period totals 3400 to 4500°C (Beijing, 4056°C), and the frost-free season lasts between 170 to 220 days. The cropping system of three crops in two years dominates, and this is one of the major winter wheat, corn, soybean, and cotton-producing areas in China. However, owing to the limitations of the cold severe winter, subtropical crops, such as citrus fruits, tea, and tung oil cannot survive, and no vegetables can be grown in the winter (outside of greenhouses).

Annual precipitation totals 500 to 800 mm (Tianjin, 529.5 mm; Beijing, 640.6 mm). Rainfall is concentrated in summer, with a maximum daily precipitation of more than 100 to 200 mm. Winter (December to February) accounts for only 3 to 7 per cent of the annual precipitation and spring (March to May) only 10 to 14 per cent. Annual variation is generally as high as 25 per cent, and the rainiest year may be more than 10 times the precipitation of the driest year. For example, Beijing had 1406 mm of precipitation in 1959, but only 168.5 mm in 1891. Therefore, spring drought and summer floods together with salinization and aeolian sand often plague agricultural production in the region. According to incomplete documentary statistics, there occurred 407 droughts and 387 floods in the Haihe River basin from A.D. 1368–1949.

The region has been cultivated more than 7000 years, hence, practically all natural vegetation has been destroyed or transformed. Soils, being more "conservative", are a better index for reflecting the physical geographical environment. There are two major soil units in the North China Plain: the drab soil (cambisols) and the wet soil (fluvisols). The former is the so-called zonal soil, mostly distributed on the diluvial-alluvial inclining plains, with deep solum light texture and good drainage; under good management, these have been excellent agricultural soils for thousands of years. The latter are widely distributed on alluvial plains, with a heavier texture and usually impeded drainage and

hence, plagued by salinization hazards. Other locally distributed soils include Shachang soils, coastal saline soils, and meadow soils.

By the integration of all areal differentiation of the above-mentioned physical elements, four subregions might be identified:

5(1) Lower Liaohe Plain

Separated from the North China Plain roughly by Shanhaiguan (Photo III-6)—the starting point of the Great Wall, this subregion has been formed by alluvial depositions of the Liaohe River on a sinking valley for a long period, with a maximum thickness of loose sediments of nearly 2000 m. Geomorphologically, it is low and flat, with an elevation of less than 50 m. Climatically, it is the coolest part of the region, with mean January air temperature of about –10°C and a frost-free season about 180 days; nevertheless, cotton can still be grown. Annual precipitation totals about 600 mm. In its southern part, wild marshes were extensively distributed until only a few decades ago, and it was, thus, called the Southern Great Wilderness in contrast to the Northern Great Wilderness north of Harbin. But now the wilderness (together with a part of the sea) has been reclaimed as productive paddy lands.

5(2) The Haihe Plain

Delimited on the north by the Yanshan Mountains, on the west by the Taihang Mountains, on the south by the lower Huanghe River channel, and bordering the Bohai Sea on the east, this region has been formed by the huge depositions of the Haihe River and the Huanghe River. It consists of immense low plains encircled by mountains and hills on its northern and western borders. Distribution of major land types around the Beijing area is shown by Figure 9-4 and Landsat image 1. This is one of the most important grain-and cotton-producing areas in China, with double cropping in its southern part and three crops in two years in its northern part.

5(3) The Huanghe River Flooding Area

Located mainly along the Huanghe River and its flood plains, this area has better climatic conditions than the Haihe Plain. But it has been frequently scoured by the flooding of the Huanghe River, with sandy dunes extensively distributed along the river channel. The damages inflicted by the intentional breaching of the Huayuankou during 1938–1946 war efforts were devastating, and the area had not fully recovered until recently.

5(4) North Huaihe Plain

Located north of the Huaihe River, this area has the best climatic conditions in the region, yet owing to frequent

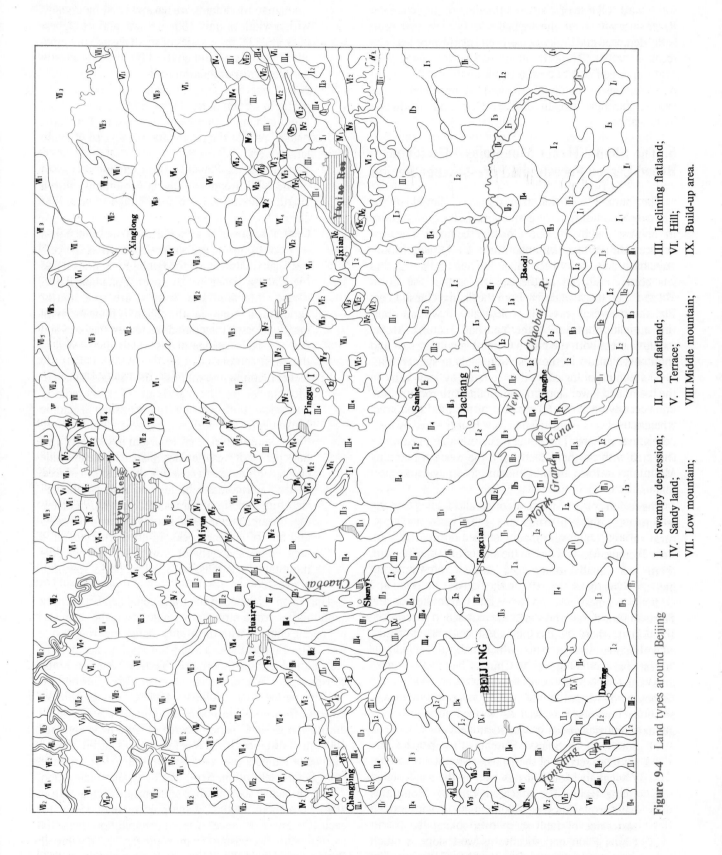

Figure 9-4 Land types around Beijing

I. Swampy depression;   II. Low flatland;   III. Inclining flatland;

IV. Sandy land;   V. Terrace;   VI. Hill;

VII. Low mountain;   VIII. Middle mountain;   IX. Build-up area.

southward collapse of dikes and flooding of the Huanghe River since A.D. 1194, the original river systems have been badly deranged and drainage has been greatly impeded. The result is that "small rains result in small hazards; large rains result in large floods; no rains result in drought". Recently, conditions in the Huaihe River have been greatly improved through stream management and landscape modification.

6. The Shanxi-Hebei Mountains—Deciduous Broad-Leaved Forests and Forest-Steppe Region

The Shanxi-Hebei Mountains region is delimited by the Taihang Mountains in the east, by the Lüliang Mountains in the west, and by the Great Wall in the north. The area between the Taihang Mountains and the Lüliang Mountains has traditionally been called the Shanxi Plateau and has been considered as the eastern part of the Loess Plateau. Yet, this area differs from the Shanxi-Gansu Loess Plateau in several ways: (1) geologically, it was an uplifting anticlinorium during the Yanshan Tectonic Movement, whereas both the North China Plain in the east and the Loess Plateau in the west were sinking synclinoria;

(2) Geomorphologically, it is covered by loess only in the basins and lower slopes of mountains with rocky mountains occupying about 60 per cent of the total land area, whereas the Loess Plateau is extensively distributed by loess nearly everywhere. Therefore, it is preferable to combine this area with northern Hebei and the western Liaoning Mountains to form an independent physical geographical region.

The region consists of a series of parallel folding-faulting mountains as well as intermontane graben basins and structural valleys. They were mainly shaped during the Yanshan Tectonic Movement. On the axis of the anticlinorium, there are usually outcropped ancient pre-Cambrian metamorphic rocks, whereas on the flanks, Cambrian and Ordovician limestones and shales and Carboniferous and Permian sandstones and shales are exposed in succession. In the latter, there are also thin limestone layers interbedded with numerous coal seams, which makes this region the largest coal producing area in China. The relief is rugged and varied, with many peaks towering above 2000 m (the highest being Mount Wutai at 3058 m), resulting in conspicuous horizontal as well as vertical zonation. Five geomorphological units can be identified:

1) Southeastern Shanxi Plateau—this includes the southern section of the Taihang Mountains and the Zhongtiao Mountains as well as the sandwiched Qinhe Basin. The southern Taihang Mountains have an elevation of 1500 to 2000 m, with a steep east slope of fault scarp overlooking the North China Plain majestically; its west slope is much more gentle, merging into the Shanxi Plateau. The Zhongtiao Mountains are narrow horst mountains, with a width of only 10 to 15 km, and its highest peak at 2322 m. A series of small basins surrounded by mountains and covered by loess are mostly dissected into undulating hills.

2) Northeastern Shanxi Plateau—this includes the northern section of the Taihang Mountains, Wutai Mountain, and other block mountains. This is the highest part of the region and the source of many large rivers.

3) Western Shanxi Plateau—this is mainly composed of the NNE-trending Lüliang Mountains, with a length of about 400 km and its highest peak is at 2831 m.

4) Central basins of Shanxi Plateau—these include a series of Cenozoic graben basins, such as the central Shanxi Basin along the Fenhe River graben and the Datong Basin in the upper Yongding River. These intermontane basins are the major agricultural areas in the Shanxi Plateau, with marginal loessic materials dissected into terraces and flat alluvial-lacustrine plains widely distributed in the basin centers.

5) Hebei-Liaoning mountains—the famous Yanshan Mountains are the backbone of this geomorphological unit. During the Yanshan Tectonic Movement, the Paralleling range and valley landforms were shaped. Uplifted again in the early Tertiary period, it was then peneplaned in the Pliocene epoch, with red earth covering the erosion surfaces. In the intermontane basins and valleys, alluvial-lacustrine deposits are distributed. In western Liaoning, the area is composed mainly of low mountains and hills of igneous rocks, in northern Hebei, the area consists chiefly of low mountains, middle mountains and hills, with a few peaks towering above 2000 m.

The region is located at the transitional belt between the subhumid warm-temperate climate and the semiarid temperate climate, with the aridity increasing from east to west in the Shanxi Plateau and from south to north in the Hebei and Liaoning mountains. Vertical zonation is also conspicuous. Accumulative temperature during the $\geq 10°C$ period totals 3000 to 4500°C. On the Shanxi Plateau, from south to north, accumulative temperature during the $\geq 10°C$ period decreases by 340°C for each 100 km of distance. If the relief element is added, the difference in temperature is even more sharp. For example, from Beijing (elevation, 51m) to Zhangjiakou (elevation, 770m), the active accumulative temperature decreases from 4046°C to about 3000°C, a sharp decrease of 550°C for each 100 km of distance. Annual precipitation totals 400 to 700 mm, decreasing from southeast to northwest. Besides, windward slopes are much more abundant than leeward slopes and mountainous areas are much more numerous than the basins. The seasonal proportion of precipitation is roughly: summer, more than 60 per cent; autumn, 20 per cent; spring, 15 per cent; winter, only 2 to 4 per cent.

Rivers consist mainly of the upper reaches of the Haihe River such as the Chaobai River, the Yongding River, and the Ziya River. The Fenhe River and the Qinhe River are tributaries of the Huanghe River, whereas the Luanhe River and the Liaohe River flow into the sea independently. These rivers are generally characterized by: (1) low discharge—the largest river of the region, the Luanhe River has an annual discharge of only 3.2 billion cu m and an annual runoff depth of 80.3 mm; (2) uneven seasonal distribution of discharge—generally the first peak is in August and the second peak is in March. Annually, the maximum discharge is usually much greater than the minimum discharge. The Luanhe River had its maximum record of 9.5 billion cu m in 1959, and its minimum record of only 1.05 billion cu m in 1951; (3) heavy silt load—the heaviest silt load is in the Yongding River, as stated above.

Distribution of vegetation and soils is also transitional. Horizontally, zonal vegetation changes from deciduous broad-leaved forest in the southeast to forest-steppe in the northwest, while zonal soils change from drab soils to dark loessic soils and chestnut soils. Vertical zonation is also conspicuous. As a rule, from the bases of the mountain up to 1200 m, warm-temperate deciduous broad-leaved forests dominate with different species of oaks (*Quercus variabilis, Q. acutissima, Q. liaodongensis,* and *Q. dentata*) most commonly seen. There are also other deciduous broad-leaved trees, such as *Fraxinus chinensis, Carpinus turczaninoii,* and *Acer mono,* and some coniferous trees, such as *Pinus tabuliformis* and *P. densiflora.* From 1200 to 1800 m, the temperate deciduous broad-leaved forest dominates, with *Betula platyphylla* and *Populus davidiana* as the chief tree species. From 1800 to 2500 m, the forest becomes the cool-temperate needle-leaved type, with spruce (*Picea meyeri, P. wilsonii*) forest on the lower slopes (1800 to 2300 m) and larch (*Larix principis-ruprechtii*) on the upper slopes (2300 to 2500 m). Their upper limit represents the tree line in North China. Finally, from 2500 m up to the montane top, subalpine shrubby meadow predominates with common deciduous shrubs, such as *Caragana jubata,* and *Spiraea alpina,* and numerous grasses, such as *Festuca ovins, Koleria cristata,* and *Kobresia bellardi.*

Practically all level lands, including flood plains, terraces, and platforms have long been intensively cultivated, with winter wheat, corn, millet and sorghum as the chief crops. Cotton can also be grown in the southern Shanxi basins. The dominant cropping system is three crops in two years. But, north of the Zhanjiakou–Fengning–Chaoyang line (somewhat north of the Great Wall), the cropping system changes to one crop in one year, with spring wheat, naked

oat, millet, and Irish potato as the chief crops. By contrast, mountains and hills are rather extensively used or even misused; forest cover has been badly removed, resulting in severe soil erosion, although on hills and some lower gentle slopes of low mountains, fruit trees have been extensively planted and subalpine shrubby meadows have been well used as summer-autumn pasture. Overall planning for comprehensive and better use of the land, especially the extensively distributed mountainous lands, is urgently needed for the region.

Eight subregions can be identified in the region:

6(1) Southern Shanxi Basins

The southern Shanxi includes basins along the southern Fenhe River graben valley. It features level landforms, deep soil, and the warm climate, with an accumulative temperature during the ≥ 10°C period of 4000 to 4500°C. The Yuncheng Basin has an elevation of 330 to 360 m, composed chiefly of alluvial-lacustrine deposits with thick salt beds in Tertiary lacustrine deposits. The famous salt-producing Xiechi Lake (324 m) is located at its southern margin.

6(2) Southeastern Shanxi Plateau

The southeastern Shanxi Plateau consists mainly of mountainous and hilly lands, with numerous small intermontane basins of between 800 to 1200 m in elevation. Annual precipitation totals 520 to 680 mm, which is richer than other parts of the Shanxi Plateau, yet, level land suitable for irrigation is rather limited. Some forests still blanket the mountains with an average coverage of 10 to 25 per cent.

6(3) Central Shanxi Basins

The central Shanxi basins includes the Taiyuan and Xinxian basins. The Taiyuan Basin has an area of 5050 sq km and an elevation of 700 to 900 m; it is bounded by steep fault scarps, both on the east and the west, with extensive alluvial fans spreading along their foothills. Extensive loessic terraces are also distributed both on the northeastern and southwestern margins. With an accumulative temperature during the ≥ 10°C period of 3300 to 3600°C and an annual precipitation of 500 to 600 mm, it is the most important agricultural and economic center in Shanxi Province. The Xinxian Basin has physical conditions similar to the Taiyuan Basin, but with a shorter growing season (by 10 to 15 days). Consequently, the three-crops-in-two-years agricultural system is rather precarious here.

6(4) Upper Yongding River Valley

Located at the transitional belt between the Shanxi and

the Nei Mongol Plateaus, the upper Yongding River valley area includes a series of basins along the upper reaches of the Yongding River from the Yanqing Basin in the east to the Datong Basin in the west. It is a transitional area, with a subhumid warm-temperate to a semiarid temperate climate as well as with deciduous broad-leaved forest to forest-steppe and steppe vegetation. It is also a stepping stone from the high Nei Mongol and Shanxi plateaus to the low North China Plain. It enjoys an accumulative temperature during the ≥ 10°C period of 2500 to 3000°C and an annual precipitation of 400 to 500 mm. It is also a transitional belt between the farming and the pastoral.

6(5) Northern Hebei Mountains

Consisting mainly of low and middle mountains, northern Hebei is essentially a mountainous area with a subhumid warm-temperate climate and an accumulative temperature during the ≥ 10°C period of 3000 to 4000°C and an annual precipitation of 500 to 700 mm. Vertical zonation is conspicuous. The Luanhe River and the Chaobai River together with their numerous tributaries cut through the northeast-trending ranges in deep gorges; many reservoirs have been built at the mouth of these gorges, which provide the chief water supplies for Beijing, Tianjin, and Tangshan.

6(6) Western Liaoning Low Mountains and Hills

Located at the eastern part of the geomorphological unit of the Hebei and Liaoning mountains, the western Liaoning subregion consists chiefly of low mountains and hills, with a climate approaching humid temperate conditions.

6(7) Western Henan Mountains

Located south of the Huanghe River and north of the Qinling and Funiu mountains, this area is composed chiefly of the Xiaoshan, Xiong'er, Songshan and other Palaeozoic mountains in western Henan Province, with foothills and river valleys extensively covered by loessic deposits. The subregion is deeply dissected by the Huanghe River and its tributaries—the Yihe River and Luohe River, and so on. The Xiaoshan Mountains together with the Zhongtiao Mountains form the famous Sanman Gorge of the Huanghe River, where a large-scale reservoir has been built. In terms of physical geographical division, this is still a part of North China, with a subhumid warm-temperate climate and deciduous broad-leaved forest as the zonal vegetation.

7. The Loess Plateau—Forest-Steppe and Steppe Region

The Loess plateau is defined on its eastern margin by

the Lüliang Mountains, on the western margin by the Helan Mountains, and on the southern margin by the Qinling Mountains. Its northern boundary, located on a transitional zone, is not distinct— for the time being, the Great Wall, which represents the landward limit of the wet summer monsoon, is used. As a whole, it has an area of about 300000 sq km. Administratively, it includes a part of Shaanxi, Shanxi, Gansu, and Ningxia. With an elevation of 1200 to 1600 m, it is covered extensively by loess deposits 30 to 60 m thick on the average (the thickest coverage reaches 200 m). This is a classic zone of a loess-deposited area and one of the most problematic areas in China. It is especially notorious for soil erosion and low agricultural production.

In Chapter 3, we discussed the chief characteristics of loess deposits in China. Geologically, the Loess Plateau might be subdivided into two parts: the first is located east of the Liupan Mountains and is composed chiefly of the Ordos Plateau; the second is west of the Liupan Mountains and constitutes the Qilian Mountains fold belt. The former is an important part of the extensive North China Platform. The loess has been deposited since the Pleistocene epoch, and three layers can be clearly identified: (1) the lowest layer, called Wuchang Loess (Q1) is distributed rather restrictedly, with a thickness of about 17.5 m; (2) the Lishi Loess (Q2) is most extensively distributed, with a thickness of 80 to 120 m; and (3) the Malan Loess (Q3) is also extensively distributed and is between 20 to 40 m thick. The Loess Plateau west of the Liupan Mountains has a different geological history, although loess has also been deposited since Pleistocene epoch. All these loess deposits are mainly composed of loose silt that is easily subjected to erosion. When it is soaked with water, serious erosion might occur even on slopes as gentle as 3 to 5°.

Geomorphologically, the Loess Plateau is subdivided into three parts: (1) the southern part is the broad Weihe River valley, which represents a faulted graben; (2) the middle part is characterized by the extensive distribution of flat loessic high plains, which denotes a better preserved level surface with a series of deep well-developed gullies— sometimes even more than 100 m deep; and (3) the northern part is a loessic, hilly area composed mainly of loessic gentle slopes and ridges dissected into rugged topography (Figure 9–5). There are also rocky island-like mountains that stand high above the Loess Plateau such as the Liupan Mountains and Lüliang Mountains. The upper limit of the loess deposits is the so-called "loess line", about 1000 m on the Lüliang Mountains rising to 1800 m on the eastern slope of the Liupan Mountain and 2400 m on the western slope. These rocky mountains are generally covered by forests or grasses and, hence, soil erosion is limited.

This is a warm-temperate region with an accumulative temperature during the ≥ 10°C period of 3200 to 3600°C. Annual precipitation totals 350 to 650 mm, of which 90

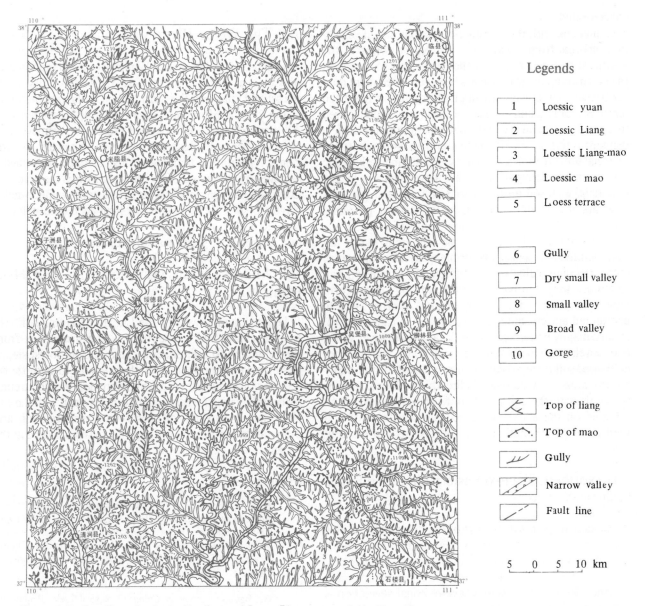

Figure 9-5 Landform in the badly dissected Loess Plateau (near Suide, Shaanxi)

Legends

1	Loessic yuan
2	Loessic Liang
3	Loessic Liang-mao
4	Loessic mao
5	Loess terrace
6	Gully
7	Dry small valley
8	Small valley
9	Broad valley
10	Gorge

	Top of liang
	Top of mao
	Gully
	Narrow valley
	Fault line

5   0   5   10 km

per cent is concentrated in the ≥ 10°C period. Rainstorms occur frequently during summer time resulting in severe soil erosion. Moisture conditions change from subhumid in the southeast to semiarid in the northwest, with the aridity index increasing from 1.5 to 4.0. To suppress strong evaporation, farmers around the Lanzhou area usually spread pebbles and sands over the farmlands, forming the so-called sand farmland.

Rivers mostly flow into the Huanghe River. The annual runoff depth decreases from 50 mm in the east to 25 mm in the west. The Weihe River is the largest tributary of the Huanghe River. It creates a broad, fertile valley with an accumulative temperature of 4500 to 5000°C during

the ≥ 10°C period, a growing season of 260 days, and an annual precipitation of 500 to 600 mm. It is one of the oldest farming areas in China and the most important commercial grain-and cotton-producing base in Northwest China. Other important tributaries are the Qianhe, the Luohe, and the Wuding rivers. All are heavily loaded with silt, especially during summer time. The Wuding River is so highly silt-laden that its channel changes frequently, and it is, thus, called Wuding which means always changing. These tributaries are also the chief agents for soil erosion and alluvial deposition. The floodplains along their channels, the so-called chuan, although limited in area, are where most of the irrigated farmlands and settlements are concentrated.

Coinciding with the moisture conditions, zonal vegetation changes from deciduous broad-leaved forest in the southeast to forest-steppe in the center and to steppe in the northwest. As the region has been cultivated more than 7000 years, practically all original vegetation has been removed. Only small patches of deciduous broad-leaved forest are left, scattered on steep northern slopes and a few coniferous forests are distributed on high mountain tops.

Zonal soils are drab soils in the southeast and dark loessic soils in the northwest. Both are developed on loessic materials, with soil profiles 2 to 3 m thick and a humus layer about 1 m thick. Dominant soil-forming processes are calcification and humification. Cultivation also plays an important role. The dark loessic soils are in reality ancient cultivated soils, occurring generally on level, cultivated high plains and not affected by severe soil erosion; they are extensively used for growing wheat, corn, millet, sorghum, and other crops. Another widely distributed ancient cultivated soil is the so-called lou tu, located mainly on terraces of the broad Weihe River valley and developed chiefly on the drab soils. Still another cultivated soil is the so-called min tu, distributed in badly eroded areas and cultivated directly on loessic parent materials with the soil profile not yet well developed.

Based on the above-mentioned physical features and their areal differentiation, four sub-regions can be identified:

7(1) Northern Shaanxi-Eastern Gansu Hills and Gullies

Located at the northern part of the Loess Plateau, this the northern Shaanxi-eastern Gansu hill and gulley subregion is mainly composed of dissected gentle slopes. The loess grains are rather coarse and are sometimes called sandy loess. Soil erosion is severe. Zonal vegetation is temperate forest-steppe, but it is now mostly destroyed. In overall regional planning, this subregion should be devoted to a comprehensive development of farming-pasture-forestry with emphasis laid on pasture. At the moment, urgent measures are needed to restore the grass cover and the crop-grass rotation system and to enforce other soil conservation measures.

7(2) Northern Shaanxi-Eastern Gansu Dissected High Plain Subregion

Located south of the above-mentioned subregion, this subregion is composed mainly of extensive level high plains. Two secondary forest covered mountainous areas are also included. Major soil types are the cultivated drab soil and the zonal vegetation is forest-steppe. This subregion should be devoted to farming and pasture.

7(3) The Weihe River Valley

The broad valley of the Weihe River is composed of floodplain as well as three terraces, the so-called first yuan, the second yuan and the third yuan. On the northern foot of the Qinling Mountains, there are still higher terraces. All these terraces are fairly well preserved with level surfaces and are badly dissected only along some large rivers. Zonal vegetation is temperate forest-steppe and is mostly removed by cultivation. The zonal soil type is the drab soil. Loess deposits here generally have fine grains—the so-called clayey loess. This subregion has long been, and will continue to be, an intensive farming area and the most important commercial grain base of Northwest China.

7(4) Middle Gansu Dissected Hills and High Plains

Located mainly west of the Liupan Mountains (Photo I–7), this subregion is essentially a transitional belt from the subhumid to the semiarid climate, from forest-steppe to steppe vegetation, and from loessic soils to siernozems. Soil erosion is quite severe. This subregion should be comprehensively used for farming-pasture-forestry, with the emphasis laid on pasture. Soil conservation, reforestation, and evaporation suppression are urgent measures needed at the present.

References

[1]   *Physical Geography of China* Compilation Committee, Chinese Academy of Sciences, 1979 –1985, *Physical Geography of China.* 12 Vols, Science   Press, Beijing. (In Chinese)

[2]   *Physical Regionalization of China* Working Committee, Chinese Academy of Sciences, 1958 – 1959, *Physical Regionalization of China,* 8 vols, Science Press, Beijing. (In Chinese)

[3]   Academy of Geological Sciences, 1978, *Main Features of Geological Structure in China,* Geological Press, Beijing. (In Chinese)

[4]   Yuan Baoyin et al., 1980, "Cenozoic Deposits and Fault Structure in North China", *Origin and Development of North China Fault Region,* Science   Press, Beijing. (In Chinese)

[5]   Liu Dongsen (Liu Tong-sen) et al., 1964, *Loess in the Middle Reaches of the Huanghe River,* Science Press, Beijing. (In Chinese)

[6]   Department of Geography, Hebei Normal University, 1974, *The Haihe River,* Hebei People's Press. (In Chinese)

[7]   Liu Chongming et al., 1978, "The Influence of Forest Cover upon Annual Runoff in the Loess Plateau", *Acta Geographica Sinica,* Vol. 33, No.   2. (In Chinese, with English abstract)

[8]   Institute of Soil and Soil Conservation, Chinese Academy of Sciences,1961, *Soils of North China Plain,* Science Press, Beijing. (In Chinese)

[9]   Chen Chun-wei et al., 1977, "Physical Environmental Changes of Southern Liaoning Province in the Last 10,000 Years", *Sienta Sinica,* 1977, No.   6. (In Chinese)

[10]   Shanxi Agricultural Regionalization Committee et al., 1981, *Interpretation and Analysis of Agricultural Physical Conditions by Landsat Imagery, Taiyun Sheet,* Science Press, Beijing. (In Chinese)

Chapter 10

Humid Subtropical Central China and South China

This is the largest and most populous division in China. Administratively, it includes all of Hubei, Hunan, Jiangxi, Zhejiang, Fujian, Taiwan, Sichuan, and Guizhou provinces, the southern parts of Jiangsu, Anhui, and Henan provinces, and the northern parts of Guangdong, Guangxi, and Yunnan provinces. The area totals approximately 2.5 million sq km, occupying about 26.1 per cent of China's total land area (Figure 10–1). It is inhabited by about 560 million people, accunting for about 56 per cent of China's total population, or about 14 per cent of the total population of mankind—more than the two superpowers, the United States and the USSR, combined.

This natural division starts from the eastern margin of the Qinghai-Xizang Plateau, coinciding roughly with the 3000 m contour line, eastward up to the Pacific Ocean. Its northern boundary is the Qinling Mountains-Huaihe River line, coinciding roughly with the 0°C isotherm of mean January air temperature or the 750 mm isohyet of total annual precipitation. This is probably the most significant geographical divide in China, showing conspicuous areal differentiation between the north (including North China, Northeast China, and Northwest China) and the south (including Central China, South China and Southwest China). The north, as discussed in Chapter Eight and Chapter Nine, is characterized by (1) extremely cold winters, with a mean January air temperature below 0°C and with freezing rivers and soils throughout; (2) no evergreen broad-leaved forest; (3) deficiency of moisture, with annual precipitation less than 750 mm; (4) dry farming dominant; (5) salinization and calcification as the chief soil-forming processes; (6) inhabitants use wheat as the staple food and traditionally travel chiefly on horse. By contrast, the south is characterized by: (1) a mean January air temperature above 0°C; (2) evergreen broad-leaved forest dominating; (3) a surplus of moisture, with annual precipitation of more than 750 mm; (4) mainly paddy rice culture; (5) leaching and laterization as the chief soil-forming processes; (6) inhabitants use rice as the staple food and formerly traveled chiefly by boat.

The division's southern boundary is not so distinct as the other three sides. We will discuss it more in detail in the next chapter. It is sufficient to point out here that it lies at about 21 to 22°N in the Guangdong and Guangxi coasts and at 23 to 25°N in southern Yunnan, whereas the southern tip of the Taiwan Island, although tropical in climate and vegetation, is combined with northern subtropic parts of the island to form an integral region of this natural division.

CHIEF PHYSICAL FEATURES

This extensive natural division is generally characterized by the following physical features.

Predominantly Montane, Hilly Topography, Partly Interwoven with Broad Basins and Valleys

Under humid, subtropic climatic control and complicated geological evolution, humid subtropic Central China and South China is a natural division distributed with numerous mountains and hills interwoven with broad basins and valleys. Chemical weathering and fluvial action are strong; red beds and karst topography are well developed. It is composed chiefly of the following seven geomorphological areas:

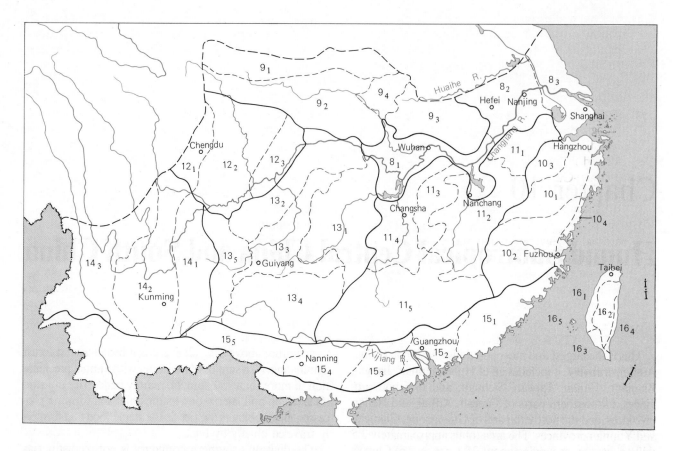

Figure 10-1   A location map of the humid subtropic Central China and South China

1) Middle and lower Changjiang Plain—the middle and lower Changjiang River has been developed along a great fault line and its extensive valley plains (the Jianghan Plain,[1] Dongting Lake Plain, and the Poyang Lake Plain) have been mostly located at dominantly submergent basins since the Cretaceous Period. Rivers and lakes are widely distributed; elevations are mostly below 200 m. The Changjiang Delta has a rather complicated geologic history. the delta plain extended far eastward into the continental shelf during the late Pleistocene recession; it then submerged and gradually silted up again, with the immense Taihu Lake transformed from a sea gulf into a freshwater lake.

2) Southeast coast mountains and hills—the southeast coast is dominated by mountains and hills, mostly with elevations of 500 to 1500 m. They are NE or NW-trending, and composed of Mesozoic granite and igneous rocks. Amid these mountains and hills, are many small graben-structured basins, deposited

with Cretaceous, Tertiary and Quaternary beds. Along the rocky and irregular shoreline, there are low, rolling granite hills and small patches of delta plains as well as many rock islands, of which Zhoushan Island (524 sq km) is the largest.

3) Taiwan Island mountains and plains—two thirds of Taiwan is made up of mountains and hills, mostly in the middle and eastern parts of the island. They are the youngest mountains in China and are still actively undergoing tectonic movement. Plains stretch chiefly along the west coast. Around Taiwan Island, there are also more than 100 small islands, of which, the basaltic Penghu Islands are the most important.

4) South Changjiang hills and basins—this extensive geomorphological area is a great amphitheater of hills surrounded by mountains around 1000 m in elevation and scattered with small basins and strips of level plains along the rivers. Geologically, these northeast-trending mountains and hills have been controlled by the Mesozoic Yanshan Tectonic Movement with anticlines as ridges and synclines as valleys. Tertiary red beds deposited in basins and valleys have been largely dissected into hills.

---

1) Jianghan Plain denotes an extensive low plain around the confluence of the Changjiang River and its largest tributary — the Hanshui River. Wuhan is located just in the center of this plain.

5) Sichuan Basin—surrounded by high mountains and plateaus, the basin itself is mainly composed of low mountains and hills, with an average elevation of about 500 m. The fertile Chengdu Plain is an outcome of downfaulting along the piedmont of western marginal high mountains.

6) Guangxi Basin—this is the best developed karst topography area in China, probably also in the world, with arc-shaped mountains around 1000 m located in the middle of the basin.

7) Yunnan-Guizhou Plateau—with an elevation around 1000 to 2000 m, this plateau is located on the second great topographic step in China. High and rugged ground surfaces cut by deep valleys and crossed by towering mountains make up most of the region. Upper Palaeozoic thick-bedded limestones are widely distributed and have been within a warm humid environment for a long time. Consequently, karst topography is well developed.

Humid Subtropic Monsoon Climate

During winter the atmosphere within an altitude of 1 to 2 km is dominated by continental northeastern monsoons, whereas the higher atmosphere is dominated by westerlies. Cold waves often occur and result in quite low temperatures. Consequently, the winter temperature of this natural division is generally lower than other parts in the world with similar latitudes. During the summer maritime southeastern monsoons predominate, bringing in high temperatures and heavy rainfall. Plum rain usually lasts about one month in early summer along the middle and lower Changjiang valley; typhoons with high winds and rainstorms occur frequently in late summer and early autumn. Spring and autumn are transitional seasons with variable weather.

The mean January air temperature ranges from 2 to 8°C, with isotherms generally paralleling the latitudes. The absolute minimum temperature may drop below −10°C north of the Changjiang River (e.g., Wuhan −18.1°C on January 30, 1977), below −7 to −10°C south of the Changjiang River, and below −4°C south of the Nanling Mountains. Summer is universally hot, with a mean July air temperature around 28°C (the Yunnan-Guizhou Plateau is an exception, it is generally below 25°C). Absolute maximum temperatures generally rise above blood heat, Chongqing (44°C), Wuhan (41.3°C), Nanjing and Changsha are called the four ovens of China. Seasons are distinct, with spring beginning in middle March, summer in early June, autumn in middle September, and winter in middle November.

Based on latitudinal location and actual temperature conditions, three subdivisions can be identified: (1) the northern subtropical zone, which is located generally north of the Changjiang River, with a mean January air temperature 0 to 5°C and an accumulated temperature during the ≥ 10°C period of 4500 to 5000°C; (2) the middle subtropical zone, which is located between the Changjiang River and the Nanling Mountains, with a mean January air temperature of 5 to 10°C and an accumulated temperature during the ≥ 10°C period of 5000 to 6500°C; (3) the southern subtropical zone, which is located south of the Nanling Mountains, with a mean January air temperature of 10 to 16°C and an accumulated temperature during the ≥ 10°C period of 5000 to 8000°C.

Annual precipitation is abundant, generally above 1000 mm. It decreases from southeast to northwest and from montane slopes to sheltered basins. The windward slope of the Wuyi Mountains has an annual precipitation above 2200 mm, whereas the Nanyang Basin and Hanzhong Basin have less than 900 mm. Seasonal distribution of precipitation is unbalanced, although not so unbalanced as in North China and Northwest China. More than 70 per cent of the total annual precipitation is concentrated in summer, although winter still enjoys more than 10 per cent of the total annual precipitation. In the western part of the division, the dry season is more conspicuous.

Bountiful Surface Water Resources

This natural division is again characterized by a dense river network (including canals) and numerous lakes (including reservoirs and ponds). In connection with mountainous and hilly topography, the dendritic drainage pattern dominates. In areas with rather homogeneous bedrocks, such as in the Sichuan Basin, the dendritic drainage pattern is typical. In the southeast Coast, owing to prominent geological control, the trellis drainage pattern is better developed, with rivers flowing independently into the sea. The density of the river network is generally more than 0.3 to 0.4 km/km^2, and even 6.4 to 6.7 km/km^2 in the Changjiang Delta where the most intricate river-network and the famous "watery country" in China lies. The annual runoff depth ranges from 500 to 1200 mm, decreasing from southeast to northwest and from higher to lower elevations. The Wuyi Mountains area with an annual runoff depth of more than 1600 mm, has the greatest, the Sichuan Basin drops to less than 450 mm, and the Nanyang Basin to less than 200 mm.

Annual runoff is consequently quite rich. Each of these rivers, the Ganjiang, Xiangjiang, Yuanjiang, and Qiantang River has a drainage basin equal to only one eighth to one fifteenth that of the Hanghe River; yet, they all have a greater annual runoff than the latter. Eight tributaries of the Changjiang River, including the Ganjiang, Xiangjiang, Minjiang, Jialing, Yuanjiang, Hanshi, Wujiang, and Yalong rivers have a greater annual runoff than the Huanghe River. The Changjiang River itself is, of course, the largest river in China. Bountiful runoff combined with mountainous topography have created abundant hydroelectric power resources and potential. According to an estimate, the hydroelectric capacity of the Sanxia (Yangtze) Gorge, or the Three Gorges alone, is 15 million kilowatts (kW); that of the Wujiang River is 8 million kW. Even along a tributary of the Mingjing River, a large hydroelectric station with a capacity of 600000 kW has recently been built.

This division contains one of the five major lake regions in China—the so-called East lake region. The most spectacular lake group is the ancient Yummeng Swamp Area; the dwindling Dongting Lake (Landsat image 3) is simply one of its numerous relic lakes. Other important lakes include Poyang Lake (now the largest freshwater lake in China), Taihu Lake, and Chaohu Lake. Most of these are being silted up or reclaimed as croplands. For the sake of flood control, freshwater fisheries, and other land and water uses, overall planning must be undertaken to protect these lakes from disappearing.

Profuse and Varied Vegetation Types

Extensive land areas together with favorable natural conditions result in profuse and varied vegetation types. Furthermore, this is the most populous natural division in China. Since the time of the Hemudu culture on the Southeast coast (Table 1-4), the human impact on the vegetation and soil-forming processes has been very great. All flat plains have been intensively cultivated, with paddy rice as the most important crop and paddy soil as the most extensively distributed fertile soil. Dry farming has also played an important role in the soil-forming process. Mountains and hills are much less intensively used. Practically all original vegetation types have been changed; instead, secondary evergreen broad-leaved forests, shrubs, grassland, and many trees of economic value dominate along with extensive terraced farming.

The horizontal zonal vegetation type is subtropical broad-leaved forest. Corresponding with the three climatic subtypes, there are three vegetation subtypes:

1) Mixed evergreen and deciduous broad-leaved forest—this is a transitional vegetation type between deciduous and evergreen broad-leaved forests, and it is widely distributed on the mountains and hills of the northern subtropical zones. The dominant evergreen tree species are *Cyclobalanopsis* spp., *Castanopsis* spp., and *Ilex chinensis*. The dominant deciduous species are *Fagus* spp. and *Platycarya strobilacea*. The zonal soil type is yellowish brown soil, a transitional type between brown soil and red earth.

2) Evergreen broad-leaved forest—this is the zonal vegetation type of the middle subtropic zone. All dominant species are evergreen and are much more profuse and varied than the above-mentioned mixed vegetation types. Bamboo groves flourish extensively (Photo IV-3) and epiphytes also appear. The zonal soil types are red earth and yellow earth.

3) Monsoon evergreen broad-leaved forest—this is a transitional vegetation type between the subtropical evergreen broad-leaved forest and the tropical monsoon forest. It is chiefly distributed on low mountains and hills of the southern subtropic zone. The flora are profuse and varied, the *Castanopsis* tree being the most important and widely distributed. There are many kinds of ferns and epiphytes. The zonal soil types are lateritic red earth and yellow earth.

Vertical zonation of vegetation and soil is also conspicuous. We will discuss it later, according to different latitudinal and altitudinal locations.

NATURAL REGIONS

This extensive natural division may be demarcated into 3 subdivisions, 9 regions, and 37 subregions (Figure 10-1), chiefly based on climatic and geomorphological areal differentiation. The divide between northern and middle subtropic zones, using a mean January air temperature of 5°C and an absolute minimum temperature of –5°C as the diversifying index, runs roughly along the Hangzhou–Jiujiang–Changde–Yichang–Guangyuan line. The boundary between the middle and southern subtropic zones, with a mean January air temperature of 10°C and an absolute minimum temperature of 0°C as the diversifying index, generally follows the Fuzhou–Yingde–Wuzhou–Bose line. A rice-wheat double-cropping agricultural system dominates in the northern subtropic zone; five agricultural crops (with two crops of rice) in two years grow in the middle subtropic zone, with tea, bamboo, orange, and tung tree as the typical economic tree crops; and triple agricultural crops (with two crops of rice) each year are produced the southern subtropic zone, with banana, lichi (Photo IV-4), longan, coffee, and the *Hevea* tree as the typical economic crops.

Northern subtropic zone includes two natural regions.

8. Middle and Lower Changjiang Plain—Mixed Forest Region

Located between 28 to 34°N and with an area of about 250000 sq km, the middle and lower Changjiang Plain is essentially a region of low, flat plains and watery country. Its northern boundary runs roughly along the Daba Mountains–Dabie Mountains–Huaihe River line. Its southern boundary, which is rather irregular, zigzags along the northern border of the South Changjiang Hills and Basins.

It is chiefly composed of extensive alluvial plains, deposited by the Changjiang River and its numerous tributaries. A series of lake basins and valley plains stretch from the Three Gorges eastward up to the sea for a total length of more than 1800 km and a total area of about 160,000 sq km. There are also numerous low mountains and hills encircling around or scattered amid these alluvial plains. Three sections may be observed.

1) The middle section is located from Yichang to Wuhan and is essentially a lake-basin plain with an area of about 80000 sq km, of which one eighth consists of lakes. It was formerly the ancient Yunmeng Swamp Area and is now mostly drained, which has resulted in extensive low plains with elevations below 200 m (Wuhan, 27 m). The Changjiang River with many braided channels and a valley plain about 2000 m wide, flows sluggishly eastward. Its largest tributary, the Hanshui River, and each of the four large tributaries of the Dongting Lake system (Xiangjiang, Zishui, Yuanjiang, Lishui), have a larger annual runoff than the Huanghe River.

2) The middle-lower section; located from Wuhan to Jiujiang a ribbon like belt of lake basins and valley plains bounded on either side by low mountains and hills. Sometimes scenic isolated peaks may border the Changjiang River, which has a much narrower channel and flows much more rapidly in this section. Tributaries are not so numerous as in the middle section, yet the Poyang Lake system, with its four large tributaries (Ganjiang, Xiushui, Pojiang, Xinjiang), enters the Changjiang River here, so that the annual discharge of the Changjiang River increases from 14400m³/sec in Yichang (Hubei Province) to 29700m³/sec in Datong (Anhui Province).

3) The lower, or delta, section is characterized by low elevation and a dense river network. Along the northern bank of the Changjiang River, the elevation may drop to 2 m at sea level. The Changjiang River has a width of about 1500 m near Nanjing, about 2000 m near Jiangyin, about 10000 m near Nantong, and even 80000 to 90000 m near its mouth. It empties into the sea with an annual discharge of about 979.35 billion cu m and a total silt content of about 450 million cu m—this pushes the coast seaward at the rate of 1 km per 60 years. The subwater delta extends eastward up to 125°30'E, about 450 km away from the present mouth of the Changjiang River. Four stages of subwater deltaic deposition are identified, with newer ones overlying older ones. There are also many lakes and scenic isolated hills inside the delta.

Climatically, the region is characterized by high temperatures and heavy precipitation that come together during the growing season, which makes agriculture very productive. With an accumulated temperature during the ≥ 10°C period of 4500 to 5000°C. It is possible to practice the highly productive rice-rice-wheat triple cropping system. Annual precipitation totals 900 to 1500 mm, with 50 to 60 per cent concentrated in summer and autumn and 30 to 40 per cent in spring, the widely distributed spring drought hazard in China is practically no problem here. Yet, owing to the east coast location and the frequent invasion of cold waves, it is rather cold during winter with a mean January air temperature of 0 to 5°C, and an absolute minimum temperature below –5°C, or even below –10°C. Therefore, the northern limit of the subtropic zone in China lies about 10° of latitude more to the south than in the Mediterranean region. Summer is quite hot; three out of four "oven" cities in China are located here. What is worse, an excess of precipitation in summer and autumn together with the above-mentioned topographical and hydrologic conditions make the region susceptible to destructive flood hazard. The flood of the Changjiang River in 1931 was probably the most disastrous one ever recorded in the world. Since 1949, a series of engineering works have been created for flood control with encouraging results.

The zonal vegetation is mixed evergreen and deciduous broad-leaved forest, the zonal soil type is yellowish brown soil. Yet, the so-called zonal vegetation and soil are restricted to low mountains and hills. The extensive plains have been intensively cultivated and turned into cultural vegetation and paddy soil. Even on low mountains and hills, original mixed forests are mostly destroyed and have been substituted by secondary growths of shrubs or grasses, as well as having been partly planted with "horse's tail pine" (Pinus massoniana) and economically valuable tree crops, such as tea and fruit. The great contrast between the intensively used and densely populated plains and the rather extensively used and sparsely populated slope lands is quite outstanding in the region.

Three subregions are identified in this natural region.

8(1) Hunan-Hubei (or Lianghu) Plain*

Coinciding roughly with the middle and middle lower sections of the Changjiang River, the Hunan-Hubei Plain is essentially an alluvial plain of numerous rivers and lakes, in particular two great rivers—the Changjiang River and the Hanshui River, and two large lakes—the Poyang Lake and the Dongting Lake. Human activity has long played

---

* Lianghu or Two Lakes denotes the Hunan and Hubei provinces, which are named after the Dongting Lake, being respectively south and north of that immense lake.

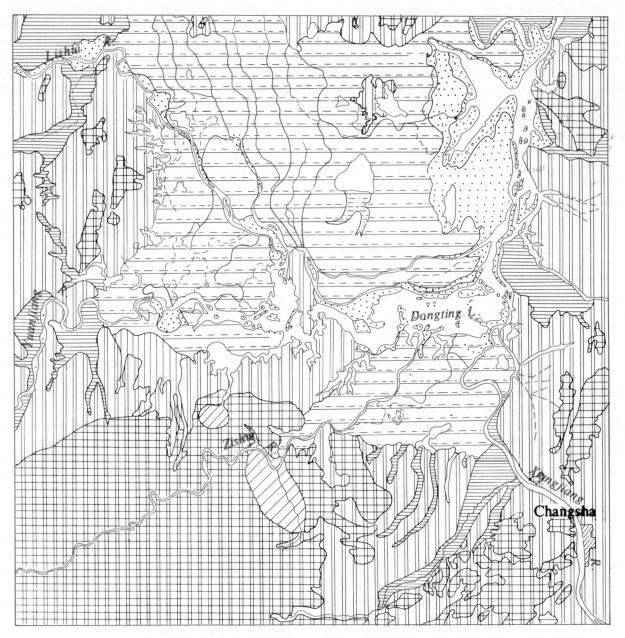

Figure 10-2 Landform types around the Dongting Lake

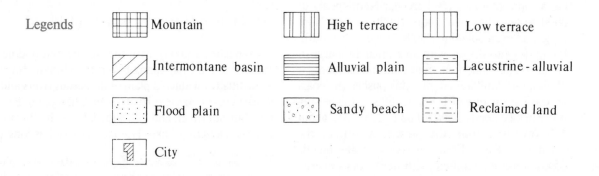

Legends

an important role in shaping the physical geographical environment. For example, the Dongting Lake (Figures 10-2, Landsat image 3)had an estimated maximum lake area of 14000 sq km during historical times. It has been gradually silted up over the last 1000 years. It was reduced to an area of about 6000 sq km in 1820, to 5400 sq km in 1890, and to 4360 sq km in 1949. Now it has an area of less than 2820 sq km. If effective measures are not taken, it will disappear entirely.

The Hunan-Hubei subregion, with its rice-wheat or cotton-wheat double-cropping agricultural system, has long been one of the greatest granaries of China. As an old saying goes, "When Lianghu has a good yield, all China has enough food". The freshwater fishery is also important. In marginal hilly lands, mulberry, tea, and other tree crops are well developed. The chief meliorating measure for this subregion is, of course, flood control; besides, care must also be taken to combat low temperatures in late spring and early summer as well as occasional drought in the growing season.

8(2) Lower Changjiang Plains and Hills

The lower Changjiang River subregion includes plains and hills on either side of the lower Changjiang River from Jiujiang to Zhenjiang. Plains are largely below 50 m in elevation, and hills between 200 to 300 m, with a few peaks towering above 500 to 1000 m. The plains are studded with lakes, such as Chaohu Lake. These are famous grain-producing areas in China. Hills and low mountains should be mainly devoted to forestry and tree crops.

8(3) The Changjiang Delta

The Changjiang Delta subregion includes broadly all plains eastward of Zhenjiang. It has been a down-warping basin since the Yanshan Tectonic Movement. Since the last submergence in the early Holocene Epoch, it has been gradually silted up by alluvial deposits of the Changjiang River and the Huaihe River (1194–1855 A.D.) as well as

by extensive lacustrine and marine deposits. This extensive plain has generally an elevation below 10 m, with some isolated scenic hills of 100 to 200 m in elevation.

This is one of the oldest cultivated areas in China. Except for a narrow strip of saline coastal beach, it has been nearly entirely reclaimed and transformed into "watery country" and a "land of rice and fish" (Photo IV-1, Landsat image 4). Probably the most characteristic feature of the landscape is the innumerable rivers and canals that are also the very arteries of life and transportation and serve as an artificial drainage system. Besides being the largest industrial center, it is now the most productive grain-producing base in China with many communes enjoying an average annual grain production above 15 tons/ha. It also prospers through fishing, silkworm raising, and other productive enterprises. In short, this is a showcase area for getting still higher production from the present already high level.

9. Qinling-Daba Mountains—Mixed Forest Region

The Qinling-Daba mountains region, with an area of about 300,000 sq km, forms a mighty mountain barrier that stretches from the lofty Qinghai-Xizang Plateau eastward almost to the Pacific Ocean. It cuts directly across Central China and separates the country into the north and the south. Its northern limit is the northern slope of the Qinling Mountains (coinciding roughly with the 700 m contour line) and the Dabie Mountains; its southern limit is the southern foothill of Micang Mountain and Daba Mountain.

The Qinling Mountains are broad latitudinal folded mountains, with a ridge line generally 2000 to 3000 m in elevation and the highest peak is Mount Taibai, at 3767 m (Photo I-9). It overlooks the Weihe River valley by a relative relief as high as 2500 to 3000 m. It extends southeastward as do the Tongbai Mountain and the Dabie Mountain, generally 500 to 1000 m in elevation. The Daba Mountain (including the Micang Mt. and the Wudang Mt.)

Table 10-1 Vertical Distribution of Temperature and Precipitation in the Qinling Mountains Area

Place	Location	Elevation (m)	Mean January temperature (°C)	Mean July temperature (°C)	Annual precipitation (mm)
Xi'an City	Northern foot hills	398	− 1.3	26.7	604.2
Taibai County	Northern slope	1543	− 5.2	19.3	736.7
Huashan County	Northern slope	2063	− 7.0	17.7	925.1
Fengxian County	Intermontane basin	970	− 1.2	22.9	644.7
Ankang County	Southern foot hills	327	3.1	27.7	779.6
Hanzhong County	Southern foot hills	509	2.0	25.9	889.7
Foping County	Southern slope	1192	0.2	22.4	938.5

runs southeastwardly, with an elevation decreasing from 2000 to 3000 m to about 1000 m; the highest peak here is the famous Mount Shennongjia, at 3105 m (Photo I-10). The Nanyan-Xiangfan Basin is the largest inland basin in the region, with terraces and hills (mostly 40 to 50 m in relative relief) dominating.

This is probably the most significant climatic barrier in China. For example, in January 1955 when a strong cold wave swept southward, Xian (located at the northern foot of the Qinling Mountains) had an absolute minimum temperature of –20.6°C; Ankang (located at the southern foot of the Qinling Mts. and with a similar longitude) had –7.6°C. Annual precipitation varies from 700 to 1550 mm, decreasing from east to west and from south to north. Vertical distribution of temperature and precipitation is conspicuous, as shown in Table 10-1.

Vertical zonation of vegetation, soil, and land use is conspicuous. Besides, as this is the transitional belt—not only between the north and the south but also between the east and the west—both flora and fauna are profuse and varied. For example, Mount Shennongjia alone has 166 families, 765 genera, and 1919 species of plants.

Four subregions are identified in this natural region:

9(1) Qinling Mountains

The Qinling Mountain subregion includes the Qinling Mts. in Shaanxi Province and its extension Funiu Mountain in Henan Province. The northern slopes of the Qinling Mts. are short and steep and their foot hills are delimited sharply by a great fault line. One of its border ranges, the Mount Huashan, is famous for its steepness; it is composed of five peaks, each with a northern slope consisting of stupendous precipices several hundred meters high. On the other hand, its southern slope is rather long and gentle; there are nine ranges running from west to east across numerous intermontane basins and valleys. The Funiu Mountains encircle the Nanyan-Xiangfan Basin, with the ridge line decreasing southeastwardly from more than 2000 m to only 300 to 1000 m.

The Qinling Mountains subregion has about 90 per cent of its total land area composed of mountains and hills. Owing to its rather sparse population and the difficulties of communication, 47 per cent of the land is still preserved as original forest, panda, the goldenhaired monkey, and other unusual and rare vertebrates survive here. Forestry should continue to be a major resource in the future, with local intermontane basins and valleys devoted to farming and economic tree crops.

9(2) Daba-Micang Mountains

The Daba-Micang Mountains subregion has a similar tree-clad montane landscape, with middle and high moun-tains occupying about 80 per cent of the total land area and low mountains and hills 10 per cent.

Another significant feature is the extensive Hanzhong Basin sandwiched between the Qinling Mountains and Daba Mountain. It comprises the upper reaches of the Han-shui River and is characterized by meandering and braid-ed channels and four terraces (3 to 5 m, 10 to 15 m, 36 to 80 m, and 100 m respectively) on either side. Agriculture has been well developed from ancient times, and this subregion has long served as a great highway be-tween the north and the south.

9(3) Tongbi-Dabie Mountains

The Tongbi-Dabie Mountains are between the Chang-jiang River and the Huaihe River and are composed chiefly of dissected low mountains and hills, with elevations generally below 1000 m. Consequently, the barrier action is not so effective as that of the Qinling Mountains and the Daba Mountain. Human intervention is also more heavy, resulting in severe soil erosion and devastation of forests. However, croplands are more extensive, with the rice-wheat double-cropping system prevailing.

9(4) Nanyan-Xiangfan Basin

The Nanyan-Xiangfan Basin is located in a Cenozoic down-warping area, deposited with Cenozoic red beds of 3000 m in thickness, and then redeposited with early Quaternary red earth about 3000 to 4000 m thick. Neotec-tonic movements uplifted the land and then dissected it into rolling terraces and hills, with an absolute relief of 100 to 150 m and a relative relief of 20 to 30 m. The Han-shui River and its tributaries flow through the basin with abundant irrigation water, although severe flood hazards are not infrequent.

This is a break or narrow pass between the otherwise continuous Qinliang-Dabie mountain system. It is the path of the southward-moving cold waves during winter, with an absolute minimum temperature below –10°C and an-nual precipitation of less than 1000 mm. It is significant as a highway between North China and South China since ancient times. Most of the basin has been cultivated with croplands occupying 60 per cent of the total land area, Soil conservation and irrigation development are two impor-tant problems here.

Middle Subtropic zone includes five natural regions.

10. Southeast Coast—Evergreen Broad-Leaved Forest Region

The Southeast coast region includes Zhejiang and Fu-jian provinces and stretches northeast-southwestwardly along the coast for more than 1700 km. Its western boundary approximates the crest line of the *Tianmu* Mountains, Xianxia Mountains, and Wuyi Mountains. Its northern border coincides broadly with the funnel-shaped Hangzhou Estuary, and its southern border with the Fuqing–Yongchun–Yangding line. With an area of about 150000 sq km, it is characterized by hilly and mountainous topography and numerous swift rivers as well as luxuriant evergreen vegetation. Zhejiang Province is noted for its "seven mountains, two waters and one cropland", that is, 70 per cent of the total land area consists of mountains and hills. Fujian Province is even more mountainous and hilly and land communication is more difficult. This is probably the main reason why there are 104 dialects of Chinese in this province alone. There are three series of NE-SW trending mountain ranges paralleling the coast. The western array is the Tianmu Mountain, the middle is the Xianxia–Wuyi mountains, and the eastern series includes Tiantai, Kuocang, and Daiyun mountains. Most of them are low mountains between 500 to 1000 m. These mountains together with their spurs and foothills are interwoven into a great network of mountains and hills, with only a small area occupied by intermontane basins and valleys where most of the inhabitants and croplands are concentrated. Along the coast, hills and terraces dominate, although alluvial and marine plains are getting broader. The coast itself and the numerous islands are also rocky, giving rise to numerous good harbors but limited hinterlands.

The climate is humid and subtropical and is quite favorable for agricultural development. Double cropping and triple cropping are generally practiced. During the cold, dry winter, isotherms run parallel to the latitudes, with a mean January air temperature of 0 to 10°C in the northern part and above 10°C in the southern part. During the warm, moist summer, isotherms parallel the coast, with a mean July temperature of 26 to 30°C. Annual precipitation totals 1100 to 2000 mm, with less precipitation in the plains and basins and along the coast. The rainy season lasts from March to June, so that the widely spread spring drought in other parts of China is nonexistent here. Typhoons are frequent from July to September, bringing downpours of rain and a temporary relief from excessively hot weather.

The density of the river network is high, usually more than 0.1 km/km², although most of rivers are short and swift and flow into the sea independently. The drainage is mostly of the trellis pattern, controlled mainly by geological structure and the alignment of the mountain ranges. The gradient is usually steep, and all basins are dotted along the river, like a string of pearls. The annual runoff depth ranges from 900 to 1400 mm, decreasing from southwest to northeast. The largest river is the Minjiang River, with a length of 539 km. It has an annual runoff of 65.5 billion cu m, greater than the Huanghe River by

a margin of 16 per cent. Many rivers have a funnel-shaped mouth, resulting in high and bore tides. The Qiantang River is especially famous for its bore; high tide may attain an amplitude of 8 m during the autumn full moon.

The zonal vegetation type is evergreen broad-leaved forest, distributed mainly on mountains and hills below 1000 m in elevation. From 1000 to 1500 m, it turns to mixed evergreen and deciduous broad-leaved forest which is also the case in the karst region. On low mountains and hills below 800 m, *Pinus massoniana* are widely distributed. Again, many commercially valuable trees, such as Chinese fir (*Cunninghamia lanceolata*) and bamboo are widely planted (Photo IV-3). Tea gardens are especially popular along the foothills, and some of the best green tea in China is produced here. In the limited intermontane basins and valley plains, the land is intensively cultivated. Zhejiang is famous for being the first province in China to have an annual grain yield above 1000 catty/mow (or 7.5 tons/ha). All this natural and cultural vegetation makes the region a vast sea of verdure, both productive and scenic.

Four subregions are identified.

10(1) Xianxia-Kuocang Mountains

Located in the northern part of the region, this subregion includes most parts of Zhejiang Province. It is mountainous and contains many famous scenic spots. It has been described in these terms: "Thousands of rocky peaks compete for beauty, tens of thousands of swift streams vie for velocity". The zonal vegetation is evergreen broad-leaved forest, yet owing to the occasional invasion of cold waves and the resulting low temperatures (a mean January air temperature of 2.5 to 6.5°C and a mean July air temperature of 27 to 28°C), mixed evergreen and deciduous broad-leaved forests appear on its northern border. Annual precipitation totals 1300 to 1800 mm.

Vertical zonation is conspicuous. Zonal vegetation (evergreen broad-leaved forest) and zonal soil (red earth) dominate on hills and valleys below 500 m in elevation. Owing to heavy human intervention, the natural zonal vegetation has long been mostly removed; there is now planted vegetation and farmland instead. From 500 to 800 (1000) m is the belt of mixed evergreen and deciduous broad-leaved forest with yellow earth dominating. From 800 (1000) to 1200 m, mixed deciduous and evergreen broad-leaved forest and montane yellow earth dominate. Above 1200 m, deciduous broad-leaved forest and montane yellowish brown soil occur. Such a vertical distribution is a basis for vertical agricultural development. Mountains and hills above 500 m should be mainly devoted to forestry, although the extensive flat erosion surface may continue to be cultivated in such a highly populated and land-hungry area. Valleys and hills below 500 m may be exploited for intensive farming and economic tree crops;

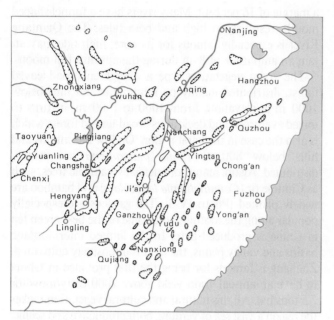

Figure 10-3   Distribution of red basins on the Southeast Coast and in the South Changjiang Hills and Basins

care must be taken to combat soil erosion, cold waves, and drought and flood hazards.

10(2) Wuyi-Daiyun Mountains

This subregion includes most montane and hilly lands of Fujian Province. Compared to the above-mentioned subregion it is even more mountainous, with plains occupying only about 5 per cent of the area. Climate and vegetation are also more tropical, with a mean January air temperature of 8 to 11°C, a mean July air temperature of 20 to 29°C, and annual precipitation generally above 1800 mm. There remain some undisturbed forests on higher mountains. Forestry and commercial tree crops (including tea and bamboo) should be the chief land-use types in this subregion.

10(3) Jinhua-Quzhou Basin

The Jinhua-Quzhou Basin subregion is one of the important red basins in Southeast China (Figure 10-3), which besides the Jinhua-Quzhou Basin includes a series of neighboring small basins in the middle Qiantang River. These basins are surrounded by low mountains and hills and, thus, combine into a large NE-SW trending basin, with a length of more than 200 km and a width of several dozen kilometers. Low hills swarm inside the basins, generally with an absolute relief of 50 to 250 m and a relative relief

of 30 to 50 m. The colorful red sandstones outcrop in many places, especially in the margins of the basins, and are interwoven with the green pine trees and bamboo groves into an immense colorful landscape.

Owing to the sheltered location, the climate is warmer than in the Xianxia-Kuocang Mountains subregion, with a mean January air temperature of 4.6 to 6.3°C, an absolute minimum temperature of –7.9°C, a mean July air temperature of 29.5°C, and an absolute maximum temperature as high as 42°C. Annual precipitation totals about 1500 mm and is mostly concentrated in June to August (June being the month of plum rain). Agriculture has been well developed in the basin since ancient times with a triple cropping system (rice-rice-winter wheat) predominant. This is also the most famous area in China for producing ham and pork. On the surrounding low mountains and hills, the zonal vegetation is evergreen broad-leaved forests; but most of this cover has been cleared, resulting in severe soil erosion.

10(4) Coastal Hills

This is a narrow strip of the Zhejiang and Fujian coast liable to the direct impact of the sea. The coastline and neighboring islands are determined by a NE-SW tronding structural line. They are usually rocky and steep, distributed with numerous hills and drowned valleys to constitute the so-called "Rias" submerged coast. At the mouths of many rivers lie a series of small delta plains where the population and farmlands are concentrated. Annual precipitation totals 1200 to 1600 mm, decreasing from the islands to inland areas. Owing to strong sea winds, evergreen broad-leaved forests are generally dwarfed. Banyan and other tropical trees are commonly seen in the southern part. In the future, this will be an area for developing fisheries and subtropical crops.

11. South Changjiang Hills and Basins— Evergreen Broad-Leaved Forest Region

This region extends from the southern boundary of the middle and lower Changjiang Plain southward up to the Nanling Mountains area, and from the Xuefeng Mountain eastward up to the Wuyi Mountains. over an area of about 300000 sq km.

The region is a great amphitheater of hills surrounded by low and middle mountains on all sides, except along its northern boundary where there are openings to the Changjiang Plain. Another outstanding feature is the extensively scattered distribution of red basins, such as the Hengyang and the Changsha-Liuyang in the Xiangjiang Basin, and the Ganzhou and the Jian-Taihe in the Ganjiang Basin (Figure 10-3). They are composed mainly of

Cretaceous-Tertiary continental deposits and have been mostly eroded into hills and terraces, with an absolute relief below 200 m and a relative relief of less than 100 m. Between these red basins are low mountains and hills generally with elevations of 300 to 600 m. They are mostly composed of pre-Devonian metamorphic rocks and granites and are badly eroded. Along larger rivers, there are ribbonlike flood plains and terraces (10 to 15 m, 25 to 30 m, and 40 to 60 m).

The climate is characterized by a rainy spring and a hot summer. Annual precipitation totals 1400 to 1700 mm, 40 to 50 per cent of which is concentrated from April to June. From July to September, the region is controlled by subtropical high pressure, with frequent drought hazards. Then follows a second high peak of precipitation in October or November. The mean January air temperature ranges from 3 to 8°C, with an absolute minimum below −6°C. The mean July air temperature generally reaches 27 to 30°C, with an absolute maximum above 38°C; this ranks as one of the hottest areas in China.

Just as in the Changjiang Delta, the river network is quite dense. In the delta, however, canals are more numerous than rivers, whereas canals are of comparatively little significance in this subregion. Two river-lake systems—the Dongting Lake and the Poyang Lake—are most important and are distributed mainly in Hunan Province and Jiangxi Province respectively. In Jiangxi Province alone, with an area of 160000 sq km, there are more than 2400 rivers, with an average river length of 115 m/km². The mean annual runoff depth totals 800 to 900 mm, with 700 to 800 mm in the Xiangjiang and Ganjiang valleys and 900 to 1200 mm in the surrounding mountains and hills. Soil erosion is severe in red sandstone and granite mountains where natural vegetation has long-since been destroyed.

The zonal vegetation type is the profuse and luxuriant evergreen broad-leaved forest, and the zonal soil is the thick, clayey, strongly acid red earth. The dominant evergreen trees are Cyclobalanopsis glauca, Castanopsis sclerophylla, C. taiwaniana, C. fardii, and Schina superba. From 27°30'N southward, there is a mingling of some tropical components; on the northern border there are some warm-temperate components. There are usually 3 to 7 stories in the physiognomy of the evergreen broad-leaved forest, with 2 to 3 stories of high trees, 1 to 2 stories of underwoods, and 1 to 2 stories of ground cover. The upper limit of such zonal vegetation rises to 1200 to 1400 m in the south, lowering down to 500 to 800 m in the north. Evergreen needle-leaved forests such as Pinus massoniana and Cunninghamia lanceolata as well as bamboo groves are also commonly seen. Economic tree crops, such as tea, oil tea (Camellia oleosa), tung oil (Aluriles fordii), peach, and orange are widely planted.

Five subregions are identified in this natural region:

11(1) Zhejiang-Anhui Low Mountains and Hills

Located in northwestern Zhejiang Province, southern Anhui Province, and northern Jiangxi Province, this area is composed chiefly of structurally eroded low mountains and hills; depositional landforms are limited to intermontane basins and broad valleys. Mountains are mainly NE-SW trending, with elevations around 1000 m. The highest and most rugged peaks are usually composed of rhyolite, granite, and other hard igneous rocks, as is the case of the famous scenic spot, Mount Huangshan (1873 m). Intermontane basins are interspersed by low hills and valleys, the latter being the most important farming areas in the subregion.

The mean January air temperature ranges from 4.0 to 5.5°C and the mean July air temperature ranges from 27 to 29°C. The frost-free season totals 230 to 260 days, and the annual precipitation is between 1600 to 1700 mm. Evergreen broad-leaved forest and red earth predominate below 800 m asl, mixed evergreen and deciduous broad-leaved forest and montane yellowish brown earth between 800 to 1500 m asl, and on a few peaks above 1500 m, dwarf forest and shrubby meadow predominate. This subregion is noted for its rich timber production and tree crops including tea.

11(2) Central and Southern Jiangxi Hills and Basins

The Central and Southern Jiangxi hills and basins subregion includes the greater part of the Jiangxi Province and the Poyang Lake drainage system. It consists of a series of NE-SW trending low mountains and hills interspersed with intermontane red basins. It is heavily dissected by numerous rivers and consequently, hills and rolling terraces predominate, with ribbonlike floodplains distributed along the rivers. Loose, easily eroded Quaternary red earths cover the foothills and intermontane basins. As a result of devastation of the vegetation cover from ancient times, heavy soil erosion, probably second only to that of the Loess Plateau, has been induced.

The mean January air temperature ranges from 4.5 to 6.1°C and the mean July air temperature ranges from 28.7 to 30°C. The frost-free season totals 281 to 292 days, and the annual precipitation is about 1400 mm. Secondary vegetation includes Pinus massoniana and oil-tea forests and graminaceous grassland in the red basins with sparse P. massoniana forest on low mountains and hills. Small patches of evergreen broad-leaved forest are restricted to narrow valleys. Paddy rice is the most important crop and other crops include cotton, wheat, and tea. Control of soil erosion and reforestation are the two most critical problems in the hilly lands; irrigation in summer and autumn is urgently needed for more intensive farming in the basins.

11(3) Hunan-Jiangxi Low Mountains and Hills

Located on the border between Hunan and Jiangxi provinces and serving as the water divide between the Dongting Lake and Poyang Lake river systems, the Hunan-Jiangxi low mountains and hill subregion is dominated by NNE-SSW or NE-SW trending mountains and intermontane valleys enechelon. The former are mostly around 1000 m in elevation with some high peaks above 1500 m, such as the Mufu Mountain (1594 m) and Wugong Mountain (1585 m). Two level erosion surfaces are conspicuous at about 1000 m and 1500 m respectively. Two steps of terraces at 15 m and 30 m respectively are also distinct.

Owing to the higher elevation, the temperature is somewhat lower than in the surrounding basins, and the precipitation is more abundant. The Lushan Mountain (1474 m) is one of the best known summer resorts in China. Vertical distribution of vegetation and soil is conspicuous. Below 800 m, large tracts of Pinus massoniana, Chinese fir (Cunninghamia lanceolata), and bamboo, with dominating red earth and yellow earth, are distributed. Between 800 to 1500 m, there is mixed forest and montane yellow earth; above 1500 m, there is shrubby meadow. This is, and will be, one of the important timber and economic tree-crop producing areas in Central China.

11(4) Middle and Southern Hunan Hills and Basins

This subregion is similar to the central and southern Jiangxi hills and basins subregion, with similar economic outlook and physical conditions. The only differences are that: it comprises the Dongting Lake drainage system instead of the Poyang Lake drainage system; it includes a greater part of Hunan Province instead of Jiangxi Province; its alluvial plains are somewhat more extensive and, hence, there is greater capability for agricultural development; and the vegetation cover is more dense so that the soil erosion hazard not as menacing.

11(5) Nanling Mountains

The Nanling Mountains, or South Ranges, were once called "Five Ranges" because of five famous ranges. It is the great geographic divide between Central China and South China, stretching from west to east for more than 600 km and from north to south for about 200 km. It is also the water divide between the Changjiang River and the Zhujiang River. Yet, it is comparatively low and discontinuous, and there exist many "breaks" where both cold waves and human traffic may pass through conveniently. Therefore, it is not so effective a barrier as the Qinling

Mountains between Central China and North China. On the whole low mountains dominate the subregion, and three sections might be identified:

1) The western section: mainly NE-SW trending, with peaks about 2000 m in elevation. Between the Yuecheng Mountain and Haiyang Mountain lies the famous Xingan Corridor, where a canal was built more than 2000 years ago to connect the Changjiang River and the Zhujiang River drainages. The corridor is also famous for its remarkable karst topography. Guilin is said to be one the most scenic spots in the world; its neighbor Yangshuo, is even better.

2) The middle section: mainly W-E trending, with peaks about 1000 m. The Beijing-Guangzhou railway traverses through one of its passes.

3) The eastern section: the lowest section, generally below 1000 m in elevation.

Winter lasts one to two months, usually with heavy frosts and occasionally with snow. The mean January air temperature ranges from 8 to 10°C. Annual precipitation totals about 1500 mm, with two peaks in the southern slopes (first peak, May–June; second peak; August), and three peaks in northern slopes (April, June, and August). Subtropical evergreen broad-leaved forest grows luxuriantly, with a considerable number of tropical components in southern slopes. The main target for agricultural development is, and will continue to be, to grow subtropical timber woods and subtropical economic crops. Better farming in intermontane basins depends on a larger supply of irrigation water.

12. Sichuan Basin—Evergreen Broad-Leaved Forest Region

The Sichuan Basin, with an area of about 260 000 sq km, is one of the largest inland basins in China. Administratively, it occupies about 46 per cent of the extensive Sichuan Province, which with a population of nearly 100 million, is certainly one of the most populous agricultural regions in the world.

This is a typical rhombic-shaped basin (Figure 10-4) that has been formed since the Indo-China Tectonic Movement during the Mesozoic era. Lithologically, reddish sandstone and purple shale of the Jurassic and Cretaceous periods are most commonly seen; hence, it is sometimes called the Red Basin. The basin, demarcated roughly by the 700 to 750 m contour line, is uninterruptedly encircled by a series of high mountains. The Micang and Daba Mountains on its northern border have been discussed in the Qinling-Daba Mountains section. The Longmen, Qionglai, Daliang, and the Emei mountains on its western border are the marginal mountains of the Qinghai-Xizang Plateau, with ridgelines above 3000 to 4000 m asl (the highest peak of the

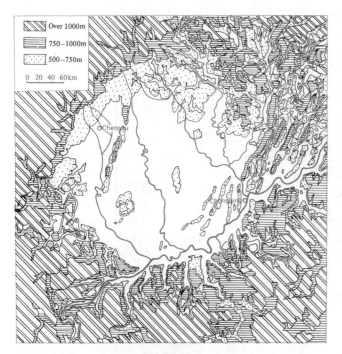

Figure 10-4 Topography of the Sichuan Basin and its surrounding areas

Longmen Mountain is 4982 m). On its eastern and southern borders, a series of ranges, mostly of 1500 to 2000 m and sometimes with well-developed karst topography are distributed. The remarkable Wushan Mountain of the Changjiang Gorges (Photo II-5) is the most famous. The basin itself has an elevation of 250 to 700 m asl and is mostly occupied by rolling hills. It stretches 395 to 455 km from west to east, and 330 to 335 km from north to south. Mountains occupy almost 50 per cent of the total land area, hills about 42 per cent, and plains only 8 per cent.

Thanks to the protection provided by the surrounding mountains, the basin has a much warmer winter than the middle and lower Changjiang Plain with a mean January air temperature of 5 to 8°C. Therefore, a pleasant green color dominates a Sichuan winter. Summer is long and hot, with a mean July air temperature of 26 to 29°C; Chongqing has experienced an absolute maximum temperature of 44°C. Annual precipitation ranges from 1500 to 1800 mm in the surrounding mountains to 900 to 1300 mm in the basin. The relative humidity is high throughout the year, and in western Sichuan Basin, there are more than 300 foggy days. There is an ancient saying, "A Sichuan dog will bark at the sun".

The above-mentioned topographic and climatic conditions make for abundant water resources and a dense concentric drainage system, with the mighty Changjiang River passing through the basin center and numerous tributaries flowing into the river from either side[1]. More than 50 tributaries have a length above 100 km, of which six (Jialing, Minjiang, Tuojiang, Qujiang, Fujiang and Wujiang) exceed 500 km. These rivers offer abundant irrigation water and transportation facilities but are also liable to flood hazards. Most of these rivers meander amid intermontane valleys and sometimes cut through hills and mountains in scenic gorges of which the Changjiang Gorges is the most famous. According to an estimate, the hydroelectric capacity totals nearly 50 million kW in the Sichuan Basin.

Both flora and fauna are profuse and varied. Zonal vegetation in the basin tends to subtropical evergreen broad-leaved forest, but this has been virtually all removed, with only a few small remnants distributed on montane slopes and hills below 1500 to 1800 m. There are also large patches of secondary oak forest and needle-leaved forest as well as bamboo groves. In combination with a vast expanse of cultural vegetation, the green verdure of all these plants interweaves with the reddish ground surface to form a brilliant mosaic. Surrounding high mountains are the most important refuge for ancient relic fauna in China, such as the panda, the goldened-hair monkey, and 40 to 50 other species of rare vertebrates.

The Sichuan Basin has long been famous as an attractive and fruitful land in China. It has been intensively cultivated for more than 2000 years. Several kingdoms were founded here. Practically all plains and hills have been developed. Numerous irrigation ponds fed by rainfall are built on tops of hills. It has now more than 6 million ha of cropland, of which about 56 per cent are paddy land. It is probably the most concentrated paddy growing area in the world.

Three subregions are identified:

12(1) Chengdu Plain

Located near the western border of the Sichuan Basin, the Chengdu Plain has an area of about 8000 sq km, which makes it the largest and most fertile plain in mountainous Southwest China. It is essentially composed of eight alluvial fans, with an elevation of 450 to 750 m inclining from NW to SE. The plain is underlain by Quaternary sand and gravel beds, with a maximum depth of more than 300 m. Annual precipitation totals 1000 to 1300 mm. Numerous rivers pass through the subregion, and ground water resources are also abundant.

This subregion has long been the grain and rapeseed oil base for Sichuan Province. The triple-cropping system (two crops of rice) predominates. The key factor for agricultural development is irrigation. The famous Dujiang Dam irrigation system (Figure 10-5, Photo IV-2) was built in 250 B.C.

---

1) Actually, the name Sichuan in Chinese denotes four rivers.

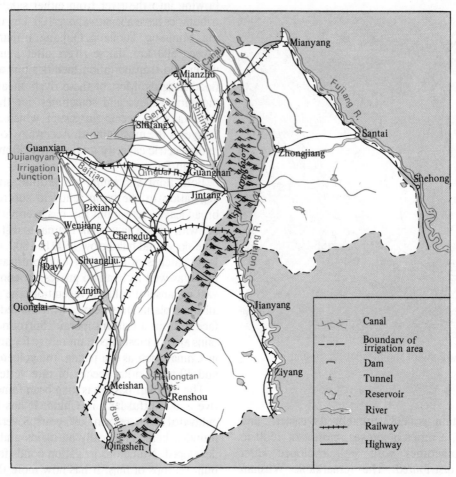

Figure 10-5 Dujiang Dam Irrigation System in the Chengdu Plain

to make use of fluvial runoff resources of the Minjiang River and Tuojiang River. It has been uninterruptedly kept in operation ever since. Recently, it has been greatly improved, with the total canal length increasing from 1100 km to 8100 km and the irrigation area from 140000 to 560000 ha.

12(2) Central Hills

Located in the central part of the Sichuan Basin and dominated by low hills, the central hills area has an elevation of 250 to 600 m. Hills are mainly composed of reddish sandstone and purple shale, with nearly horizontal strata. They have been deeply dissected by numerous running rivers, resulting in steep slopes and flat tops or the so-called mesa topography. In winter, this is the warmest part in the Sichuan Basin with a mean January air temperature of 6 to 8°C. Annual precipitation totals 900 to 1100 mm. Since ancient times, this has been a well-developed agricultural area specializing in grains, cotton, and sugarcane. Nearly all flat hill tops have been cultivated, mostly for dryfarming and partly for winter paddy land.

Consequently, the most urgent problems in this subregion are the improvement of irrigation systems and more efficient control of soil erosion.

12(3) Eastern Paralleling Ranges and Valleys

East of the Huaying Mountain lies a closely folded area with anticlines as ranges and synclines as valleys. More than 30 NE trending ridges and valleys are, thus, arranged and parallel each other. The elevation of ridges increases from 700 to 800 m in the east to about 1000 m in the west. Ridges are mainly composed of reddish sandstone and purple shale and covered by dense evergreen broad-leaved forest; the valleys are mainly cultivated. This subregion has a short and warm winter (65 to 70 days), with a mean January air temperature of about 8°C. Summer is long and hot (145 to 150 days), with a mean July air temperature above 40 to 42°C. The frost-free season lasts 320 to 340 days without any frost at all along the Changjiang valley. The annual precipitation totals 1100 to 1300 mm. Climatically, this is a transitional area between the middle and southern subtropical zones.

13. Guizhou Plateau—Evergreen Broad-leaved Forest Region

The Guizhou Plateau and its neighboring mountains, with an area of more than 400000 sq km, is located at the second great topographic step in China, with an elevation mostly between 1000 to 2000 m. It was described, rather exaggeratedly, as without 3 feet of continual level land, without 3 successive days of fine weather.

It is essentially a dissected plateau with high and rugged surfaces cut by deep canyons and crossed by towering mountains. Scattered among and encircled by these mountains and hills are high plains, the so-called Batzi which occupy probably only 5 per cent of the total land area, yet account for a lion's share of farmlands and settlements. Owing to widely outcropped Paleozoic limestone, karst topography is also well developed and extensively distributed. In the central part of the Guizhou Plateau, such as the Anshun area where intense headward erosion of large rivers has not yet occurred, the rolling plateau surface formed during the Tertiary period has been better preserved. On tops of higher mountains, with elevations between 1500 to 2000 m, there are extensive remnants of an ancient erosion surface.

Climatically, the region is characterized by the absence of excessively cold winters and excessively hot summers although there is an abundance of rainy and cloudy days. The mean January air temperature generally rises above 5°C; the mean July air temperature is below 30°C. Annual cloudy and overcast days total more than 200; Meitan, with 253 overcast days, holds the highest record in China.

Vegetation is characterized by its transitional features, both from east to west and from south to north. In the east evergreen broad-leaved forest dominates; in the west, as the dry season lasts longer, there is a considerable portion of deciduous broad-leaved trees mingled in the forest. In the south, a monsoon forest quite similar to that of the southern subtropical zone dominates on gentle slopes and in level valley plains. In the north, there is the typical evergreen broad-leaved forest of the middle subtropic zone. On the extensively distributed karst topography, certain special types of karst vegetation dominate. These consist mainly of mixed evergreen and deciduous broad-leaved forest; once the forest is cleared the vegetation usually is reduced to sparse shrubs.

Five subregions can be identified.

13(1) Eastern Low Mountains and Hills

The eastern low mountains and hills subregion is the transitional belt between the Guizhou Plateau and the South Changjiang hills and basins region. It is composed mainly of low mountains, hills, and basins with an elevation generally below 800 m. Located on the eastern margin of the plateau, numerous tributaries of the Changjiang River and the Zhujiang River extend their headward erosion deep into the subregion, resulting in rather rugged topography. Annual precipitation totals 1200 to 1300 mm. Vertical zonation of vegetation is conspicuous. Taking the Fanjin Mountain (elevation, 2494 m) as an example, evergreen broad-leaved forest is restricted to elevations below 1400 m, mixed evergreen and deciduous broad-leaved forest to between 1400 to 2200 m, and alpine shrubby meadow predominates above 2200 m.

13(2) Northern Mountains and Gorges

The norther mountains and gorges subregion is the transitional belt between the Guizhou Plateau and the Sichuan Basin. The NE-SW trending Dalou Mountain is uplifted on its northern border, with an absolute relief of 1300 to 1500 m (highest peak, 1800 m) and a relative relief of 500 to 700 m. Many gorges and canyons have been formed. Karst topography is also well developed. The Dalou Mountain serves as a climatic barrier, with wet, windward southeastern slopes and drier, leeward northwestern slopes. Some intermontane basins, are also distributed and this is where both farmlands and population are concentrated.

13(3) Central Hilly Plateau

The central part of the Guizhou Plateau, including the provincial capital Guiyang, is an undulating hilly plateau with an absolute relief of 900 to 1500 m and a relative relief mostly below 300 m. More than 80 per cent of the total area is characterized by limestone strata and the main landforms are composed of structural basins, block mountains, and karst topography. Besides the Wujiang River which cuts a deep gorge, most large rivers have rather broad valleys and sluggish water where farmlands and settlements are densely distributed. The climate is favorable for agricultural development, with a mean January temperature of about 5°C, a mean July temperature of 24°C, a frost-free season of more than 260 days and an annual precipitation between 1100 to 1200 m.

The subregion is located south of the Miaoling Mountain, which is the divide between the Changjiang River and the Zhujiang River, with ridges around 1000 m in elevation (highest peak, 2179 m). It is essentially the dissected southern slope of the Guizhou Plateau, with an absolute relief of 500 to 1400 m and a relative relief of 300 to 700 m. In its northern part, limestone strata are widely distributed and the plateau surface is deeply dissected, with an elevation ranging from 800 to 1200 m, a frost-free season of about 300 days, an annual precipitation of about 1200 to 1400 m, and a vegetation predominantly consisting of middle subtropic evergreen broad-leaved forest. In its southern part, many tributaries of the Zhujiang River cut the subregion into low mountains, hills, and deep valleys, with an elevation generally below 800 m, a frost-free season of about 335 days, and a conspicuous dry season from October to March (comprising only 20 per cent of total annual precipitation). Consequently, middle subtropical evergreen broad-leaved forest predominates, mingled with some southern subtropical components, such as the banyan tree. This is a land suitable for growing sugar-cane, bananas, and other tropical crops.

13(5) Western Mountains and Gorges

The western mountains and gorges subregion is located at the transitional belt between the Guizhou Plateau and the higher Yunnan Plateau. The plateau surface is preserved only in small patches on water divides. Mountains are mostly above 2000 m in elevation (highest peak, 2900 m), with slopes at more than 20 to 25°. Intermontane valleys generally have an absolute relief of 900 to 1700 m and a relative relief of 400 to 600 m (deepest gorge is > 1000 m). On the whole, it is a subalpine area. Although the evergreen broad-leaved forest is still dominant, montane grassland is also widespread; there is a high potential for developing forestry and animal husbandry in this subregion.

14. Yunnan Plateau—Evergreen Broad-Leaved Forest Region

The region is located roughly between 24 to 28°N and 98 to 105°E, with an area of about 300000 sq km. It is the transitional belt between two great natural realms in China—the Qinghai-Xizang Frigid Plateau and Eastern Monsoon China. Its dominant landscape is an extensive level and reddish plateau interspersed with greenish patches of well-farmed basins and tree-clad mountains.

The region is first of all characterized by a rather well preserved undulating plateau surface and broad lake basins, mostly with elevations ranging from 1500 to 3000 m. Second, it is the southern part of the spectacular Hengduan Mountains area, a series of nearly north-south-trending, lofty mountain ridges running parallel to a series of nearly north-south-trending mighty river gorges. This ridge and gorge topography is particularly prominent north of 25°N, with a relative relief of more than 2000 to 3000 m and sometimes even more than 5000 to 6000 m. From west to east, they are: the Gaoligong Mt., the Nujiang R., the Nushan Mt., the Lancang R. the Daxue Mt. (Yunling Mt.), the Jinsha R., the Mianmian Mt., the Yalong R., and Lunan Mt. Third, karst topography is extensively distributed, especially in eastern Yunnan where Shilin in located, an example of a particularily striking stone forest (Photo IV-6). Finally, volcanic landforms have also developed in the western Yunnan Plateau of which the Tengchong area (Figure 10-6) with its numerous volcanic cones and hot springs are the most significant.

The area is dominated by the so-called plateau monsoon—the southwestern (Indian) monsoon in the western part and the southeastern (Pacific) monsoon in the eastern part. Owing to its topographic conditions and latitudinal location, the weather is mild all year round, and the provincial capital Kunming, with an elevation of about 1960 m, is famous for its eternal spring. Annual precipitation totals 1000 to 1200 mm, decreasing both from southeast and southwest to middle north, with the deep valley bottoms of the Jinsha, Nujiang, and Lancang rivers having less than 1000 mm. The division between the rainy season (May to October) and the dry season (November to April) is apparent, with "winter" (December to February) accounting for only 5 per cent of total annual precipitation.

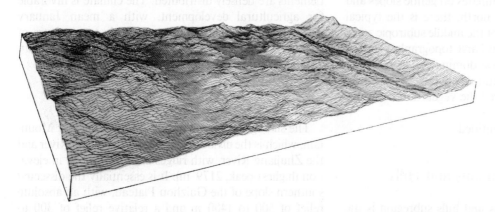

Figure 10-6   A digital landform model of the Tengchong area, western Yunnan

Vegetation is accordingly varied. The zonal vegetation type is evergreen broad-leaved forest, with *Cyclobalanopsis glaucoides, C. delavayi,* and *Castanopsis delavayi* as dominant trees. After heavy human intervention, the drought-tolerant Yunnan Pine (*Pinus yunnanensis*) forest now prevails, intermingled also with *P. armandii* and Keteleeria evelyniana. Vertical distribution of vegetation is conspicuous. One famous example is the snowcapped Yulong Mountain (5596 m) overlooking the Jinsha River valley, with a relative relief of more than 4000 m. Five vertical zones are demarcated: (1) the Jinsha River valley below 2000 m is distributed with the semiarid subtropical shrubby savanna; (2) the basic vegetation belt (2000 to 3100 m) is mainly composed of Yunnan Pine forest; (3) fir forest occurs between 3100 to 3800 m; (4) alpine meadow occurs between 3800 to 4500 m; (5) continual snow appears above 4500 m.

Such a conspicuous distribution is mirrored by the vertical agriculture in the region. Semiarid, subtropical valleys below 1000 m might be used as farmland with double-cropping of rice if irrigation water is available. Farmland between 1000 to 24000 m is generally under a rice-wheat double cropping system. Farmland above 2400 m is characterized by one-crop dryfarming. Such vertical zonation is even reflected in the distribution of races, with Dai, Bai, and Hani mainly in the low valley; Tibetan, Kachin, Yi, and Lisu on the high plateau, and Han and Hui in the middle.

Three subregions can be identified.

14(1) Eastern Yunnan Karst Plateau

With carbonate rock occupying more than 50 per cent of the total land area at a thickness of more than 3300 m, karst topography is widely distributed in the eastern Yunnan karst plateau subregion. The ground surface is generally rather dry, rivers are few, and farmlands are mostly restricted to dryfarming. On limestone hills and mountains, once the original vegetation has been destroyed, shrubby grassland or even bare rock tends to emerge. The conditions on sandstone and shale outcrops are much better; they are usually covered by a thin veneer of weathered materials and Yunnan Pine or oak forest. The distinction between rock mountain and earth mountain is quite significant in the subregion, and soil conservation and reforestation are the most important measures to transform the present poverty-stricken landscape.

14(2) Central Yunnan Plateau and Lake Basins

Mainly located in the water-divide area, the plateau surface is comparatively well preserved. However, near river channels the plateau is dissected into four erosion surfaces of different levels: 4000 to 4100 m, 3600 to 3700 m, 2400

to 2500 m, and 1800 to 2100 m. There are also numerous faulted lake basins of which Dianchi Lake near Kunming is the most famous. These lake basins are rich areas of rice and fish in Yunnan Province, with a larger proportion of paddy land than dry farming. Climatically, this is a land of eternal spring, with one of the warmest mean January temperatures (8 to 10°C) and the coolest mean July temperatures (19 to 22°C) in China. Annual precipitation totals 700 to 1200 mm; spring drought, however, is a problem for agricultural development.

14(3) Hengduan Mountains

Located mainly in western Yunnan, this area is composed of parallel high mountain ranges and deep river gorges. North of the Baoshan-Xiaguan line (about 25°30'N), they are closely crowded together and there are practically no broad intermontane basins nor large patches of plateau surface preserved. South of that line, the mountain ridges become lower and intermontane basins broader. Two climatic factors stand out quite conspicuously. One is the barrier action of these lofty mountains. For example, Baoshan, which is located at the eastern foot of the Gaoligong Mountain, has an annual precipitation of only 903.2 mm; while Longling, on the western windward slope, is as high as 2595.7 mm. The second factor is the conspicuous vertical zonation. On the western slopes of the Gaoligong Mountain below 2500 m, secondary vegetation and farmlands lie; between 2500 to 2700 m, there is luxuriant evergreen broad-leaved forest; between 2700 to 2960 m there is mixed needle-and broad-leaved forest; between 2960 to 3500 m, needle-leaved forest predominates; and between 3500 to 3680 m, there are mainly alpine shrubs. Mountainous land makes up about 96 per cent of the total land area (Photo IV-5) and different forest types about 25 per cent. Forestry and tree crops certainly are, and will be, two major resources in this subregion.

Southern subtropic zone includes Two natural regions.

15. Lingnan Hills—Evergreen Broad-Leaved Forest Region

This region, with an area of about 350000 sq km, is extensively distributed south of the Nanling Mountains, which are traditionally regarded as the divide between Central China and South China. On the southeastern side, it faces the Taiwan Strait and South China Sea. On the Southwestern side, a series of coastal mountains and hills separate it from the humid tropic division.

The region is dominated by hills below 500 m in elevation. There are also a few mountains above 1000 m and a series of ribbonlike alluvial plains located along rivers. Mountains and hills are largely controlled by the NE-SW trending structural lines composed chiefly of granites and metamorphic rocks in Guangdong and southeastern Fujian and of thick limestone beds in Guangxi, where typical karst topography has developed. In the structural basins, such as Guangzhou and Meixian County, middle Cenozoic red beds are distributed. The coastline is generally rugged and irregular. There are many promontories and protected bays suitable for harbors. The alluvial plains are recently deposited, and the deltaic plains of the Zhujiang River and other large rivers still continue to grow.

The climate is characterized by a long, rainy summer and a short, dry winter. Summer lasts more than 6 to 7 months, with frequent typhoons accompanied by torrential rain. Winter has a mean January air temperature of 10 to 15°C. Nevertheless, when a strong cold wave passes southward through the low and broken Nanling Mountains the absolute minimum temperature may drop below −3°C; such a cold wave generally occurs once a year. Annual precipitation totals 1400 to 1800 mm, with some windward slopes recording more than 2000 mm and some secluded valleys less than 1000 mm. The seasonal variability of rainfall is rather great; spring and autumn drought as well as summer floods usually put restrictions on agricultural production. Soil erosion is severe on montane slopes where natural vegetation has been destroyed.

Abundant rainfall feeds numerous rivers paralleling mountain ranges with broad valleys or cutting through them in gorges. The most important river system is the Zhujiang River, which is composed of three main tributaries. The largest one, the Xi Jiang (West River), rises in the eastern Yunnan Plateau and southern Guizhou Plateau and flows eastward across Guangxi Province and western Guangdong Province to meet two other tributaries—the Beijiang River (North River) and the Dongjiang River (East River) near the head of the delta at Guangzhou. It then splits up into a ramifying system of channels and distributaries and finally empties into the South China Sea. Hong Kong and Macao are located on either side of the main river mouth. There are also numerous independent rivers, of which the Hanjiang River (flowing into the sea at Shantou) and the Jiulong River (flowing into the sea at Xiamen) are the most important.

The dominant zonal vegetation is the so-called evergreen monsoon forest, which is different from both the evergreen broad-leaved forest of the middle subtropical zone and the tropical rain forest or monsoon forest of the tropics. Inside the region, tropical components increase from north to south, and subtropical components decrease accordingly. Owing to heavy human intervention, the original forests have long-since been cleared and replaced; secondary forests, shrubs, and grasslands are widely distributed instead. Little use has been made of secondary growth. Soil conservation and reforestation are urgent problems on mountains and hills. The low-lying alluvial plains, on the other hand, have been intensively used, mostly for double-cropping paddy rice and highly productive economic crops, such as sugarcane, mulberry trees, and hemp. There are about 4 million ha of cropland in the region, accounting for about 10 to 11 per cent of the total land area. The Zhujiang Delta, the Shantou-Chaozhou Plain, and the Zhangzhou Plain, are among the most productive and densely populated areas in China.

Five subregions can be identified.

15(1) Eastern Guangdong and Fujian Coastal Hills

Owing to the meliorating influence of the sea, the eastern Guangdong and Fujian coastal hills subregion extends northeastward along the coast up to about 25°N with practically a year-round growing season. A series of mountains and hills, paralleling the coast, dominate the landscape. The coastline is rugged and irregular and contains many excellent harbors, such as Shantou, Xiamen, and the ancient Quanzhou. Numerous southeastward-flowing rivers cut through the mountains and hills in gorges or narrow valleys. There are also highly fertile deltaic plains, such as the Shantou-Chaozhou Plain, the Zhangzhou Plain, and the Quanzhou Plain. The hard-working inhabitants are fighting against both the sea and the mountainous topography. They have traditionally made efforts to reclaim land from the sea with considerable success. They are also making better use of mountains and hills by means of reforestation and soil conservation.

15(2) The Zhujiang Delta Plain

With an area of about 10000 sq km, the Zhujiang Delta Plain is the most important farming area in South China and probably one of the most productive agricultural ecosystems in the world (Landsat image 5) . The United Nations University in cooperation with the Guangzhou Institute of Geography is now making a close quantitative study of its mulberry-dike-fish–pond-ecosystem. In this subregion, the growing season lasts all year long, with a mean January air temperature of 13 to 15°C. Annual precipitation totals more than 1600 mm.

15(3) Western Guangdong-Eastern Guangxi Mountains and Valleys

This subregion is dominated by a series of NE-SW trending low mountains and hills, with the highest peak at 1704 m in elevation. The Xijiang River has cut through these mountains in deep gorges; its tributaries on either side form broader valleys and intermontane basins. Annual precipitation totals 1800 to 2000 mm on windward slopes and 1400 to 1600 mm on the leeward side. Evergreen broad-leaved forests and evergreen monsoon forests are luxuriant; this is one of the major timber-producing areas in South China. Soil conservation is now a growing problem.

15(4) Central Guangxi Broad Valleys and Karst Basins

Karst topography is well developed in western and northern parts of this subregion, and broad valleys occur extensively in the eastern and southern parts. This area, which produces sugarcane and hemp, is the most important grain and economic crop producing section of Guangxi. It needs more intensive management and a better irrigation system for further development.

15(5) Northwestern Guangxi Mountains and Karst Topography

The northwestern Guangxi mountains and karst topography subregion is the most western and the highest part of the region, with an elevation of 500 to 1000 m. Only the valleys of the NW-SE stretching Youjiang River

(Right R.) and the SW-NE flowing Zuojiang River (Left R.) lie below 200 m, where they enjoy abundant sunshine and a mean daily temperature of above 10°C nearly year round. Annual precipitation totals 1000 to 2000 mm, although spring drought is sometimes still a problem. Agriculture is well developed. On either side of these two rivers, there are shrubby or grassy terraces and hills, and soil erosion is a great problem. The extensive mountainous karst topography between these two rivers is deeply dissected.

16. Taiwan Island—Evergreen Broad-Leaved Forest and Monsoon Forest Region

The region is composed of the largest island of China, Taiwan Island, and its more than 100 neighboring islands. Taiwan Island itself has an area of 35760 sq km, with a north-south length of about 384 km, and a west-east width of 144 km at its widest point. All neighboring islands are rather small; the largest one, Penghu Island has an area of 64 sq km. They are clustered around Taiwan Island at

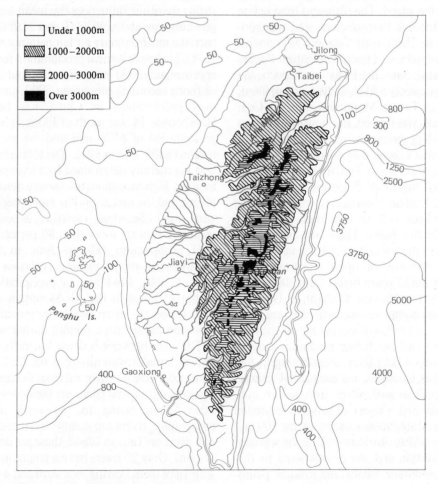

Figure 10-7 Topography of Taiwan Island

a distance usually not greater than 70 to 80 km.

Owing to the unique distribution of land and sea and the related physical geographical features (Photo IV-7) Taiwan Island and its neighboring islands are integrated into one single natural region. Climatically, however, a larger part of the island north of the Tropic of Cancer pertains to the southern subtropical zone, whereas a smaller part south of the Tropic of Cancer is in the tropical zone. The region is part of the great marginal island arc of the northwestern Pacific Ocean and marks the divide between the shallow continental shelf and the deep sea basin. Taiwan's east coast, with an immediate offshore water depth of more than 2500 m, is the Chinese "window" directly facing the immense Pacific Ocean. The Taiwan Strait, with a length of about 300 km and a width of 150 to 250 km (the narrowest part is only 130 km), is a part of the continental shelf, with a water depth generally less than 100 m. The Taiwan Warm Current (Kuro Shio) flows northward and embraces both the east and west coasts of Taiwan.

Taiwan Island is dominated by Cenozoic mountains and hills that occupy about two thirds of the total land area. The Central Range runs from north to south for nearly the entire length of the island. The Taiwan Geosyncline was formed during the late Paleozoic era, and was concluded by the Pliocene Himalayan Tectonic Movement. The great marginal island-arc of the northwestern Pacific Ocean was then formed. Since the Pleistocene epoch, the island-arc has been repeatedly folded, faulted, and uplifted, and the Taiwan Arc or Taiwan Mountains, with the Central Range as backbone, was formed. It is now mainly composed of four paralleling mountain ranges running from south to north; from east to west, they are the East Coastal Range (composed mainly of upper Tertiary beds), the Central Range (composed mainly of Paleozoic to Mesozoic metamorphic rocks), Yushan Mountain (composed mostly of lower Tertiary beds) and Ali Mountain (composed mainly of upper Tertiary beds). These are among the youngest mountains in the world, and tectonic movement is still in progress (the rate of uplift is estimated at 18 cm/100yr). There are 62 peaks higher than 3000 m in elevation, the highest one (Mount Yushan), is 3997 m. These lofty mountains contrast strongly with deep sea trenches (more than 3000 to 5999 m deep) not far off the east coast. Between these four paralleling mountain ranges, there are great paralleling fault lines, contributing to the presence of earthquakes, landslides, and geothermal springs. Mineral resources, such as gold, silver, iron, sulfur, and coal, are widely distributed. The divide between montane and hilly areas is represented approximately by the 500 m. contourline. Hills are widely distributed along the western foot of the Ali Mountain and merge westward to the coastal plains. Intermontane basins and coastal plains together account for 20 per cent of the island's total land

area. They are densely covered with farmlands and settlements.

A long but rather monotonous coastline is another geomorphological feature of the region. In total there are about 1500 km of coastline, of which Taiwan Island itself accounts for 1100 km. Except for the northeastern corner where the Central Range meets the coast and results in a rugged coastline and abundant promontories— including the famous Jilong (Keelung) port— all Taiwan's coastlines are rather straight.

The western coast, which borders the shallow Taiwan Strait, has a well-developed alluvial plain; the coastal beach quickly advances seaward. Near Tainan the mean annual advance of the coastal beach has been 35.4 m during the last 200 years. Therefore, natural good harbors are few; Gaoxiang (Kaohsiung) is the only exception. The eastern coast is largely given character by the paralleling coastal range and the straight fault line; the offshore deep sea makes the formation of depositional coastal beach difficult. Penghu Island has the most varied and irregular coastline, and its Magong (Makung) is the best port in the region.

The region is extremely wet and rather hot year round. Owing to the low latitudinal location (22 to 25°N) and strong maritime influences the mean annual temperature generally rises above 20°C (Taibei 21.7°C). In the northern part, the mean January air temperature may drop to about 15°C. The mean annual precipitation for the whole island approximates 2600 mm, with the mountainous area south of Jilong recording more than 5000 mm. The highest annual precipitation in China has been recorded in Haoshaoliao 14 km south of Jilong; it has a mean annual precipitation of 6576 mm and the annual precipitation reached 8408 mm in 1912. The total amount of precipitation is generally determined by elevation and geographic location; high mountains are usually much more rainy than plains, and the east coast has more precipitation than the west coast. Seasonal variation is also significant. In southern Taiwan, more than 80 per cent of the total annual precipitation falls in June to September when southwestern monsoons and typhoons dominate. In 30 August 1940, 1164 mm of precipitation poured down within 24 hours in the Ali Mountain area. In northern Taiwan, on the other hand, the prevailing northeastern monsoon brings more rainfall during winter.

The river network is dense but rather asymmetric. As Taiwan's major waterdivide is the Central Range, which is located a little bit eastward from the central line, the river network and alluvial plains are better developed on western side. Again, owing to the small area and rugged topography, rivers are mostly small and swift. There are 151 rivers on Taiwan Island, these are divided into 48 river systems. Only 20 rivers have a length in excess of 50 km, with 16 of them flowing westward and 4 eastward. Gorges, usually formed in their upper and middle reaches, have a great hydroelectic potential. According to an estimate, the output could reach 5 million kW. Extensive alluvial fans are formed as these rivers flow suddenly from mountainous areas and drop into level plains. Extensive alluvial plains are also deposited in their lower reaches. Both alluvial fans and alluvial plains are intensively cultivated.

Vegetation is luxuriant and varied. Taiwan Island has long been famous for its forests and timber production. Formerly, forests occupied about 70 per cent of the total land area; now the coverage is 50 to 55 per cent. Large patches of original forests and numerous economically valuable, trees (more than 20 species) grow extensively up to 1600 m in elevation. In the northern and central parts, the dominant zonal vegetation type is subtropical evergreen broad-leaved forest mingled with some tropical components. The most spectacular species is the banyan tree, which is widely distributed on low mountains, hills, and basins below 500 m in elevation. The southern tip is characterized by tropical monsoon forest and rain forest. Dominant species in the tropical rain forest are *Myristica cagayanensis, Pterospermum niveum,* and *Artocarpus lenceolatus.* Vertical distribution of climate and vegetation is particularly conspicuous in the mountains of Taiwan Island—from low plains to lofty mountains—tropical (subtropical), temperate, and alpine climates and vegetation types appear in succession. Broadly speaking, horizontal zonal subtropical or tropical vegetation types are distributed below 500 m in elevation, evergreen broad-leaved forest and needle-leaved, forest from 500 to 2000 m; mixed evergreen broad-leaved deciduous broad-leaved, and needle-leaved forest from 2000 to 3000 m and alpine shrubby meadow predominates from 3600 to 3900 m.

Areal differentiation in Taiwan Island is prominent. First, it might be distinguished horizontally by the northern subtropic zone and southern tropic zone. Second, vertical zonation is quite conspicuous, from subtropical (or tropical) to temperate and alpine. In addition, there is divergence between eastern and western coasts. Larger, neighboring islands, such as Penghu Island, should be separately delimited.

16(1) Northern Subtropic Hills and Plains

The northern subtropic hills subregion includes extensive coastal hills and plains. The mean annual precipitation totals 1500 to 2000 mm. The mean annual temperature ranges from 15 to 16°C, with an absolute minimum temperature above 4°C. Alluvial soils are thick and fertile and chiefly support paddy rice and sugarcane. Natural vegetation below 300 m in elevation is oak forest, which is now mostly removed.

16(2) Central Subtropic Mountains

The central subtropic mountains are the highest subregion in Taiwan Island including most lofty peaks above 3000 m. Consequently, this subregion has the heaviest rainfall, with annual precipitation generally above 4000 to 5000. It is also the most important timber-producing area. Vertical zonation of vegetation and soil is conspicuous: (1) below 500 m—subtropical monsoon evergreen broad-leaved forest and red earth; (2) from 500 to 2000 m—evergreen broad-leaved forest and yellow earth; (3) from 2000 to 3000 m—Mixed broad-and needle-leaved forest and yellowish brown earth; (4) from 3000 to 3600 m—coniferous forest; (5) from 3600 to 3900 m—alpine shrubby meadow.

16(3) Southern Tropic Hills and Plains

Located generally south of the Tropic of Cancer and below the 500 m contour line, the southern tropic hills and plains area is composed of two alluvial plains—the Tainan Plain and the Pingdong Plain. Both have been extensively cultivated for paddy rice and sugarcane. On the southern margin of Tainan Plain there is a hilly range that contains coral reef, which shows the violence of the neotectonic movement and the great amplitude of its uplift. The climate is much warmer but less humid than other subregions, with an annual temperature above 24°C, a mean January temperature above 16°C, and an annual precipitation around 1500 mm. The zonal vegetation is the luxuriant monsoon rain forest, with Myritia spp, Sideroxylon ducliton, Diospyros utilis, and Artocarpus lanceolata dominating. Numerous epiphytes are interwoven with the trees. Except for small patches of tropical rain forest still surviving on foothills or valley slopes, most of the original forests have been removed.

16(4) Tropical East Coast

The tropical east coat area is restricted to a narrow coastal strip east of the Central Range. A longitudinal valley is sandwiched between the Central Range and the East Coastal Range, with a length of about 150 km and a minimum width of only 4 to 5 km. On the valley floor, the elevation ranges from 50 to 250 m. Alluvial fans are well developed on either side of the valley. On its eastern slope, the East Coastal Range virtually plunges into the deep Pacific Ocean, with a precipitous fault scarp several hundred meters high; little coastal plain has been formed. Climatic and vegetation conditions are similar to 16(3) subregion, but annual precipitation, however, is somewhat more abundant (1500 to 2000 mm).

16(5) Penghu Islands

The archipelago stretches longitudinally for more than 60 km and latitudinally for more than 40 km; however, it has a total land area of only 127 sq km. Penghu Island with an area of 64 sq km, is the largest island. The archipelago is mainly composed of rocky basalt platforms, there are frequent earthquakes. The elevation nowhere exceeds 80 m. Located on the transitional belt between alternating monsoons, the area is windy year round. Annual precipitation totals around 1000 mm. Originally, there was no forest, but luxuriant banyan trees have been planted by the local inhabitants.

References

[1] *Physical Regionalization of China* Working Committee, Chinese Academy of Sciences, 1958 – 1959, *Physical Regionalization of China,* 8 vols, Science Press, Beijing. (In Chinese)

[2] Chu Ko-chin (Zhu Kezhen), 1958, "Subtropic Zone of China", *Science Bulletin,* 1958, No. 17. (In Chinese)

[3] Kiang Ai-liang, 1960, "On the Delimitation of the Subtropic Zone in China", *Acta Geographica Sinica.* Vol. 26, No. 2. (In Chinese, with English abstract)

[4] Liu Yinhan, 1980, "On Natural Zonation in Shaanxi Province", *Acta Geographic Sinica,* Vol. 35, No. 3. (In Chinese, with English abstract)

[5] Wuhan Institute of Botany, Chinese Academy of Sciences, 1980, *Plants of the Shennongjia.* Hubei People's Press, Wuhan. (In Chinese)

[6] Tan Vanyi et al., 1980, *Physical Geography of Hubei Province,* Hubei People's Press. Wuhan. (In Chinese)

[7] Department of Geography, Hangzhou University, 1959, *Physical Geography of Zhejiang Province,* Zhejiang People's Press, Hangzhou. (In Chinese)

[8] Guo Guanmin et al., 1980, *Physical Geography of Hunan Province,* Hunan People's Press, Changsha. (In Chinese)

[9] *Sichuan Vegetation* Working Group, 1980, *Sichuan Vegetation,* Sichuan People's Press, Chengdu. (In Chinese)

[10] Department of Geography, Kunming Teachers' College, 1978, *Geography of Yunnan,* Yunnan People's Press. Kunming. (In Chinese)

[11] Xu Xianghao, 1981, *Plant Ecology and Plant Geography of Guangdong Province,* Guangdong Science and Technology Press, Guangzhou. (In Chinese)

[12] Zhong Gongfu, 1980, "Mulberry-Dyke-Fish-Pond on the Zhujiang Delta — A Complete Artificial Ecosystem of Land-Water Interaction", *Acta Geographica Sinica,* Vol. 35, No. 3. (In Chinese, with English abstract)

[13] Zhang Hunan, 1980, "The Fault Delta", *Acta Geographica Sinica,* Vol. 35, No. 1. (In Chinese, with English abstract)

[14] Chen Chin-siang, 1959 – 1960, *Taiwan,* 3 vols, Fumin Institute of Geography, Taibei (Taipei). (In Chinese)

[15] Hsieh Chiao-min, 1964, *Taiwan: A Geography in Perspective,* London: Butterworths.

Chapter 11

Humid Tropical South China

China is essentially a country with temperate and sub-tropic climate. The tropical area is not extensive, occupying only about 1.6 per cent of the total land area. Yet, it stretches along the Chinese southern border through more than 15 degrees of longtitude from the Sino-Burmese international boundary in southern Yunnan up to the southern tip of Taiwan Island. From north to south, it spreads out in the form of numerous islands dotted amid the immense South China Sea to the Zengmu Shoal, through more than 20 degrees of latitude. It enjoys bountiful climatic and biologic resources and is especially famous both for its tropical crops and strategic value.

There still exists some disagreement about the northern boundary of this natural division. The controversy extends over a width of 21 to 25°N. Owing to inadequate scientific data, we will refrain from making a final judgment on this matter at present. Here we will adopt, for the time being, the line drawn by the Physical Regionalization of China Working Committee of Chinese Academy of Sciences in 1959, that is a line running from the southwestern margin of the Zhujiang Delta (about 22°N), westward through Maoming, Hepu, Chongzuo, Gejiu, Simao, and Mangshi up to the Sino-Burmese international boundary northwest of Wanding. This line coincides roughly with the 8000°C isotherm of accumulated temperature during the ≥10°C period and the 16°C isotherm of mean January air temperature. The major factor shaping the physical geographical environment of Taiwan Island is the distribution of land and sea. The sub-tropical zone of its northern part and the tropical zone of its southern tip may be combined together into one single natural region as discussed in Chapter 10. Hence, an analysis of the southern tip of the Taiwan Island will not be included in this chapter.

HUMID TROPICAL LANDSCAPE

This natural division is first of all characterized by the humid tropical landscape that contains the following common features.

Humid Tropical Climate

Humid tropical climate has an accumulated temperature during the ≥10°C period above 8000°C (above 9000°C in the southern part), a mean January air temperature >16°C, an annual precipitation ranging from 990 to 2500 mm (Figure 11-2) and an aridity index of <1.0. Hence, the growing season lasts year round and paddy rice can be triple cropped. Yet, from the global point of view, most parts of this natural division lie at the northern margin of the tropical zone. Its tropical landscape is not as typical as that of Malaysia or Indonesia because the temperature is somewhat lower and there is occasional frost occurring in the northern parts of this region. Furthermore, this division is dominated by alternating monsoons, with a marked distinction between the rainy season and the dry season, quite different from the year-round humid conditions in Malaysia or Indonesia.

This natural division can be again subdivided into two parts: the western part and the eastern part. The former is mainly distributed in low valleys or the intermontane basins of southern Yunnan, with precipitation deriving chiefly from the Indian Ocean without any direct influence of the Pacific typhoons. Owing to the barrier action of the

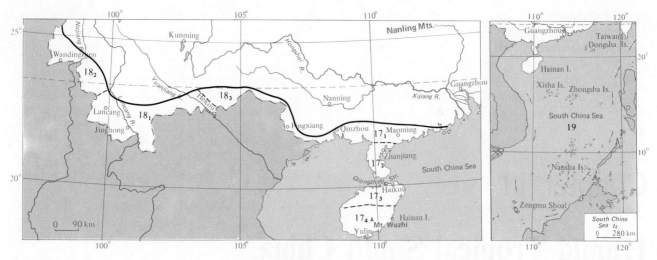

Figure 11-1   A location map of humid tropical South China

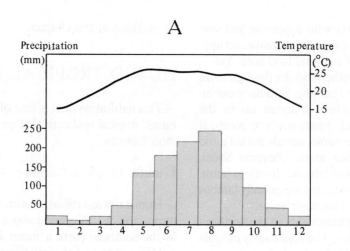

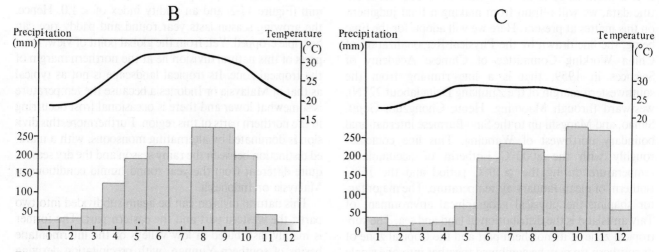

Figure 11-2   Climate of humid tropical South China
A. Jinghong (in Yunnan); B. Zhanjiang (in Guangdong); C. Xisha Archipelago.

lofty Qinghai-Xizang Plateau on the north as well as its dissected plateau topography, the winter is much warmer, temperature inversions happens frequently, and the limit of the tropical zone stretches deeply northward along river valleys as far as 23 to 25°N, with elevations as high as 700 to 900 m asl. The eastern part of the division is located along the sea-coasts of Guangdong and Guangxi, as well as on numerous South China Sea islands, with precipitation deriving mainly from the Pacific Ocean. Typhoons occur frequently in summer and autumn. Owing to the rather low and dissected topography of the Nanling Mountains, its barrier action is not very effective and strong cold waves may penetrate southward up to northern Hainan Island in the winter. Consequently, the winter temperature is much lower and the annual variation of temperature much greater. The northern limit of the tropical zone is, thus, pushed southward along cold wave tracks up to about 22°N.

There also exist conspicuous climatic distinctions between the southern and northern parts. The mean temperature and absolute minimum temperature of the cold est month rise gradually from north to south. For example, Jinghong (22°52′N) has a mean January air temperature of 15.7°C and an absolute minimum temperature of 2.7°C; whereas Zhanjiang (20°13′N) records 15.6°C and 2.8°C respectively, Haikou (20°2′N) 23.8°C and 2.8°C respectively, Yaxian (18°.24′N) 20.8°C and 5.7°C respectively, Xisha (16°50′N) 22.8°C and 15.3°C respectively, and Nansha (10°23′N) 26.5°C and 21.1°C respectively. If we use a mean monthly temperature ⩾22°C as an index for the summer season, then there will be summer year round south of the Xisha Archipelago and two distinct seasons (summer and spring-autumn) occur north of it. Precipitation increases also from north to south. If we use a monthly precipitation of ⩾50 mm as an index for the dry season, then there is no dry season in the southern part of the Nansha Archipelago but a dry season of about three to four months in the northern part of the Nansha and in the Xisha, and one of about three to six months in Hainan Island and in the Leizhou Peninsula. Therefore, as a whole, this natural division may be divided into two natural zones: the southern typical tropical zone (south of the Xisha Archipelago) and the northern marginal tropical zone (north of the Xisha Archipelago).

Dynamic Lateritic Soils

A humid tropical climate, such as that described above, exerts a great influence on the soil-forming process. In humid tropical South China, chemical weathering and leaching are intensive, and decomposition of organic materials is rapid. Consequently, the organic content of the soil is generally low and the laterization soil-forming process predominates. The endproduct of laterization may be called laterite, a porous, crumbly textured, bright reddish soil often hardening on exposure to the air and composed almost exclusively of clays and iron compounds; it is much the same whatever the parent materials. Yet, only in a limited area is this zonal soil type (laterite) fully developed. Most of the region's soils are at an intermediate stage of laterization. The stage may be assessed by measuring the proportion of silica to aluminum; the higher the silica content, the less laterization has taken place.

One feature of lateritic soils is their dynamic nature and the delicate balance of their fertility. We should not hastily call them infertile because — even if compared to the well-known fertile chernozem, which has a much higher humus content — the much quicker recycling of nutrients in the tropical zone makes up the humus content deficiency. Therefore, if the land is well managed, quick disintegration and absorption will be met by the quick supplement of organic nutrients. Thus, soil fertility and productivity will remain high. On the other hand, if the land is poorly managed or overused, soil erosion, leaching, and other depletion processes will soon dominate. In these cases, soil fertility is quickly degraded or even completely destroyed.

Bountiful Biological Resources

Heavy precipitation and constant high temperature generate continuous plant growth and a profuse, varied flora. The luxuriant tropical monsoon forest is the typical zonal vegetation type here. Two subtypes may be distinguished: evergreen tropical monsoon forest and deciduous tropical monsoon forest. The former is distributed in areas with annual precipitation above 1600 mm and with a shorter dry season; it is largely located in the eastern part of this natural division. The latter is mainly distributed in the western part of this natural division, with a longer dry season and annual precipitation of less than 1600 mm. However, both subtypes have similar forest physiognomy, with quite varied species and numerous tall trees as well as high annual productivity in biomass. As high temperatures usually coincide with the rainy season, and somewhat lower temperatures with the dry season, there appear corresponding periods with different growth rates—the luxuriant, rapid-growth rainy season and the less luxuriant, slow growth dry season. The distinction between the greenish phase in summer and the yellowish phase in winter is particularly prominent.

Within such a humid tropical habitat, the faunal resources are also bountiful and are characterized by complicated faunal composition and many endemic species. Arboreal and fruit-eating species of vertebrates are particularly abundant, such as many species of primates (Chiropheus and Carnivora in animals), Hylidae and Rhacophoridae (in amphibians) and *Draco* spp. (in reptiles). The most outstanding marine fauna is coral, whose innumerable skeletons

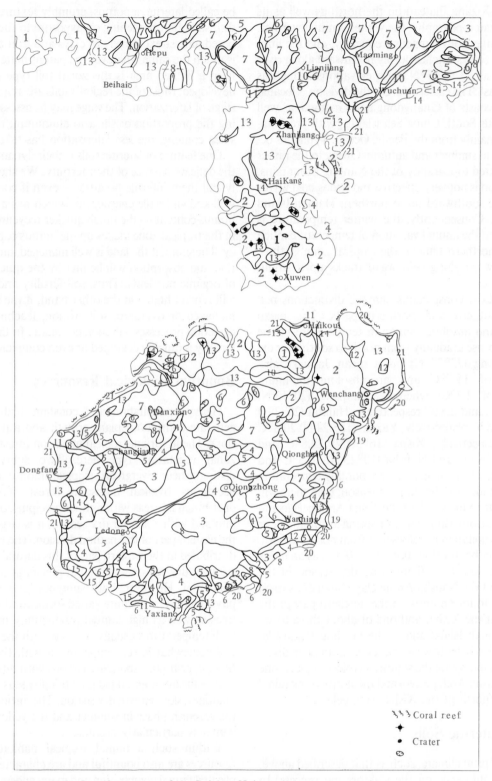

Figure 11-3 Landform types of the Leizhou-Hainan region

1. Basalt hill; 2. Basalt platform; 3. Middle mountain; 4. Low mountain; 5. Low hill; 6. Hill; 7. Terrace; 8. River terrace; 9. and 10. Alluvial plain; 11. Delta plain; 12. Marine eroded terrace; 13. Marine deposited terrace; 14. Marine deposited plain; 15. Alluvial-maritime deposited plain; 16. L agoon plain; 17, 18. and 19, 20. Karst hill and mountain; 21. Unstable sand dune.

have built up multitudes of coral reefs and islands in the South China Sea (Photo IV-10) Coral flourishes best in sea water with a temperature of 25–30°C and perishes when the minmum water temperature drops below 13°C. Hence, it is mainly restricted to the tropical sea, with the exception of the eastern coast of Taiwan Island where, owing to the beneficient influence of the warm Kuro Shio, coral is distributed along the coast northward into the subtropical zone (about 25°N).

Montane and Hilly Topography

The humid tropical South China natural division is overwhelmed by montane and hilly topography, to a degree much greater than the average condition in China. For example, on Hainan Island, with an area of 33 891 sq km, low mountains (elevation above 500 m) occupy about 25.4 per cent of the area; hills (100 to 500 m), 13.2 per cent; rolling terraces (below 100 m), 49.1 per cent; flat plains, 9.6 per cent; miscellaneous, 2.7 per cent. In the Xishuangbanna area of southern Yunnan, with an area of 19200 sq km, flat plains occupy only 5 per cent, low mountains and middle mountains 65 per cent, and hills and terraces 30 per cent.

Absolute relief exerts a great influence on the vertical zonation of climate and vegetation. In southern Yunnan especially, there often occur inversions of temperature in areas below 400 to 500 m in elevation, with a temperature increase rate of + 1.0° for every increase of 100 m (maximum record, + 2.7°C/100 m). Consequently, tropical monsoon forest may be distributed up to 800 to 900 m, sometimes even above 1000 m in elevation. This is the northernmost (22.5°N) and highest (900 m) *Hevea* plantation limit in the world. In the Xishuangbanna area, the lower tropical forest occupies about 28.5 per cent of the total land area, whereas the upper subtropic forest type accounts for 41.5 per cent. Relative relief also plays an important role. Low, flat plains are usually devoted to grain crops, hill and terrace land to Hevea and other tropical crops, whereas low and middle mountains are mainly forested.

NATURAL REGIONS AND SUBREGIONS

Based on the integration of areal differentiation of all the above-mentioned zonal and azonal factors, three natural regions may be demarcated in this natural division.

17. Leizhou-Hainan — Tropical Monsoon Forest Region

The Leizhou-Hainan region is located south of 21°31′N, including the narrow coast of the Guangdong and Guangxi as well as the broad Leizhou Peninsula and Hainan Island. Its land area totals 46451 sq km of which Hainan Island occupies about 73 per cent (Landsat image 6).

The Leizhou Peninsula and Hainan Island, although separated by the narrow Qiongzhou Strait (only 15 to 30 km in width) — which was formed during the Quaternary period by a differential fault — have essentially similar climatic and geomorphological features. Three major landform types stand out (Figure 11-3).

1) The extensive basalt platform and terrace, with an elevation generally below 150 m, occupies about 46 per cent and 90 per cent of the total land area of Hainan Island (Photo IV-9) and the Leizhou Peninsula respectively. The dominant basalt and associated igneous rocks erupted during the Quaternary period in different stages, consequently, a series of stepped surfaces (10–15 m, 25–35 m, 45–55 m, 60–80 m, and 100–150 m in elevation) were formed. The terrain is usually combined with many shield landform types, with a volcanic cone uplifted in the center of each shield that then slopes radially to the border. There are several dozen volcanic cones distributed in this region, with elevations ranging from several dozen to several hundred meters. Sometimes crater lakes have been formed. The basalt platform and terrace of earlier stages have been subjected to intense weathering for a long time; the reddish weathering crust is thick, the ground surface gentle and rolling. These conditions are quite favorable to the agricultural use of land, consequently, farmlands are extensive. In younger stages, which are only slightly weathered and eroded, rugged topography usually dominates. The farmers laboriously use rocks and gravels to build terrace walls and small patches of farmland here, so as to grow crops on the terraces and plant fruit trees along the walls. This is referred to as the rock agriculture.

2) The dome mountains, with elevations generally above 500 m, are restricted to middle southern Hainan Island. There are 81 peaks towering above 1000 m, the highest one is Mount Wuzhi (Mount Five Fingers) at 1867 m. This system is mainly composed of granite and other igneous rocks. They were first formed and uplifted during the Indo-Chinese Tectonic Movement. Then, they were repeatedly uplifted and peneplaned, resulting in many NE-SW trending fault lines and a number of erosion surfaces, of which, those of 300 m, 500 m, and 800 m are the most important.

3) Coral reefs are distributed intermittently along the coast. There are more than 120 species of reef-building corals in the Hainan Island area alone. Their optimum living conditions include a water

temperature of 25 to 30°C, a salt content of 2.7 to 2.8 per cent, a water depth of less than 50 m, and a rocky coast not strongly attacked by waves. The coral reefs create an excellent habitat for fish and shell. They are also an excellent wave barrier. If they are allowed to spread without control, however, they can limit access to sea ports.

Climatically, the region is controlled by alternating continental northeastern monsoons, maritime southeastern and southwestern monsoons, as well as typhoons originating from the Pacific Ocean and from the South China Sea. The temperature is high year round, with an accumulated temperature during the ≥ 10°C period of 8200 to 9200°C, and a summer season lasting 7 to 12 months. Therefore, greenness dominates the landscape. Paddy rice and other food crops can be cropped three times per year, and *Hevea* (rubber tree), coconut, coffee, pepper and other tropical economic crops grow luxuriantly. Owing to the marine influence, summer is not so stiflingly hot as in the subtropical middle Changjiang Valley. The maximum high temperature of the region never rises above 40°C; the minimum low temperature generally ranges from 5 to 8°C. When strong cold waves move far southward, however, the temperature may drop below 3°C. Sometimes, frost may occur; within 24 hours the temperature may drop more than 10 to 15°C. According to statistics, such severe cold waves have invaded this region 7 times in the past 50 years. Therefore, the planting of *Hevea* and other tropical crops in the Leizhou Peninsula and northern Hainan Island means that great care must be taken to control or to avoid low temperatures during the winter.

Precipitation is abundant, but there is a conspicuous dry season and a marked areal variation. Mean annual precipitation in Hainan Island totals 1700 mm, decreasing from more than 2000 mm along the east coast (Wanning, 2151 mm) to less than 1000 mm along the west coast (Dongfang, 997.8 mm). Annual variation is also great, for example, the rainiest year in Haikou amounted to an annual precipitation of 2282.3 mm, whereas in the driest year, only 1120 mm was recorded. Precipitation is concentrated in the rainy season (May to October) when maritime monsoons and typhoons prevail, accounting for 80 to 90 per cent of total annual precipitation. The rainiest month has more than 400 mm of precipitation; one single day may even have maximum precipitation of 250 to 400 mm, one single hour 80 to 100 mm. The driest month may record less than 10 mm. Hence, both spring drought and summer flood set limitation to agricultural development.

This is also the region most often plagued by typhoon. From April through November, especially from August to October, typhoons are familiar visitors. The strongest typhoons usually occur in September. According to statistics, there were, from 1949 to 1978, 38 typhoons landed on Hainan Island of which 21 had wind velocities of

more than 33 m/sec, and the strongest one (Typhoon No. 14 in September 1973), reached 61.2 m/sec. Typhoons land on Hainan Island most frequently along the central east coast, from Wanning to Wenchang. This area accounts for about 55 per cent of the total landed typhoons. These landed typhoons have presented a grave danger to human life and tropical crops; for example, Typhoon No. 14 destroyed virtually all the houses and uprooted about 80 to 90 per cent of the *Hevea* trees in the Qionghai area. Therefore, typhoon forecasting is of great significance in this region.

Abundant precipitation feeds numerous rivers in this hilly land. There are three relatively large rivers on Hainan Island. Their hydrologic features are briefly described in Table 11-1. There are also numerous small rivers originating in the foothills. Many of them are intermittent rivers and have empty channels during the dry season.

The zonal lateritic soils, under different geological, topographical, climatic, and vegetation conditions, are further developed into varied subtypes. The typical laterite is distributed on low hills and terraces that have different characteristics during the rainy and dry seasons; they undergo a strong weathering-leaching process. Laterite developed on basalt is characterized by an abundance of iron. In eastern Hainan Island, with high temperatures and heavy precipitation as well as a short dry season, the laterite is yellowish. In southwestern Hainan Island where the climate is somewhat drier and is dominated by deciduous monsoon forest, the laterite is brownish. On sandy land or the marine terraces of southwestern Hainan Island, where the soil-forming process is rather young, the soil color is reddish brown. These soil subtypes are again distributed in conspicuous vertical zones. For example, in eastern Hainan Island, yellowish laterite is generally distributed in areas below 300 m; lateritic red earth, from 300 to 600 (700) m; montane yellow earth, from 600 (700) to 900 (1000) m, and small patches of montane meadow, above 1300 m. In southwestern Hainan Island, owing to the drier conditions, brownish laterite appears below 400 m, lateritic red earth from 600 to 700 m, and montane yellow earth, from 800 to 1000 m.

Zonal vegetation is the luxuriant, evergreen tropical monsoon forest. There are more than 1400 species of woody plants here, of which more than 800 species are tall trees and about 50 species rank as precious timbers. Shrubs and epiphytes are also profuse and varied, but grasses are comparatively few. The physiognomy of the evergreen tropical monsoon forest is characterized by multi-stories, with tall (usually 20 to 40 m.), uneven tree crowns within the first 1 to 3 stories; luxuriant shrubs as stories 3 to 5; and sparse grasses as stories 5 or 6; all these stories are interwoven by numerous epiphytes. Under special conditions, some particular vegetation types have also developed. Tropical rainforest is restricted to low hills and the coast of eastern Hainan Island where annual precipitation is

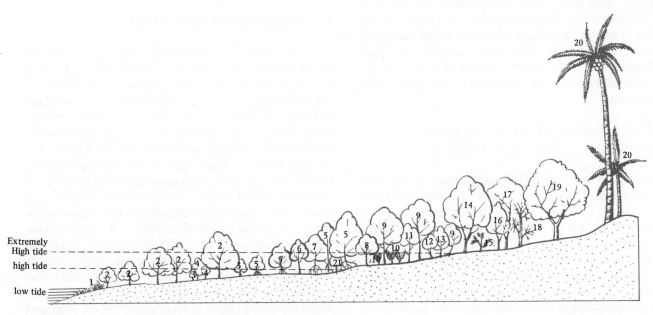

Extremely
High tide
high tide
low tide

Figure 11-4 (Quoted from "Vegetation of Guongdong")

1. *Ruppia rostellate;* 2. *Sonneratia caseolaris;* 3. *Aegiceras corniculatum;* 4. *Avicennia marina;* 5. *Rhizophora apiculata;* 6. *Bruguiera gymnorrhiza;* 7. *Rhizophora stylosai;* 8. *Kandelia candel;* 9. *Bruguiera setangula;* 10. *Acrostichum aureum;* 11. *Ceriops tagal;* 12. *Scyphiphora hydrophylacea;* 13. *Lumnitzera racemosa;* 14. *Excoecaria agallocha;* 15. *Clerodendron inerme;* 16. *Barringtonia racemosa;* 17. *Heritiera littoralis;* 18. *Pendanus tectorius;* 19. *Hibiscus tiliaceus;* 20. *Cocos nucifera* (Coconut tree); 21. *Acanthus ilicifolius.*

Table 11-1 Hydrologic Features of the Three Major Rivers in Hainan Island

Rivers	Area (km²)	Length (km)	Number of tributaries above 100 km²	Mean annual runoff (m³/sec)	Total annual runoff (10⁹m³)	Annual runoff depth (mm)
Nandu	7177	311	19	209	61.2	894
Changhua	5070	230	10	122	37.1	801
Wanquan	3683	196	8	166	51.0	1575

more than 2000 mm and where there is no conspicuous dry season. Deciduous tropical monsoon forest, park savanna, and sandy vegetation are located in low hills and intermontane basins of southwestern Hainan Island where the climate is both hot and comparatively dry. Coastal mangrove forests are distributed intermittently along the coast (Figure 11-4). Vertical vegetation zones include: (1) montane rainforest — located on lower slopes (400 to 800 m), with annual precipitation of more than 2000 mm; (2) montane evergreen broad-leaved forest — located above 1000 to 1200 m; and (3) montane top dwarf forest — distributed on windy montane tops above 1300 m, with dwarf trees about 5 m in height and abundant mosses and lichens.

As a whole, this natural region is rich in natural resources; yet, they are far from being adequately used. For example, Hainan Island has an agriculturally usable land area of about 20000 sq km of which only 45 per cent is being exploited. A primitive shifting agricultural system still exists widely in montane areas. With a view to conserving these natural resources and using them in a more productive way, every land type should be allotted and used according to its specific suitability and capability. Low, flat plains should be devoted to intensive farming, rolling terraces and hills below 500 m should be devoted chiefly to *Hevea* and other tropical crops, and montane areas above 500 m could be better developed for tropical and subtropical forestry. From the national point of view, *Hevea* and other tropical crops have a special significance, and this is the best place in China for planting them. At present, Hainan Island contains 53 per cent of the *Hevea* plantation area and 70 per cent of the rubber production in China. These advantages might be further capitalized in the future.

Four subregions of the Leizhou-Hainan region can be identified.

17 (1) Coastal Plain of Southwestern Guangdong and Southern Guangxi

The first subregion of the Leizhou-Hainan region is mainly composed of plain and terrace below 100 m in elevation interspersed with low hills. It is located on the northern margin of the humid tropical zone, with an accumulated temperature during the $\geqslant 10°C$ period of about 8200°C and an annual precipitation of 1450 to 1600 mm. A series of natural hazards, such as frost, spring drought, summer floods, and typhoons, play havoc in this subregion; owing to the devastation of natural vegetation since ancient times, the most serious hazard is soil erosion, which should be controlled as soon as possible.

17 (2) The Leizhou Peninsula

The Leizhou Peninsula is mainly composed of basalt terrace (occupying about 43% of the total land area), marine terrace (27%), and alluvial plain (17%); it is also characterized by scattered diluvial plains, sandy beaches, and volcanic cones, all with elevations below 100 m. It has an accumulated temperature during the $\geqslant 10°C$ period of about 8000 to 8500°C and an annual precipitation of 1400 to 1700 mm. It is also limited by a series of natural hazards similar to the coastal plain of Southestern Guangdong and southern Guangxi. Yet, owing to better human use patterns and efforts at afforestation as well as irrigation engineering, this subregion has a potentially better and the prospect for future development.

17 (3) Northern Hainan Island

Located north of the Chanjiang-Qionghai line, 85 per cent of the northern Haina Island subregion is composed of extensive basalt and granite platforms below 300 m in elevation; the fertile low alluvial plain is mainly restricted to the middle and lower reaches of the Nandu River. It differs from the Leizhou Peninsula subregion in that it has a much larger area of basalt platform but a smaller area of terraces and plains. It is also a little bit warmer, with an accumulated temperature during the $\geqslant 10°C$ period of about 8400°C and an annual precipitation of 1500 to 1700 mm. Both in natural hazards and economic development, it is similar to the Leizhou Peninsula, although they are separated by the Qiongzhou Strait.

17 (4) Southern Hainan Island

Located south of the Chanjiang-Qionghai line, southern Hainan Island occupies about 55 per cent of the total land area of the island. It is mostly composed of middle mountains (elevation above 800 m), low mountains (500–800 m), and hills (250–500 m), with intermontane basins (mostly below 250 m) along the foothills. These land types overlap each other like a huge pyramid, with Mount Wuzhi (1867 m) as its pinnacle. This is the warmest subregion in the Leizhou-Hainan region, with an accumulated temperature during the $\geqslant 10°C$ period of 8700 to 9200°C (increasing from north to south) and with a minimum temperature of above 5°C, so that frost is not a problem. Annual precipitation decreases from more than 2000 mm in the east to less than 1000 mm in the west, and the aridity index increases from about 0.8 (subhumid) to about 1.5 (semiarid) correspondingly. Vertical zonation is conspicuous, generally with a decreasing rate of temperature of 0.6°C/100 m, and an increasing rate of precipitation of 140 mm/100 m. Owing to this area's topography, the influence of typhoons is much restricted. Therefore, the intermontane basins in this subregion are probably the best place for developing *Hevea* and other tropical crops in China; the extensive montane slopes are the major tropical rain forest and montane rain forest areas in China. But the long-time clearing of natural vegetation has virtually eliminated forests below 500 m in elevation; only about 400000 ha of montane forest survive in this extensive subregion. Natural conservation and reforestation are regarded to be urgently necessary here.

18. Southern Yunnan — Tropical Monsoon Forest Region

The southern Yunnan region is located south of 23 to 25°N, with its typical zonal vegetation (tropical monsoon forest) limited to intermontane basins and low valleys below 700 to 900 m in elevation. It is the southward extension of the Hengduan Mountains area, with longitudinal high mountains and intermontane deep valleys paralleling each other. The area broadens from two degrees of longitude in the northern Hengduan Mountains area to about six degrees at the southern Yunnan border. In addition, the mountains become lower and the intermontane valleys widen. This natural region is composed chiefly of dissected plateau and broad intermontane valleys of 700 to 1500 m in elevation, interspersed with mountains above 1500 m and low valleys below 700 to 900 m. A few peaks tower above 2500 m, and the lowest point is Hekou (84 m) along the Chinese-Vietnam border. On the whole, the topography is rather rugged; the so-called *Batzi* is restricted to intermontane basins and valleys. There are about 300 *Batzi* with

Table 11-2 Hydrologic Features of Three Major Rivers in Southwestern Yunnan

Rivers	Area (km²)	Length (km)	Mean annual runoff (m³/sec)	Total annual runoff (10⁹m³)	Annual runoff depth (mm)
Nujiang (Salween)	129876	1775	1650	522.1	439.6
Lancang (Mekong)	154920	2138	1840	580.1	420.5
Yuanjiang (Red R.)	37297	692	319	100.5	313.7

an area larger than 1 sq km (the three largest are the Mengzi, 369 sq km; the Yingjiang, 339 sq km; and the Longchuan, 242 sq km). They occupy about 6 per cent of the total land area. These *Batzi,* just like oases in the desert zone, contain most of the population and farmlands in the region.

Owing to higher elevation and higher latitudes, summer temperatures (generally 21–25°C) are lower than in the Leizhou-Hainan region, consequently, there is a lower mean annual temperature (mostly 18–22°C). However, winter is slightly warmer here than in similar elevations of the Leizhou-Hainan region, with a mean January air temperature around 11 to 16°C. In areas with an elevation between 1300 to 1500 m there is eternal spring; in areas of around 900 to 1300 m, there is summer and spring (autumn) but no winter. Another climatic feature is the abundant precipitation and sharp distinction between rainy and dry seasons. Annual precipitation totals 1200 to 1600 mm, decreasing from both southwest and southeast to middle north. The windward slopes facing the Indian monsoons in the southwestern part and the Pacific monsoons in the southeastern part of the region have the heaviest precipitation in the region, with annual rainfall of 1500 to 2500 mm and 1750 to 2100 mm respectively. Ximeng, with an annual precipitation of 2812.9 mm, is one of the rainiest spots in China. 80 to 90 per cent of annual precipitation is concentrated in the rainy season (May to October), contrasting markedly with the dry season (November to October). The rainy season usually bursts in early May, and it will suddenly disappear in late October. Precipitation comes mostly in sudden and brief showers.

Except in the most southeastern part, there is no direct influence of typhoons. In tropical low valleys, fogs occur frequently, with about 120 foggy days annually. Mengla, with 186 foggy days, is one of the foggiest spots in China. Fogs occur mostly in the dry season and provide a reasonable compensation for moisture deficiency at that time.

Because of the above-mentioned topographical and climatic conditions, surface runoff is abundant. Besides the Jinsha River, there are three mighty international rivers flowing southward into the Indo-China Peninsula. Their hydrologic features are outlined in Table 11-2.

This natural region is the "treasure garden" of plants in China. In the Xishuangbanna area alone, about 4000 to 5000 species of higher plants and 539 species of vertebrates are distributed. The upper limit of the tropical monsoon forest rises up to 900 to 1000 m. West of the Yuanjiang Valley, owing to a longer dry season, there are a number of deciduous trees amid the evergreen tropical monsoon forest; east of the Yuanjiang Valley, as the dry season is shorter, evergreen tropical monsoon forest dominates. In low valleys below 500 m, rain forest appears. Evergreen subtropical broad-leaved forest occurs above 900 to 1000 m. All these forests have been subjected to burning and cutting for shifting agriculture since ancient times. Hence, many parts of the region are now dominated by secondary growth — park savanna and grassland. Only about 23 to 31 per cent of the original forest coverage still remains. Conservation of vegetation (especially those precious rare species) and afforestation are two urgent meliorating measures in this region.

This is also a region with rich natural resources; yet, they are far from adequately used. For example, Dai Autonomous Prefecture of Xishuangbanna[1], with an area of 19200 sq km and an elevation ranging from 540 m to 2400 m is one of the best sites for developing tropical crops in China (Photo IV-8) and has been cultivated and developed by the Dai and other peoples since ancient times. Since 1949, croplands have been increased from 36700 ha to 92700 ha and *Hevea* plantation from nil to nearly 30000 ha. Yet, a series of acute problems occur in regional development here. The most critical one is the struggle between natural conservation and *Hevea* plantation below 900 m. From the national point of view, *Hevea* and other tropical crops should have priority. According to an estimate, lands suitable for *Hevea* plantation in this area total about 127000 ha; they should be planted with *Hevea*

---

1) Xishuangbanna, in the Dai language, denotes twelve flat plains.

and other tropical crops as much as possible. Besides, current productivity of rubber reaches only 810 kg/ha, and it should be increased as much as possible. However, a balanced and planned economy should also be maintained in conjunction with a high quality of natural environment. The restricted, precious flat plains are better devoted to intensive farming, whereas slopes and hills below 900 m can be used chiefly for *Hevea* and other tropical crops. Montane areas above 900 m should be mainly set aside for conservation and forestry.

Three subregions are identified in this natural region.

18 (1) Southern Yunnan Low Valleys

The southern Yunnan subregion includes low valleys below 700 to 900 m and their surrounding mountains in southern Yunnan. It is characterized by relatively low latitudes (mostly south of the Tropic of Cancer) and low altitudes. Consequently, the climate is hot and humid, with a mean January temperature above 16°C, an absolute minimum temperature above 10°C, and an annual precipitation above 1500 mm. During summertime, the southwestern monsoons dominate in its western part the southeastern monsoons in its eastern part. In low valleys below 700 to 900 m, evergreen tropical monsoon forests dominate, with patches of tropical rain forest in low valleys below 500 m and deciduous tropical monsoon forest in some drier sheltered valleys. Montane areas above 700 to 900 m are distributed chiefly with subtropic evergreen broad-leaved forest.

18 (2) Southwestern Yunnan Broad Valleys

The southwestern Yunnan subregion is characterized by dissected plateau and broad valley topography, mostly with an elevation of 700 to 1000 m. There are also longitudinal mountains, with an elevation of around 2000 m and along the Longchuan River and other rivers, there are low valleys below 700 m. During summer time, maritime southwestern monsoons dominate, with 85 to 90 per cent of the annual precipitation concentrated from May to October. By contrast, the dry season (November to April) is conspicuous. Owing to relatively higher latitudes (mostly 23 to 25°N) and higher altitudes, winter is somewhat cooler than in the southern Yunnan low valleys subregion, with a mean January temperature of 12 to 16°C and an absolute minimum temperature below 0°C. Therefore, the area is generally unfit for *Hevea* and other tropical crops, except on sunny slopes of low valleys below 700 m; tea and other subtropical crops, however, are at their best.

18 (3) Southeastern Yunnan Karst Plateau

The southeastern Yunnan subregion is located on the southeastern slopes of the Yunnan Plateau at elevations of between 500 to 1500 m interspersed with limestone mountains of around 2000 m. During summer, the maritime southeastern monsoons dominate, with 80 to 85 per cent of annual precipitation concentrated from May to October. This subregion enjoys about the same latitudinal and altitudinal locations as the southwestern Yunnan broad valleys subregion; yet, it is sometimes subject to invasion of cold waves during winter and consequently its mean January temperature is generally lower by 1 to 2°C and, thus, rather less appropriate for tropical crops. Here, the upper limit for *Hevea* plantation is lowered to about 400 m asl. The distinction between windward and leeward slopes is quite conspicuous, with evergreen and deciduous tropical monsoon forests dominating respectively. Owing to heavy human intervention, large tracts have been reduced to graminaceous grassland or even barren slopes.

19. South China Sea Islands — Tropical Rain Forest Region

The immense South China Sea is a deep sea basin, with an area of 3.5 million sq km and an average depth of 1212 m. Scattered amid this immense area are four archipelagos: Dongsha (East Sand), Xisha (West Sand), Zhongsha (Middle Sand), and Nansha (South Sand). These archipelagos are composed of numerous islands, beaches, reefs, and shoals — all built up by coral skeletons (Photo IV-10). They are small in area. The largest island, Yongxing Island of the Xisha Archipelago has an area of only 1.85 sq km. They are also low in elevation. The highest, Shidao Island of the Xisha Archipelago has an altitude of only 15 m asl. Furthermore, more than 300 reefs and shoals are normally submerged and only 34 islands and 8 beaches stand above high tides for a total land area of about 12 sq km. All these islands, beaches, reefs, and shoals are bound together by the immense South China Sea, to form this natural region.

The South China Sea Islands are developed on seabottom terraces along northern, western, and southern continental slopes of the South China Sea Basin. The continental shelf and continental slopes of the South China Basin are composed of sial crust, with Paleozoic and Mesozoic granite and metamorphic rocks as a base. The coral reefs at Yongxing Island have a thickness of more than 1000 m, underlain by a reddish weathering crust of 28 m. The latter might represent an ancient land surface of the Mesozoic era that was repeatedly peneplaned and submerged, resulting finally in the present sea-bottom terrace. The deep sea bottom of the South China Sea is composed of sima crust consisting mainly of basalts and volcanic eruptions. A series of sea mounts divides the deep sea basin into a northeastern shallower part and a southwestern deeper part. On either border of the deep sea basin, there is a great fault line. Some scholars argue that the deep sea basin of the South China Sea has been formed by sea-bottom spreading since the Mesozoic era.

The South China Sea Islands are located from 3°50′N up to 20°42′N. They are characterized by an equatorial climate in the Nansha Archipelago and a tropical monsoon climate in all other parts. During the summer, the sun shines vertically overhead twice. The mean annual air temperature ranges from 25 to 27°C, the mean January temperature is from 21 to 26°C, with an absolute minimum temperature of 11.2°C in Dongsha Island, 13.9°C in Yongxing Island (in the Xisha Archipelago) and 22.9°C in Taiping Island (in the Nansha Archipelago). Accumulated temperature during the $\geq 10°C$ period totals 9150 to 10000°C, which results in a year-round growing season. The mean annual sea surface water temperature ranges from 23 to 28°C, with a minimum sea surface water temperature of 20 to 26.5°C, which is suitable for all tropical marine life.

Owing to the low and flat topography, the annual precipitation (1100–1500 mm) is not so plentiful as might be expected. From May to October, southeastern and southwestern monsoons prevail in Dongsha Archipelago accounting for 87 per cent of the total annual precipitation. In Xisha Archipelago the precipitation is mainly derived from the southwestern monsoons that prevail from June to November. Nansha Archipelago has no dry season and has an annual precipitation of 1842 mm.

Again, owing to the smallness in land area and the porous nature of coral reefs, which are the basis for all South China Sea Islands, there is no surface runoff at all on these islands. Surface water exists mainly in the form of enclosed lagoons, which are encircled by coral reefs and fed by rainfall. In Yongxing Island, the Chinese inhabitants excavate ponds and store rain water so as to develop freshwater irrigation and fisheries. The first apparatus for converting seawater into fresh water in China has also been set up in the Xisha Archipelago. Again, there is some fresh ground water at a ground level of about 1.5 to 3.5 m, which may be conveniently used.

Vertical zonation of the surrounding seawater is conspicuous. Just as in the Pacific Ocean, five vertical strata of sea water may be identified: (1) tropical surface water: < 0.75 m in depth, seawater temperature > 20°C, salinity > 34.4‰, dissolved oxygen > 4.0 ml/l; (2) subtropical underlying water: 0.75–300 m in depth, seawater temperature 20–12°C, salinity 34.4–34.55‰, dissolved oxygen 2.5–1.6 ml/l; (3) middle strata water: 300–900 m in depth, seawater temperature 12–5°C, salinity 34.4–34.5‰, dissolved oxygen 2.5–1.6 ml/l; (4) Deep strata water: 900–2500 m in depth, seawater temperature

5–2.4°C, salinity 34.5–34.6‰, dissolved oxygen 1.6–2.5 ml/l. (5) Basin bottom water: > 2500 m in depth, seawater temperature < 2.4°C, salinity > 34.6‰, dissolved oxygen > 2.3 ml/l.

The South China Sea Islands are characterized by phosphorus-rich limy soils. The parent materials of the soil are mainly corals and shells, and the soil-forming process is greatly accelerated by the humid tropical climate, luxuriant tropical rain forest, and abundant bird's quano. The last component speedily disintegrates into phosphate under the humid tropic climate, and then it combines with limy materials to form phosphorous-rich limy soil. With the help of luxuriant vegetation, the humus-accumulation process is usually greater than the humus-disintegration process. According to an analysis conducted by the Nanjing Institute of Soil Science, the soil surface layer contains 8 to 10 per cent organic matter (sometimes as high as 57 per cent), 17 to 30 per cent phosphorus, and 40 per cent calcium oxide. The soil is good enough to serve as fertilizer.

Biological resources are also rich. According to a recent investigation, there are 213 species of plants (including cultivated crops) in the Xisha Achipelago alone, of which 146 species are also common to Hainan Island. The major vegetation types are evergreen tropical forests, shrubs, and psammophytes on the coastal beaches. The dominant species are *Pisonia grandis* (the so-called avoid-frost-flower), *Guettarde specieose, Calophyllum inopllum* (in trees), *Scaevola sericea, Messerschmidia argentes, Morinda citrifolia, Pemphis acidula, Clerodendron inerne* (in shrubs), and *Ipomoes peacaprae, Canavalia maritina,* and *Tribulus terrestris* (in grasses). The most commonly seen cultivated crops are coconut palms, with many kinds of vegetables grown underneath.

This area is a paradise for birds that feed on rich marine life. According to a recent investigation there are 103 species of birds living on the Xisha Archipelago alone, of which more than 60 species are commonly seen. The *Sula sula,* a bird with ducklike shape and white abdomen and red feet, is present in great numbers. It is said that such colorful birds formerly lived in the Xisha Archipelago in the tens of thousands and covered the islands like an immense "carpet of blossoming cotton". Their quano has been accumulated to a thickness of more than 1 m, playing a significant role in the soil-forming process; formerly, it was exported in large quantities as fertilizer. Another rich faunal resource is marine life. One important economic fish is *Caesio chrysozona,* which often appears in large schools. Sea turtles (*Chelonia mydas, Eretmochelys imbricata, Caretta olivacea*), *Stichopus* spp. and many species of algae are also valuable. The corals not only build up all the South China Sea islands, but also offer excellent habitats for different kinds of marine life, resulting in innumerable and remarkable sea-bottom gardens.

The South China Sea Islands include four archipelagos,

thus, each archipelago may be considered as a subregion.

19 (1) The Dongsha Archipelago

Located at about 20°N, the Dongsha Islands Archipelago are developed on the Dongsha sea-bottom terrace, which is about 300 m deep. The archipelago is composed of Dongsha Island and several submerged coral reefs. Dongsha Island, with an area of 1.8 sq km, is located in a lagoon that is surrounded by the horseshoe shaped Dongsha circular coral reef. This archipelago is the northernmost and coolest island group in the region. It is also the most windy, with a mean annual wind velocity of 6.4 m/sec. It has been Chinese territory for a long time, and numerous houses and temples as well as a lighthouse and a meteorological station have been established here.

19 (2) The Xisha Archipelago

The Xisha Archipelago are the westernmost island group in the region; they also have the largest number of visible islands (22) and beaches (4), as well as submerged reefs (5) and shoals (6). They are developed on the Xisha sea-bottom terrace, which is about 1000 m deep. Two subgroups can be recognized. The eastern subgroup is called the Xuande Islands, of which, Yongxing and Shidao are the largest and the highest islands in the region respectively. The western subgroup the Yongle Islands are composed mostly of submerged reefs and shoals. Xisha was formerly called Qizhou-yang (literally: Seven Continents' Ocean); when the famous imperial messenger Zheng He, together with his large fleet, made seven "West Oceans" travels in the early Ming dynasty (early 16th century). This was one of their important stops.

19 (3) The Zhongsha Archipelago

Located about 100 km southeast of the Xisha Archipelago, the Zhongsha Archipelago are developed on the Zhongsha sea-bottom terrace, which is separated from the Xisha sea-bottom terrace by a trough 2500 m deep. It includes 26 submerged reefs and shoals, of which the Zhongsha Great Circular Reef (140 km long and 60 km wide), is the largest. Huangyan Island located about 300 km southeast of Zhongsha Great Circular Reef and developed on a sea mount, is the only visible reef in this subregion.

19 (4) The Nansha Archipelago

Stretching far into the equatorial zone, this is the southernmost and the warmest island group in the region. It includes 11 visible islands and 4 visible beaches, with a total land area of 2 sq km; Taiping Island is the largest island. There are also numerous submerged reefs and shoals. The Zhenghe Reefs are the largest circular reefs in the whole region, and the *Zengmu* Shoal is the southernmost boundary of Chinese territory.

References

[1]   *Physical Regionalization of China* Working Committee, Chinese Academy of Sciences 1958 – 1959, *Physical Regionalization of China*, 8 Vols, Science Press, Beijing. (In Chinese)

[2]   Qiu Baojian et al., 1963, *Agricultural climate of tropical and southern subtropical zones in China*, Science Press, Beijing. (In Chinese)

[3]   Xu Xianghao, 1981, *Plant Ecology and Plant Geography of Guangdong Province*, Guangdong Science and Technology Press, Guangzhou. (In Chinese)

[4]   Han Yuanfeng, 1976, *Hainan Island*, Guangdong People's Press, Guangzhou. (In Chinese)

[5]   Zhong Gongfu et al., 1981, "Conservation and Development of Tropical Resources in Hainan Island",*Economic Geography*, 1981, No. 2. (In Chinese)

[6]   Xu Chunmin, 1978, "A Brief Remark on the Physical Geography of South China Sea Islands", *Selected Papers on Oceanography*, 1978, No. 2. (In Chinese)

[7]   Huang Jingsun et al., 1978. "Geological and Geomorphological Features of the Xisha Archipelago", *Selected Papers on Oceanography*, 1978, No. 2. (In Chinese)

[8]   Nanjing Institute of Soil Science, 1977, *Soils and Phosphate mines in the Xisha Archipelago*, Science Press, Beijing. (In Chinese)

[9]   W.J. Ludwing et al., 1977, "Profiler-Sonobuoy in the South China Sea Basin", *Journal of Geophysical Research*, Vol. 84, No. B7.

[10]   Geography Department, Kunming Teachers's College, 1978, *Geography of Yunnan*, Yunnan People's Press, Kunming. (In Chinese)

[11]   Shi Yulin, 1980, "Conservation and Utilization of Natural Resources in the Xishuangbanna Area", *Natural Resources*, 1980, No. 2.(In Chinese)

Chapter 12

Temperate Grassland of Nei Mongol

The natural division, Nei Mongol (Inner Mongolia), is located in the northernmost part of China, with its northern boundary being the Sino-Mongolian international boundary. Its eastern boundary corresponds with the western boundary of the temperate humid and subhumid division of Northeast China, coinciding roughly with the 1.2 isopleth of the aridity index. Its southern boundary, located on the southern margins of the Nei Mongol Plateau and the Ordos Plateau, coincides roughly with the 3200°C isotherm of accumulated temperature during the ⩾ 10°C period. Its western boundary is located at the western foothills of the Helan Mountains and the northern foothills of the Wushao Mountain, roughly coinciding with the 4.0 isopleth of the aridity index. As a whole, this natural division has an area of about 710000 sq km and a population of about 14 million. Administratively, it is distributed essentially in two autonomous regions — Nei Mongol and Ningxia (Figure 12-1).

In terms of climatic-biological-soil conditions, it is characterized by temperate grassland (photos V-1, V-2) — temperate steppes in the eastern part and temperate desert-steppe in the western part, with chestnut soil and brown semidesert soil as zonal soils respectively. Geologically and geomorphologically, it is mainly composed of two plateaus, with elevations of 1000 to 1500 m. These are the Nei Mongol Plateau and the Ordos Plateau together with their surrounding mountains (chiefly the Helan, the Yinshan, and the southern Da Hinggan) and river valleys (chiefly the Huanghe and Xi Liaohe). This natural division is also noted for its transitional feature: from north to south it shifts from cool-temperate to temperate and warm temperate and from east to west the climate moves from subhumid to semiarid and then to arid.

These physical features play an important role in the historical, economic, and social development of the division. Since ancient times, the Nei Mongol Plateau has been a homeland and pasture for different pastoral peoples. In the late fifth and early sixth centuries, folklore vividly described its landscape: "The sky is blue and wide, the earth is flat and vast; the grasses bow before the sweeping winds, then innumerable sheep and cattle are seen". In the past 100 to 200 years (a little earlier than in Northeast China), agriculture has been rapidly developed along the plateau borders and in surrounding river valleys, so that a broad mixed agricultural-pastoral belt as well as an intensive irrigated agricultural belt have been developed. Recently, numerous industrial and mining enterprises have also been developed. Just like Northeast China, this is seen as a setting with a rich potential for expanded development.

GENERAL PHYSICAL FEATURES

Nei Mongol features a temperate grassland landscape. It is also characterized, from southeast to northwest, by a series of arc-shaped belts both in zonal (climatic-biological-soil) and azonal (geologic-geomorphological) features.

Immense and Undulating Plateau landforms

The Nei Mongol division is mainly composed of the immense and undulating Nei Mongol Plateau and the Ordos Plateau, with elevations between 1000 to 1500 m. Since the Yanshan Tectonic Movement, which uplifted these two plateaus, this natural division has been comparatively stable and subjected to denudation and peneplanation for long

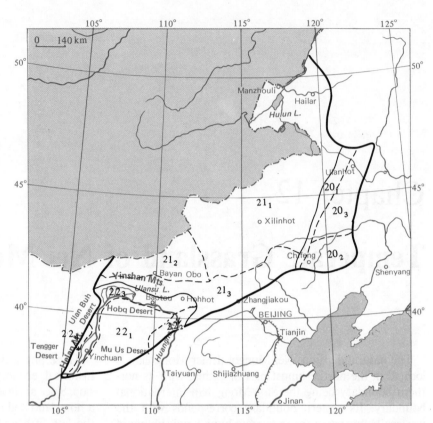

Figure 12-1   A location map of temperate grassland of Nei Mongol

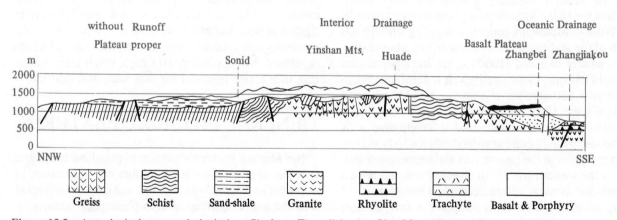

Figure 12-2   A geological-geomorphological profile from Zhangjiakou to Sino-Mongolian border

periods that resulted in the pre-Cretaceous Mongolian and Gobi erosion surfaces. The extensive, repeated eruption of basalt during the Cretaceous, Tertiary, and Quaternary periods again enlarged the flat plateau surface and smoothed out any former irregularities. The arid and semiarid climates since the early Tertiary period have preserved all these undulating, rather undissected landforms to the present.

The middle Huanghe Valley includes the Yinchuan Plain (West Elbow Plain), the Back Elbow Plain, and the Front Elbow Plain. These are high plains approximately 1000 m in elevation. Surrounding mountains that usually merge gradually into the plateaus are comparatively limited in area and below 3000 m in elevation. The Xi Liaohe

Plain, which is located east of the Da Hinggan Mountains and generally has an elevation below 500 m.

Another geomorphological feature is the belted structure of land forms. For example, from Zhangjiakou (Kalgan) northwestward up to the Sino-Mongolian international boundary (Figure 12-2), basalt platform (plateau margin), denudational middle and low mountains (Yinshan Mountains), peneplaned Mongolian erosion surfaces, and *gobi* erosion surfaces (plateau proper) up to denudational low hills with volcanic cones and basalt platforms along the international boundary appear in a sequence.

Temperate Semiarid and Arid Monsoon Climate

The above-mentioned geomorphological features together with rather high latitudinal location (mostly from 38 to 50°N) result in the dominance of westerlies in the upper atmosphere year round; dry and cold continental monsoons prevail in the lower atmosphere for most of the year. Warm and moist maritime monsoons, on the other hand penetrate only a short distance into the southeastern margin of the Nei Mongol Plateau and bring in only a small amount of precipitation. Therefore, this natural division is characterized by temperate, semiarid, and arid climates, with the aridity index increasing from 1.2 in the southeastern margin to about 2.0 along the Ondormiao – Bailingmiao–Otog–Yanchi line and to about 4.0 at the northwestern border.

Winter is long and severe and is frequented by cold waves. From November to March the weather is generally bitterly cold, with a mean January temperature ranging from −9°C in the south to −28°C in the north, with a minimum temperature from −30°C to below −50°C respectively. Spring comes late and lasts about two months, beginning in April in the south and May in the north. Summer is short and warm, lasting generally from early or middle June to middle or late August with a mean July temperature ranging from 24°C in the south to 18°C in the north. Autumn is a little bit shorter than spring and may come and depart suddenly. Sometimes, heavy snow can fall in early or middle September. The ≥10°C period lasts about 120 to 160 days, with an accumulated temperature of 2000 to 3200°C. Together with abundant sunshine (2800 to 3300 hours annually), it is adequate for growing one crop per year.

Annual precipitation decreases from 400 to 450 mm in the east to 150 to 200 mm in the west, with 60 to 70 per cent concentrated in the summer (June to August). Rain usually falls suddenly in the afternoon and lasts only a short time. Rainstorms may sometimes be very strong. For example, on August 1, 1977, a single rainstorm poured down continually for 11 hours in Uxin, with a total precipitation of 1850 mm, one of the highest records in the world. Spring and winter are usually dry, sometimes without any precipitation at all. Annual variation of precipitation is also great; sometimes, there may be no precipitation at all for more than 12 successive months. Hence, in the western part of this natural division, there is no farming without irrigation, and pastoralism remains the chief occupation for nonirrigated areas. The eastern part was formerly the best and most extensive grassland in China. Now it is chiefly a transitional belt of pastoralism and dry farming; in 9 out of 10 years, drought hazards play havoc with the dry farming yields.

Transitional Belt Between Oceanic and Interior Drainage

Drainage that is determined by the belted distribution of precipitation decreasing from southeast to northwest as well as the belted structure of land forms is apparently divisible into two systems: the oceanic system in the southeastern peripheral area and the interior system in the plateau proper. In the innermost part of the Nei Mongol Plateau, there exists no runoff at all. The great divide between oceanic and interior drainage systems runs southwestward from the western slopes of the middle and southern Da Hinggan Mountains through the northern slopes of the Yinshan Mountains up to the western slopes of the Helan Mountains.

The oceanic drainage has a more completely developed river network, mostly consisting of the upper reaches of large rivers such as the Hailar River (upper reaches of the Heilong River) and the Xar Moron River (upper reaches of the Xi Liaohe River). The largest river is the middle reaches of·the Huanghe River which encircles the Ordos Plateau in a great horseshoe-shaped arc. It is characterized by a string of broad alluvial basins connected through gorges or rapids. The most important alluvial basins are the three Elbow plains, all of them well watered and intensively cultivated. This gives rise to the Chinese saying, "The Huanghe River has done a hundred bad things, but at least one good thing; it has created the flourishing Elbow plains". Water from the Huanghe River is heavily drawn for irrigation, so that its mean annual discharge decreases from 1,019 m³/sec at Lanzhou to 830 m³/sec at Baotou, although the drainage area increases by 18.7 per cent. Another unfavorable condition is the river's northward flow of 4°27' of latitude (upstream from Baotou) and then its sudden eastward turning. It is frozen over much earlier at the Back Elbow Plain than upstream, and it melts much later as well. A high ice dam is often formed in this northern elbow of the river, and heavy flood damage ensues; sometimes even bombs are needed to break away these ice dams.

Interior drainage characterizes an immense area. It has an annual runoff depth of less than 25 mm. Perennial rivers are few in number and small in discharge. They usually originate in the surrounding mountains and then dwindle away as they approach the plateau proper and disappear amid shallow depressions or flow into inland lakes. From Hulun Buir southwestward to the Ordos Plateau more than 1,000 inland lakes are distributed. They are usually small and shallow, with depths of less than 4 m. Most of them are saline or sodic, with pH-values around 8.5, and they are widely used as sources of salt or soda. A few structural lake basins are exceptionally deep, with a freshwater depth of more than 10 m. There exist several higher, older shorelines, showing the larger dimensions of ancient lakes.

Along the Sino-Mongolian international border, which represents the central and lowest part of the Nei Mongol Plateau, and in the middle west of the Ordos Plateau, practically no runoff exists. Ground water is also scarce and deeply buried. Here, pasturing can be conducted only during winter when some snow cover is available to provide drinking water.

Temperate Grassland Landscape

The Nei Mongol division is essentially dominated by temperate grassland landscape: temperate steppe east of the Ondormiao –Bailingmiao–Otog–Yanch line and temperate desert-steppe west of that line. It can be again divided into several subtypes that are distributed in NE-SW trending belts from the southeast to the northwest in a sequence as distance from the sea increases and, consequently, annual precipitation decreases (Figure 12-3). East of the Da Hinggan Mountains lies the Xi Liaohe Plain and its luxuriant temperate steppe vegetation, dominated by *Aneurolepidium pseudoagropyrum, Stipa grandis, S. krylonii,* and other graminaceous grasses. This is the famous Kolshin Grassland suitable for both agriculture and pasture. The Da Hinggan Mountains mark the eastern margin of the Nei Mongol Plateau; its middle and southern parts are characterized by montane forest-steppe. On the eastern Nei Mongol Plateau and the Ordos Plateau, luxuriant temperate steppe vegetation appears again, of which the Hulun Buir and the *Ujimqin* grasslands are most famous. These have always provided the best pasture in China. The zonal chestnut soil is quite fertile, although supplemently irrgation is necessary if it is cultivated.

The temperate desert-steppe vegetation is located on the western Nei Mongol and Ordos Plateaus. It is composed chiefly of drought-tolerant small bunch grasses and half shrubs. The dominant grass and shrub species are *Stipa capilata, Artemesia frigida,* and *Caragana microphylla.* They are shorter, sparser, and less nutritious than temperate steppe species. The zonal soil is brown semidesert soil, which is also less fertile than chestnut soil. Irrigation is an absolute necessity for cultivation.

Bountiful Animal Resources

Owing to the above-mentioned physical conditions, the animal resources, especially hoofed animals, are quite bountiful. The dominant animals are herbivorous rodents and hoofed animals. The former cause great havoc with grasslands, just as wolves do with domesticated animals. The rodents, however, have been recently effectively controlled. The Mongolian gazelle (yellow goat, *Procapra gutturosa*) is the most commonly seen large animal. By the late 1950s and early 1960s, these animals had been overhunted[1], but their population has recently begun to recover. On the other hand, birds, reptiles and amphibians are rather few in species and in number.

NATURAL REGIONS AND SUBREGIONS

By the integration of areal differentiation of all the above-mentioned zonal and azonal factors, the Nei Mongol natural division can be subdivided into the following 3 natural regions and 10 subregions (Figure 12-1).

20. Xi Liaohe Basin — Steppe Region

Located east of the Da Hinggan Mountains, the Xi Liahe Basin is the only region in this natural division to be characterized by low plains and widely distributed farmlands. It is surrounded by mountains and hills on its northern, western, and southern sides, with its eastern side opening to the southern part of the Northeast China Plain. Consequently, it has a horseshoe geomorphic configuration, similar to the Songhua-Nenjiang Plain (i.e., the northern part of the Northeast China Plain), but its general topography dips eastward instead of southward, and its gradient is much steeper (from more than 2,000 m asl that decreases to about 120 m asl). Landform types are also much more complicated and include middle mountains, low mountains, hills, terraces, plains, depressions, and sandy dunes.

As a whole, this natural region enjoys the best temperature and moisture conditions in this natural division, with accumulated temperature during the ≥10°C period of 1400 to 3200°C (higher in the plains) and annual precipitation of 300 to 600 mm (higher in the surroun-

---

1) When the Mongolians hunted yellow goats on a large scale, they made good use of microrelief. The flocks on the immense plateau surface (Mongolian erosion surface) were cut off by the hunters on three sides and then purposefully driven to the margins of local deflated depressions (Gobi erosion surface) where steep scarps of several dozen meters caused serious injury to the animals in flight.

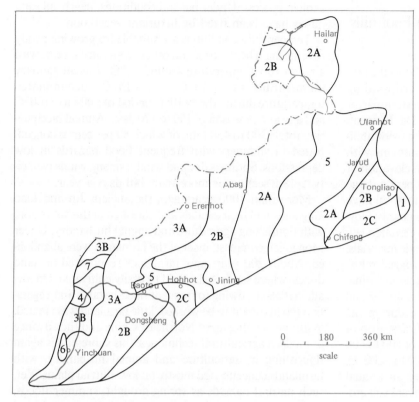

1. Meadow-steppe.
2. Steppe
   2A. *Aneurolepidium Peudoagropyrum, Stipa grandis;*
   2B. *Stipa krylonii;*
   2C. *Stipa baicalensis.*
3. Desert-steppe
   3A. *Stipa capilata, Artemisia frigida;*
   3B. *Reumuria soogarica.*
4. Desert
5. Montane forest-steppe
6. Montane forest-steppe, steppe
7. Montane desert-steppe

Figure 12-3 Grassland types of Nei Mongol

ding mountains). Summer rainfall provides about 80 per cent of total annual precipitation, which is favorable for agricultural development; however, there is poor drainage and a risk of flood hazards in low plains and depressions. Spring is dry and windy. In conjunction with widely distributed sand beds, large patches of fixed and half-fixed sand dunes have been formed. When natural vegetation cover is destroyed, shifting sands result; these ocupying now about 10 per cent of the total sandy areas. Farmers on the lands neighboring these white sands are often forced into resowing five or six times. Since 1949, a large-scale natural transformation project has been conducted here, with emphasis on forest-shelter planting. Its success has been remarkable.

Three subregions can be distinctly delimited.

20 (1) Southern Da Hinggan Mountains

The southern section of the Da Hinggan Mountains lies between the Chaor River and the Xar Moron River, with an absolute relief of 1000 to 1500m (highest peak, 2034m) and a relative relief of 100 to 500 m. Its bedrocks are composed mainly of Mesozoic rhyolite and trachyte. Its eastern slopes are rather steep, overlook the Northeast China Plain and are drained by many tributaries of the Songhua River

and Xi Liaohe River. Its western slopes are rather gentle, merge into the Nei Mongol Plateau, and are drained by small interior rivers. There are also many intermontane basins and a broad belt of piedmont plain with thick, fertile soils.

Accumulated temperature during the ≥ 10°C period increases from west to east (corresponding with descending elevation) from 1400 to 3000°C. Annual precipitation totals between 400 to 500 mm, concentrated mostly from July to September. Higher montane slopes (1300 – 1700 m asl) comprise the needle-leaved forest belt, with patches of spruce (elevation, 1300 – 1500 m) and larch (elevation, 1500 – 1700 m) on the shady slopes and montane grassland on the sunny slopes. Lower montane slopes (below 1300 m) contain deciduous broad-leaved forest, with birch, poplar, and oak predominating. The intermontane basins and piedmont plains are occupied by luxuriant graminaceous grassland, of corn and other crops.

As a whole, this subregion should be mainly devoted to forestry and pasture. The protection of natural vegetation and reforestation are two urgent measures. Local favorable sites with better climatic and topographic conditions, such as intermontane basins and piedmont plains, may be cultivated in grains or be utillized as a feeding ground.

20(2) Yanshan Mountains Northern Foot-hills and Terraces

The Yanshan Mountains subregion is located on the northern slopes of these Yanshan Mountains northward up to the Xar Moron River. Geologically, its western part is the junction between the NEE trending Da Hinggan Mountains and the W-E trending Yinshan Mountain systems, resulting in a mosaic of middle mountains with divergent directions. They generally have an elevation of between 1500 to 1800 m. Since the Tertiary period, large areas of basalt platform (elevation: 1000 to 1800 m) have been extruded. Its eastern part is mainly composed of low mountains, hills, and terraces, mostly with an elevation between 500 to 700 m and covered with thin loessic materials. They have been eroded and dissected into rugged relief. This is one of the most serious soil erosion areas in China.

As a result of the relief, the western part is cooler and less arid, with an accumulated temperature during the $\geqslant 10°C$ period of 1500 to 3000°C, a frost-free season of 125 to 150 days, and an annual precipitation of 500 to 600 mm. the eastern part records a pattern of 3000 to 3200°C, a frost-free season of 150 to 160 days and an annual precipitation of 400 to 500 mm. Owing to severe soil erosion and heavy human intervention, natural grassland vegetation has been mostly removed. Bare basalt platforms and shifting sand dunes are also widely present. In future development programs, protection of vegetation and other soil conservation measures should be greatly emphasized.

20(3) Xi Liaohe Plain

The middle reaches of the Xi Liaohe River[1] consist of a broad, triangular-shaped plain that inclines and opens eastward, with elevation decreasing from 500 to 120 m. It is composed mostly of loose Quaternary alluvial-diluvial sand and gravel beds that are covered with a thin veneer of a eolian silt. Along the Xi Liaohe River trunk channel lies a broad belt of low, flat alluvial plains composed mainly of floodplain and terraces. It is characterized by many abandoned channels and oxbow lakes, poor drainage, saline soils, and large tracts of meadow and swamp. On either side of this broad belt of alluvial plain, extensive sandy dunes are distributed. This is the famous Kolshin Sandy Land. Here, sand dunes above 2 m in height are called *tuo*, sandy gentle or rolling slopes are called *zhao*, and the low depressions among them are known as *dian*; they occupy 50%, 20%, and 30% of the total sandy land area respectively. Sands are mostly derived from local alluvial deposits, then reworked (deflating, depositing, sculpturing, etc.) by

aeolian process. Under natural conditions, nearly all sand dunes have been fixed by luxuriant vegetation.

Temperature conditions are suitable for growing paddy rice, spring wheat, corn, and other crops once a year, with a mean July temperature around 24°C, a mean January temperature of $-13°C$ to $-14°C$, accumulated temperature during the $\geqslant 10°C$ period of 3000 to 3100°C and a frost-free season of 150 to 170 days. Annual precipitation totals 300 to 450 mm, of which 80 per cent is concentrated in summer, with frequent flood hazards in low depressions. Spring is dry and windy; strong winds (velocity $> 18m/sec$) occur more than 140 days a year.

More than 1000 years ago, the ancient Jin and Liao kingdoms had their capitals and core area in this subregion, with flourishing agriculture and animal husbandry. Under Mongolian conquest, most of the farmlands were abandoned. About 200 years ago, farmlands reappeared on sand dunes, where sandy soils could be easily cultivated. On low alluvial plains, owing to frequent flood hazards and vegetation too luxuriant to be worked, little farmland was created. With the opening up of Northeast China and the advance of modern agricultural techniques, this subregion is again flourishing in agriculture and animal husbandry, with farmlands concentrated mostly on low alluvial plains. Yet, such natural hazards as spring drought, summer floods, shifting sand, and saline soil, are still playing great havoc in this subregion. Overall planning together with other effective measures must be taken to combat these forces of nature. Large-scale forest shelterbelt planting is certainly one excellent measure. Another suggested device is to use sand dunes as natural dams and low depressions as flood reservoirs to control flood, drought, aeolian, and salinization hazards and to improve the sandy soil simultaneously.

21. The Nei Mongol Plateau — Steppe and Desert-Steppe Region

The immense region of Nei Mongol Plateau stretches from west to east through 14° of longitude. It runs from the western foothills of the Da Hinggan Mountains up to the eastern border of the Ulan Buh Desert, then from south to north through 10° of latitude from the Yinshan Mountains up to the Sino-Mongolian international boundary. It includes, essentially, the Nei Mongol Plateau and the arc of the Yinshan Mountains, both uplifted since the Quaternary period, with elevations ranging from 2200 to 1500 m (Yinshan Mountains.) to 1500 to 1000 m (Nei Mongol Plateau). The Yinshan Mountains has a steeper southern slope overlooking the Elbow plains and a much gentler northern slope merging into the Nei Mongol Plateau, whereas the latter generally has a rolling plateau surface and inclines from the southern margin northward to the plateau proper. There are also many scattered local deflated depressions.

---

1) The upper reaches of the Xi Liaohe River comprise the Xar Moron. The Mongolian word "moron" means "river".

Owing to its rather high latitudinal and altitudinal location as well as to its openness on the northern side, this area becomes the highway of innumerable cold waves during the winter with much lower temperatures than other parts of the world at similar latitudes. The mean annual air temperature ranges from $-2°C$ to $-6°C$, becoming colder from southwest to northeast. Moisture conditions are more important in causing areal differntiation. Annual precipitation ranges from 400 mm on the eastern border to 150 mm on the western border. The Ondormiao – Bailingmiao–Otog–Yangchi line divides the eastern steppe zone from the western desert-steppe zone.

The region can, thus, be subdivided into the following three subregions.

21(1) Hulun Buir-Xilin Gol Plateau

The Hulun Buir-Xilin Gol Plateau subregion, the best grassland in China, (Photo V-2), has its western limit coinciding with the boundary between steppe and desertsteppe zones and its southern limit demarcated by the northern foothills of the Yinshan Mountains. It is subdivided into Hulun Buir and Xilin Gol. Geologically, it is mainly controlled by the NE-trending Cathaysian Structure system. The Hulun Buir is chiefly a graben basin with a vast rolling plateau surface of 550 to 700 m in elevation. Quaternary sediments are well developed. The so-called Hailar bed (Q_3) of fine and middle sands is especially widespread, resulting in a series of prominant eolian landforms. The Xilin Gol is chiefly composed of volcanic hills and basalt platforms along the Sino-Mongolian border, an immense peneplaned plateau surface in the middle, and the Hunshandake Sandy Land in the south. The latter has an area of about 20000 sq km, with sands mostly derived from local Quaternary lacustrine and alluvial beds. The sand dunes are mostly fixed by luxuriant vegetation.

This area enjoys a typical temperate semiarid climate, with annual precipitation ranging from 250 to 400 mm, of which 70 per cent is concentrated in the June to August period. Winter is long (5 to 7 months) and severe, with a mean January temperature that ranges from $-20°C$ to $-28°C$ and an absolute minimum temperature below $-50°C$. Under traditional pastoral management, "black hazard" can occur if there is not enough snowfall in winter (because of the shortage of drinking water, a good part of the grassland cannot be used), whereas "white hazard" (the snow-burying of pasture) might happen if there is too much snow. Summer is short and characterized by very large diurnal variation of temperature. There is a popular saying that goes, "Fur in the morning, shirt at noon; watermelon is eaten while the stove is welcome". The frost-free season totals 100 to 130 days annually; the accumulated temperature during the $\geq 10°C$ period is 2000 to 2400°C,

thus, spring wheat, spring barley, and Irish potato can be grown without too much difficulty. In addition, this is one of the windiest areas in China, with an annual mean wind velocity of 3.5 to 5.7 m/sec. This might be profitably harnessed for mechanical and electrical power in the future.

On the whole the vegetation of this subregion is luxuriant and the soil not infertile. The horizontal zonation is characterized by temperate steppe and chestnut soil. Dominant grass species are *Aneurolepidium, Stipa,* and other bunch grasses. The annual productivity of grassland reaches 900 to 2250 kg/ha of fresh grass. It has been the habitat of different pastoral peoples ever since neolithic times. In the thirteenth century, when the Mongolian hordes swept over most of Eurasia, the Hulun Buir steppe was their reliable base. Today, the pastoral Mongolians still occupy most of this subregion and an effort to modernize traditional pastoralism is now in full swing. The first measure is to use grassland by rotation and to build *kulun* (enclosures) to protect and to improve natural grassland. The next step is to improve natural grassland by sowing high quality grasses and crops. Third, it is necessary to establish cultural feedlots and grain bases. In addition, all natural hazards and limiting factors, such as the low temperature, the black and white hazards, as well as the rats and wolves, need to be controlled.

21(2) Ulanqab Plateau

Located between the Yinshan Mountains in the south and the Sino-Mongolian international boundary in the north, the Ulanqab Plateau subregion descends northward from about 2200 m to 900 m. It is essentially an uplifted, peneplaned plateau (the so-called Mongolian erosion surface), having four or five stepped erosion surfaces (mostly distributed with stony gobi) that are composed mainly of Tertiary red sandstones. Hence the area bears the name Ulanqab, which means red cleavage in Mongolian. There are also many northeast trending low hills or *ula* distributed on the plateau surface as inselbergs, which denote a pre-Cretaceous erosion surface, and many shallow aeolian depressions, which represent the gobi erosion surface.

Owing to the deep inland location and the Yinshan Mountains barrier, the climate is arid, with annual precipitation ranging from 150 to 250 mm (decreasing westward) and an aridity index of 2.0 to 4.0 (increasing westward). Many luxuriant grasslands lie unused because of the lack of drinking water. Temperature conditions are somewhat better, with a mean July temperature of 22 to 24°C, an accumulated temperature during the $\geq 10°C$ period of 2500 to 3000°C, and annual sunshine of 3200 to 3300 hours.

Zonal vegetation is a transitional type between steppe and desert and is, thus, called desert-steppe. It is much less productive than temperate steppe, with an annual productivity of 600 to 1500 kg/ha of fresh grass. Zonal soil is also a transitional type — the so-called brown semidesert soil. Irrigation is necessary for cultivation.

21(3) Yinshan Mountain

The Yinshan Mountains comprise one of great geographic divides in China; the areal differentiation between "up mountain" (Nei Mongol Plateau) and "down mountain" (surrounding plains) is quite conspicuous. The mountain itself is chiefly controlled by the so-called Giant Latitudinal Structural System (see Chapter One) and dominated by desiccation and denudation processes. The crest lines have generally an elevation of 1800 to 2000 m, with the highest peak above 2300 m. The southern slope of the western section is a majestic fault scarp that overlooks the Front Elbow Plain, with a relative relief of more than 1000 m. On the other hand, the northern slope of the western section is much less steep and merges into the Nei Mongol Plateau through a broad belt of hills and intermontane basins. Large tracts of which have been cultivated for dry farming. The eastern section of the Yinshan Mountains system is much lower and less rugged, the dominant landscapes are large basins studded with small basins and barren basalt platforms.

Vertical zonation is evident. The dominant vegetation in the foothills is temperate semiarid steppe. Accumulated temperature during the $\geq 10°C$ period totals 1700 to 2300°C, the frost-free season is between 90 to 120 days and annual precipitation is 250 to 420 mm. Uncultivated grasslands are still widely distributed, with annual productivity of fresh grass 1500 to 2250 kg/ha. Farmlands have been developed in large tracts during the last 100 years and are located mostly on gentle slopes and intermontane basins and sometimes even on steep slopes and hilltops, which results in severe soil erosion. Overall planning and better management for both grasslands and farmlands are much needed.

22. The Ordos Plateau — Steppe and Desert-Steppe Region

The Ordos Plateau is a natural region that is encircled on the north by the Yinshan Mountains, on the west by the Helan Mountains, and in the south by the Great Wall. It is essentially composed of the rolling Ordos Plateau together with its surrounding middle mountains and high plains. Climatically and biologically, it is a transitional zone between the semiarid steppe and the arid desert-steppe. It is widely interspersed with widespread aeolian sands. The mighty Huanghe River encircles and traverses the region in a great arc, furnishing bountiful irrigation water and fertile soils. Yet, owing to meager rainfall, other perennial rivers are few and the lakes are mostly saline or sodic. Areal differentiation inside this region is prominent; four subregions may be identified.

22(1) Ordos Plateau

The Ordos Plateau subregion is roughly demarcated by the Great Wall on the south and by the Huanghe River on the other three sides. Geologically, it has been stable for a very long time, with undisturbed level strata and without any igneous activities. During the Mesozoic submergence, a broad synclinic basin was formed, deposited with thick Mesozoic sandstone, conglomerate, and shale, as well as with Tertiary red sandstone and sandy clay in the western part. Since the Quaternary, it has been uplifted to its present elevation of 1200 to 1600 m, with the Mount Table (2149 m) of the northwestern margin as the highest peak. As a whole, it is an undulating denudational plateau that is covered by wide aeolian landforms. The west-eastward Hobq Desert with an area of 16000 sq km is distributed on its northern arid border. The Mu-Us Sandy Land is located on the semiarid southeastern border. It has an area of 25000 sq km, of which fixed, half-fixed, and shifting sand each comprise about one third. There is also a series of deflated low depressions between these two great sandy areas that have much better water and other more favorable physical conditions.

Winter is severe and long, controlled by cold and dry northwestern continental monsoons, with a mean January temperature of −10 to −13°C. Summer is short but warm, with a mean July temperature between 20 to 23°C. Annual precipitation totals 300 to 400 mm in the eastern part, decreasing to 200 to 300 mm in the west. Both seasonal and yearly variations' are great, resulting in marked divergence in the productivity of the grassland. East of Otok, zonal vegetation is temperate steppe, whereas west of it, the vegetation is temperate desert-steppe. Ecologically, both are mainly composed of xerophytes and psammophytes. Under the great impact of human intervention, they have seriously degenerated in the last 1000 years.

According to historical documents, the Ordos Plateau, especially the Mu Us Sandy Land, was famous for its luxuriant grassland. In A.D.413, the Xia Kingdom founded Tungfan City (now the White City ruin) deep inside the Mu-Us Sandy Land as its capital, with a population of about 200000. Yet, owing to excessive expansion of farmlands and overgrazing, the process of desertification in the Mu-Us Sandy Land has continued for more than 1000 years since the Tang dynasty. During the last 300 years, after the rebuilding of the modern Great Wall, a belt of shifting sands 60 km wide along the Great Wall has been formed. This is China's most notorious example of the southward advance of sandy desert. Other parts of the Ordos Plateau have also suffered badly from the process of desertification. Therefore, to control shifting sands and to reestablish luxuriant grassland should henceforth be the two most important meliorating measures for this subregion.

22(2) Loess Hills

The Loess Hills subregion is restricted to a small area in the northereastern Ordos Plateau. It is a transitional belt between the Loess Plateau and the Ordos Plateau, with an elevation of between 1200 to 1500 m, an accumulated temperature during the $\geq 10°C$ period about 3000°C and an annual precipitation of about 400 mm. In the last 100 or 200 years, it has been actively cultivated and overgrazed, resulting in severe soil erosion as well as low agricultural production. What is worse, the area has been permanently dissected into numerous barren hills, with a relative relief of 100 to 200 m, a length of gullies of 5000 to 7000 m/km², and a rate of surface erosion of 1 to 2 cm/yr. Therefore, soil conservation measures are of paramount importance here. Farmlands on steep slopes should be abandoned, terraced and transformed into productive grassland and woodland; fertile valley bottoms may continue to be used as farmlands.

22(3) The Elbow Plains

This subregion is sandwiched between the Ordos Plateau and the Helan -Yinshan mountain systems. It includes three fertile high plains along the middle Huanghe River — the West Elbow Plain (Yinchuan Plain), the Back Elbow plain, and the Front Elbow Plain. Geologically and geomorphologically, these plains on plateau are a series of grabens filled by diluvial-alluvial deposits with a thickness of about 1000 m in the western and northern Back Elbow Plain. This subregion inclines generally from southwest to northeast.

Protected by the Helan Mountains and Yinshan Mountains on the north and west, as well as being lower in elevation, this area is warmer than the Nei Mongol Plateau and the Ordos Plateau to either side. It has a mean January temperature of -9 to $-14°C$, a mean July air temperature of 21 to 24°C, and an accumulated temperature during the $\geq 10°C$ period of 2800 to 3200°C. In the Yinchuan Plain even cotton can be grown under irrigation. The annual precipitation ranges from 300 to 450 mm in the Front Elbow Plain to 150 to 300 mm in the Back and West Elbow plains. Dry farming is possible in the Front Elbow Plain, but yield is rather precarious. In the other two plains irrigation is an absolute necessity for agricultural development; the minimal precipitation is not only far from adequate for crops, but also is harmful because it induces salinization and soil crust.

The Huanghe River furnishes bountiful and excellent irrigation water; the three Elbow plains themselves are

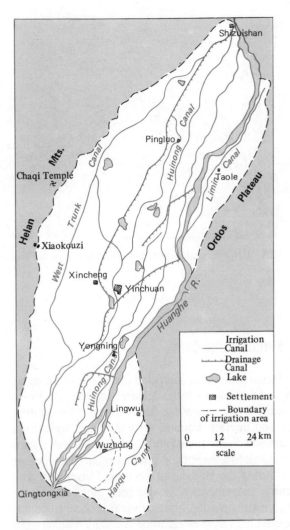

Figure 12-4 Irrigation canal system in the west Elbow Plain

(Yinchuan Plain)

A. Irrigation canal; B. Drainage canal.

chiefly created by alluvial deposits of the Huanghe River. As early as the third to second centuries B.C., the emperors of the Qin and Han dynasties sent armies and colonists first to the West Elbow Plain and shortly afterwards to the Front and Back Elbow plains. Extensive farmlands were cultivated and productive irrigation canal systems were built (Figure 12-4, 12-5; Photo V-3). The area was turned into the "land of rice and fish," comparable to the fertile and rich lower Changjiang Valley. With the improvement of irrigation and drainage systems as well as the control of drought, salinization, aeolian hazards and other limiting factors, this subregion could be further built up as a new commercial grain base in China.

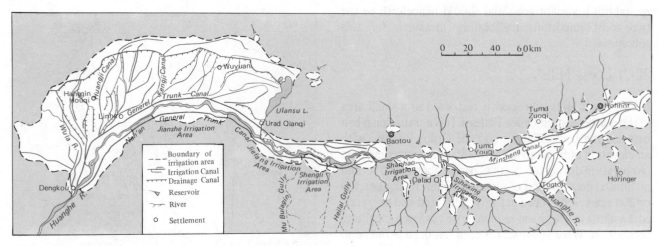

Figure 12-5   Irrigation canal system in the Back Elbow Plain and Front Elbow Plain

22(4) Helan Mountains

The Helan Mountains subregion is also one of the great geographic divides in China, demarcating the temperate grassland from the temperate desert as well as delimiting oceanic and interior drainage systems. The longitudinal mountain was firstly folded and uplifted during the Yanshan Tectonic Movement and uplifted again to its present altitude during the Himalayan Tectonic Movement. The neotectonic movements have also been quite violent; up to now, it is still uplifting, and earthquakes occur frequently. The crest lines have an elevation generally ranging from 2000 to 3000 m asl (highest peak, 3556 m). Like other marginal mountains that surround the Nei Mongol Plateau, the Helan Mountains have an asymmetric profile, with steep eastern slopes and much gentler western slopes. When viewed from the Yinchuan Plain, the silhouette of these middle mountains looks like a galloping horse, hence the Mongolian name, Helan.

Modern geomorphic processes are dominated mainly by desiccation and denudation, with only a small patch of alpine shrubby meadow on the mountain crests. Patches of needle-leaved forest also survive on upper slopes above 1500 m asl. Montane desert-steppe and montane grassland, suitable only for pasturing goats, prevail on the lower slopes. Protection of natural vegetation and reforestation are certainly two most important meliorating measures that need to be undertaken in this subregion.

References

[1]   Department of Geography, Teachers' College of Nei Mongol, 1965, *Physical Geography of Nei Mongol,* Nei Mongol People's Press, Hohhot, (In Chinese)

[2]   Integrated Investigation Team of Nei Mongol and Ningxia, Chinese Academy of Sciences, 1980, *Geomorphology of Nei Mongol,* Science Press, Beijing. (In Chinese)

[3]   ———, 1980, *Climate and its Relationship with Agriculture and Animal Husbandry in Nei Mongol,* Science Press, Beijing. (In Chinese)

[4]   ———, 1980, *Water Resources and Their Utilization in Nei Mongol,* Science Press, Beijing. (In Chinese)

[5]   ———, 1980, *Soil Geography of Nei Mongol,* Science Press, Beijing. (In Chinese)

[6]   ———, 1980, *Vegetation of Nei Mongol,* Science Press, Beijing. (In Chinese)

[7]   Chao Sung-chiao (Zhao Souggiao), 1958, "Landforms of Nei Mongol and Their Evaluation in Land Use", *Acta Geographica Sinica,* Vol. 24, No. 3. (In Chinese, with English abstract)

[8]   ———,1981, "Desertification and De-desertification in China", *The Threatened Dryland: Regional and Systematic Studies of Desertification,* Working Group on Desertification in and Around Arid Lands, IGU.

[9]   ———, et al., 1958, *A Preliminary Study on the Location Problem of Agriculture and Animal Husbandry in Nei Mongol,* Science Press, Beijing. (In Chinese)

[10] Sun Jinzhu, 1976, *Physical Conditions and Their Transformation in Elbow Plains,* Nei Mongol People's Press, Hohhot. (In Chinese)

[11] Sand-Contol Team, Chinese Academy of Sciences, 1984, *Report on Integrated Investigation of Mu Us Sandy Land,* Science Press, Beijing. (In Chinese)

[12] *Agricaltural Geography of Nei Mongol* Compilation Committee, 1982, *Agricultural Geography of Nei Mongol,* Nei Mongol People's Press, Hohhot.   (In Chinese)

Chapter 13

Temperate and Warm-temperate Desert of Northwest China

The extremely arid desert division in China extends eastward up to the Helan Mountains (about 106°E) and southward up to the marginal mountains of the Qinghai-Xizang Plateau—the Kunlun, Altun, and Qilian mountains. On the northern and western sides, it is defined by the Sino-Mongolian and Sino-Russian borders respectively. It has an area of about 2.1 million sq km, occupying 22 per cent of the total land area of China, yet it contains only 2 per cent of total population (Figure 13-1).

This arid desert division is located in the innermost part of the largest continent in the world, Eurasia. Distances to the sea are very great on all sides. Ürümqi (the capital city of the Xinjiang Uigur Autonomous Region) is about 4400 km from the nearest point of the Pacific Ocean, about 4300 km from the Atlantic Ocean, about 3400 km from the Arctic Sea, and about 2500 km from the Indian Ocean. Furthermore, it is surrounded and traversed by a series of lofty mountains that serve as effective climatic barriers and bar nearly all moist maritime monsoons. Consequently, it is characterized by and dominated by an extremely arid desert landscape. The wiid expanse of yellowish sandy desert (shamo) and greyish stony desert (gobi) together occupy about 45 per cent of the total land area of this natural division.

As a whole, this area is unfavorable to agricultural development and habitation, with drought, salinization, shifting sand, and other natural hazards frequently inflicting heavy losses on farms, pastures, transportation links, and settlements. On the other hand, there are quite rich land, mineral, and solar-energy resources in the area and a considerable amount of water and biological resources that might be profitably tapped. Besides, this is an important strategic borderland of China that has been the home of millions of Chinese and many minority peoples since ancient times. Therefore, a basic study on this immense, yet so-far-little-known natural division, is certainly of great significance, not only scientifically and economically, but also politically and socially. The major economic and social goals have two aspects: (1) to protect and to make better use of all natural resources according to specific conditions of different natural regions and different land types and (2) to control and to transform all unfavorable natural conditions and natural hazards for the greatest benefit of all inhabitants of this natural division.

ORIGIN AND EVOLUTION OF CHINA'S DESERT LANDSCAPE

The desert landscape of China is the end product of an array of interrelated geological, atmospheric, and biological processes since late Cretaceous and early Tertiary times.

According to recent paleogeographical studies in China, the dry climate in Northwest China developed as long ago as the late Cretaceous and early Tertiary periods. The Chinese mainland, after a long, relatively quiet post-Yanshan Tectonic Movement period, was then mostly denudated and reduced to low hills and peneplanes, with several depositional basins interspersed among them (see

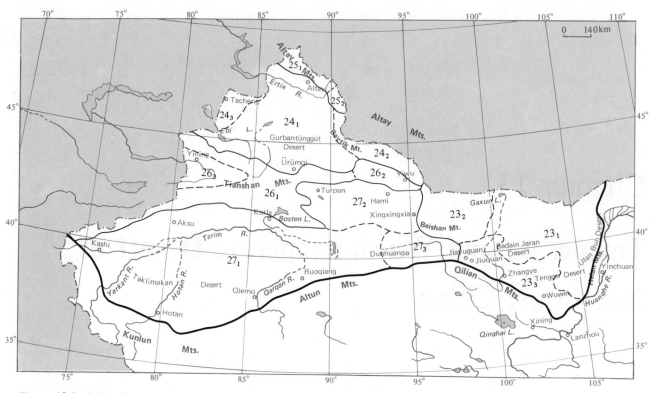

Figure 13-1 A location map of the temperate and warm-temperate desert of Northwest China

Figure 1-5). The formerly extensive submerged areas had then only one remnant — the southwestern margin of the Tarim Basin which was a part of the ancient Tethys Sea. In Northwest China, the so-called Mongolian Peneplane was widely distributed over the Nei Mongol Plateau; a reddish weathered crust was well developed in the Tianshan, Altay, and many other montane areas; and depositional basins were also extensively distributed in the Tarim, Junggar, and other inland low-lying areas. At that time, the Chinese monsoon system was not yet formed; most of the Chinese territory was under a subtropical high pressure system with northeasterly trade winds prevailing. Hence a broad belt of arid and semiarid climate characterized by park savanna and semidesert vegetation extended from northwest China southeasterly to the lower Changjiang Valley.

The great age of Chinese arid and semiarid lands might be reaffirmed by recent botanical studies. A xerophilous flora existed in northwest China as early as the late Cretaceous or early Tertiary times. Most of the flora perished at the end of the Tertiary period, yet a small number have survived, including *Gympnocarpos, Potaninia, Zygophyllum-sacrozygium, Tetraena,* and *Nitraria.* The following dominant or widely distributed species in modern Chinese deserts might also be considered as remnants of ancient flora: *Ephedra przewalski, Sympegma regelii, Il-*

jinia regelii, Reaumuria soongolica, R. trigyma, R. kaschgarica, Brachanthemum gobicam, Convolvulus tragacanthoides, Ceratoides laters, Asterothamnus centraliasiaticus, Anabasis brevifolia, Kalidium gracile, and *Psammochloa villosa.* Again, modern riparian forests with their chief components — *Populus diversifolia, P. pruinosa, Elaeagnus* spp., and *Ulmus pumila* might represent the degeneration of ancient park savanna.

Since Pliocene times, the violent Himalayan Tectonic Movement has caused great changes in the geographical environment of China (Figure 1-6). It has resulted in the complete disappearance of the ancient Tethys Sea in western China and the gigantic uplift of the Himalaya Mountains and the Qinghai-Xizang Plateau, which has been uplifted about 3500 to 4000 m since the Pliocene and continues to rise even now at a rate of about 13 to 14 cm per 100 years. Other great ranges in northwest China — the Kunlun, Qilian, Tianshan, and Altay mountains have also undergone differential fault uplift, resulting in a series of block mountains with elevations of more than 5000 m as well as a series of relatively low graben basins. Consequently, these inland basins have become even more isolated, more desiccated, and deposited with thick beds of sands and gravels, which supply rich source materials to the sandy and stony deserts. For example, according to a geophysical investigation, the Quaternary deposits in the alluvial plain along the upper reaches of the Tarim River have a depth of 400 to 500 m. In addition, the Mongolian peneplane, after uplifting and warping, has been shaped into a rolling plateau with an elevation of around 1000 m. In short, all Northwest China and the western part of North China assumed in the Quaternary period broad land forms similar to the present (i.e., extensive inland basins and high plateaus surrounded by lofty mountains).

Based on recent research of the Integrated Scientific Investigation Team of the Qinghai-Xizang Plateau, as well as by many archeological excavations in Northwest and North China, the formation and acceleration of the dry climates are closely correlated with the rapid uplifting of the Qinghai-Xizang Plateau. Fossils of the three-toed horse (*Hipparian*) of the late Pliocene Epoch that have been newly excavated in many places in the Qinghai-Xizang Plateau are analogous to those excavated in North China. Both regions might have had a subhumid forest-steppe environment at that time, whereas Northwest China had a semiarid steppe landscape. In the late Pliocene, the Qinghai-Xizang Plateau attained an uplifted elevation of about 1000 m. The Siberia-Mongolian high-pressure system was then not yet formed, and there existed only a weak high-pressure belt near Lhasa (about 30°N). At the end of the Tertiary period the Qinghai-Xizang Plateau together with its neighboring regions was uplifted violently. The plateau surface attained an elevation of about 3000 m. There were no longer any forests in the plateau proper, and the weak high pressure belt near Lhasa was strengthened and pushed northward to about 40°N. Thus, the desiccation of Northwest China was greatly accelerated. The ancient lakes in the Tarim and other inland basins diminished or dried out entirely, and the extent of the Taklimakan and other sandy deserts was considerably enlarged at that time.

From the late Pleistocene to the early Holocene the Qinghai-Xizang Plateau and its neighboring areas again underwent violent mass uplifting. Since then, the Qinghai-Xizang Plateau has attained an elevation of about 4000 m and the Chinese monsoon system has been well established. The winter high pressure belt has been greatly strengthened, and it has pushed northward to its present position (about 55°N). From this belt, dry, cold winds blow to the surrounding area. At about 97°E, they are blocked by the lofty Qinghai-Xizang Plateau, resulting in two great divergent wind systems the northwesterly and the northeasterly. The former dominates a greater part of North China and even South China during the winter whereas the latter prevails up to the southwestern margin of the Taklimakan Desert, where it is superseded by the northwesterly. Thus, the air is cold and dry over Northwest and North China in the winter. However, during summer, the maritime southeastern monsoon predominates, bringing rainfall to South China and the eastern part of North China. In Northwest China and the western part of North China, it becomes increasing difficult for the moist maritime monsoon to penetrate to the northwest because of the distance from the sea and the barrier action of a series of lofty mountains and high plateaus. Hence, from the Da Hingan Mountains westward, the annual rainfall decreases sharply as distance from the sea increases, and there is a complete array of steppe, semidesert and, desert landscapes.

It should be emphasized that the general trend of desiccation in Northwest and North China since the late Cretaceous and early Tertiary periods and the acceleration of this trend since the Pliocene Epoch does not rule out fluctuations and changes of paleoclimates. There were certainly relatively moist periods both in Tertiary and Quaternary times, especially during the glacial epochs, although so far there is no convincing evidence of the existence of Quaternary continental glaciation in Northwest and North China. An outstanding example is in the southern part of the Taklimakan Desert, where late Quaternary fluvial delta deposits are well developed and extensively distributed and ancient drainage networks are conspicuous on the ancient delta, which stretches deep into the desert interior for about 200 to 250 km. This shows that the rivers flowing out of the Kunlun Mountains had much larger discharges at that time.

CHIEF PHYSICAL FEATURES OF CHINA'S DESERT

The desert landscape of Northwest China is chiefly characterized by the following common physical features.

Climate

First, the climate is extremely dry, and as a Chinese proverb points out, "No irrigation, no agriculture". Annual precipitation decreases westward from about 200 mm along the Helan Mountains piedmont plain to the driest part in China — the eastern Tarim Basin, with annual precipitation of less than 25 mm (only 3.9 mm at Toksun in the Turpan Basin). There is then a precipitation increase wastward up to 70 mm along the Sino-Russian border. Most of the Junggar Basin records 100 to 200 mm. Rainfall is also quite sporadic; usually, there is none for more than six successive months, then suddenly one storm might produce one half or even two thirds of the total annual precipitation. Consequently, sunshine and solar radiation are abundant with generally more than 3400 hours of sunshine and 130 to 155 kcal/cm^2 of solar radiation annually.

The Alashan Plateau and the Junggar Basin have a temperate climate, with a frost-free season of 150 to 200 days and an accumulated temperature during the ≥ 10°C period of 3000 to 3500°C. Hence, spring wheat, millet and Irish potato can be grown once a year. The Tarim Basin has warm-temperate climate, with a frost-free season of 200 to 230 days and an accumulated temperature of 4000 to 4500°C. Two crops in three years might be reaped and long-fiber cotton can be grown here. All over these regions, winters are long and severe and thermal variations, both annual and diurnal, are very great. Here the highest temperature in China was recorded — 47.6°C at Turpan — as well as one of the lowest −51.5°C Fuyun (not far from Turpan). On the surface of a sand dune, the Gurbantünggüt Desert, the extreme temperature of 83°C has been recorded.

Strong winds blow frequently. In winter, cold and dry northwestern and northeastern winds predominate. In summer, there are also frequent atmospheric disturbances and "hot easters" as well as dry, hot Foehns. Consequently, Anxi and Toksun, both called the "wind reservoir of the world", have strong winds (wind velocity above 18 m/s) that blow 35 and 72 days per year respectively. The Junggar Gate, which is located in northwestern Junggar Basin, has 155 strong windy days each year.

Geomorphology

The desert division of China is essentially composed of the Alashan Plateau (including the Beishan Mountains and the Hexi Corridor), two great inland basins (the Junggar and the Tarim) and their surrounding high mountains (mainly the Tianshan and the Altay). On the whole, the ground surface is level or undulating and consists of rather coarse materials (chiefly sands and gravels). Yet, the surrounding mountains stand out conspicuously and exert a great influence on the desert environment. The highest peak of the Tianshan Mountains, Mount Tomul (Photo 1-5), towers up to 7435 m asl. The snowcapped Mount Bogda rises to 5445 m, whereas to the immediate south lies the Turpan Basin, which contains the lowest point of elevation in China (−155 m).

The Junggar and Tarim basins have generally an elevation of 500 to 1000 m and dip gradually from the surrounding mountains and their piedmonts to basin centers where the largest and second-largest sandy deserts in China, the Taklimakan and the Gurbantünggüt, are located. From the surrounding mountains to the basin centers, a series of distinctive land types appear in succession as the elevation decreases (Figure 13-2A):

1) Extremely high mountains, more than 5000 m asl, mainly under continous snow and the process of nivation.
2) High mountains, generally 3000 to 5000 m. asl, mainly under the process of erosion.
3) Denudational mountains and hills, generally less

than 3000 m asl.
4) Peneplaned stony level land (denudational stony gobi).
5) Diluvial fan (gravel gobi).
6) Diluvial-alluvial piedmont plain (sandy gravel gobi).
7) Alluvial plain (clay and silt level land)
8) Alluvial-aeolian plain (fixed, half-fixed, and shifting sand dunes).
9) Alluvial-lacustrine plain (chiefly salt marsh).

The Alashan Plateau (which might be considered as the western part of the Nei Mongol Plateau), the Beishan Mountains (Mazong Mountain), and the Hexi Corridor, generally have an elevation of 1000 to 1500 m. The highest peak of the Beishan Mountains rise to 2791 m, and is still under the process of desiccation. The region is chiefly composed of a series of NE-SW or E-W trending denudated, peneplaned mountains and hills, interspersed with depositional intermontane basins. Peneplaned hills, Inselbergs, and stony gobi predominate in the former; diluvial gravel gobi and diluvial-alluvial sandy gravel gobi chiefly comprise the latter (Figure 13-2B). In some large marginal basins extensive sandy desert are distributed, including the famous Badain Jaran, the Tengger, and the Ulan-Buh.

Rivers

There are practically no perennial rivers fed by local runoff. Only intermittent streams and a few larger rivers that originate in the surrounding high mountains flow into the zone. Even these larger rivers belong to inland drainage with but two exceptions — the Huanghe River and the Ertix River. The upper reaches of rivers are located in the surrounding montane areas with numerous tributaries and rich discharges. As soon as they flow out of the montane area, both their tributaries and discharges diminish quickly, with water quality also deteriorating. Finally, they die out in the lower reaches, either flowing into salty inland lakes or disappearing under sand or salt marsh. The largest inland river in China, the Tarim River, has had its lower reaches and terminal lakes — the Lop Lake (Lop Nor) and the Taitima Lake completely dry since 1972.

The distribution of ground water resources is quite uneven and unbalanced. There are rich resources along perennial river channels and high mountain piedmont plains, but they are very poor in most other parts of the zone. On the frontal margin of large piedmont plains, rich and good underground water often emerges in springs. This is sometimes called the "spring line" and served as the site for cities and highways; the famous ancient Silk Road was also located along this "spring line".

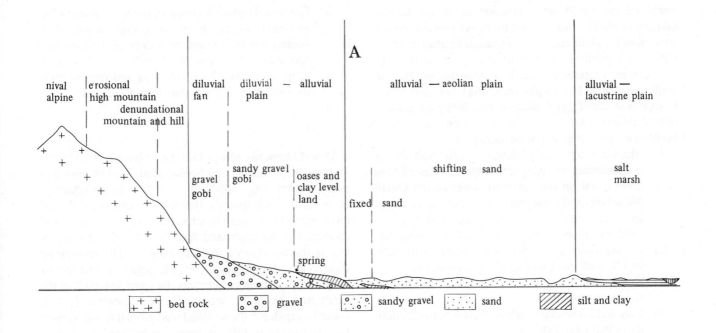

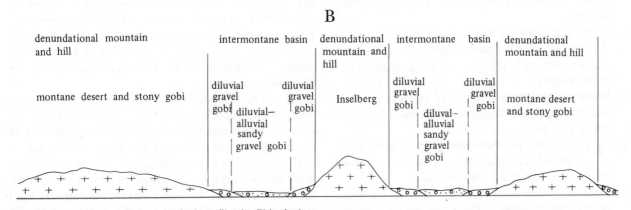

Figure 13-2 Comprehensive physical profiles in China's desert zone

A. Inland basin type; B. Plateau type.

Soil

The most widely distributed soils are yermosols, xerosols, and solonchaks, all with poorly developed soil profiles, very low soil moisture and humus content, coarse parent materials, and high salt content. Consequently, they are generally unfavorable for agricultural use, and a series of improvement measures, such as irrigation, drainage, fertilization, and salinization control, must be taken if they are to be used as farmland.

Yet, along modern and ancient rivers as well as in the lower part of diluvial-alluvial plains, there exists a considerable amount of rather fertile soils, chiefly azonal soils, such as fluvisols and gleysols. Under proper management these might be cultivated. There are now about 4 million ha of cropland in the desert zone of China, with about 12 million ha of potentially arable land, which accounts for nearly 30 per cent of the total unused arable land in China.

Vegetation

The desert vegetation is sparse — generally less than 20 to 30 per cent in coverage, with large tracts less than 5 per cent. In certain areas, such as in the gravel gobi along the upper part of the Kunlun Mountains piedmont plain, the shifting sand sea in central Taklimakan Desert, and the peneplaned stony gobi in the Gaxun Desert, there might be practically no vegetation at all for tens of kilometers. Furthermore, owing to the unfavorable natural conditions for plant growth, it is most difficult or even impossible for plants to recover once they are destroyed.

The desert vegetation is mainly composed of shrubs, half-shrubs, and small half-shrubs, including Chenopodiaceae, Compositae, and Zygophyllaceae. Paralleling the correlation of different land types, there are five major desert vegetation types that might be identified:

1) Montane and hilly desert: small half-shrubs predominate; its upper limit approximates 1800 to 2000 m asl on the northern slopes of the Qilian Mountains and the southern slopes of the Tianshan Mountains and as high as 3200 to 3600 m asl on northern slopes of the eastern Kunlun Mountains.

2) Gravel desert (gobi): distributed widely; with half-shrubs and small half-shrubs predominate.

3) Sandy desert: distributed widely; half-shrubs and shrubs predominate.

4) Clay and silt desert: rather restricted in distribution, half-shrubs predominate.

5) Saline desert: distributed widely in low depressions, composed chiefly of fleshy, saline small half-shrubs, all of them are species of the Chenopodiaceae family.

There are also several azonal vegetation types:

1) Riverine forest: distributed mostly along modern and ancient river channels; *Populus diversifolia* predominates.

2) Saline shrub: composed mostly of *Tamarix* spp and *Nitraria* pp, usually dotted on the periphery of shifting sandy areas.

3) Saline meadow: distributed mainly on the lower part of alluvial-diluvial plain, with a shallow ground water table and luxuriant grasses.

4) Cultivated vegetation or oases, which in Chinese denote "green areas", flourishing with luxuriant cultural vegetation and standing out quite conspicuously amid the vast expanse of greyish or yellowish deserts.[1].

In the surrounding high mountains, vertical zonation is outstanding. From the montane base and hilly desert zone upward the following vertical zones appear in sequence:

1) Montane grassland: composed chiefly of graminaceous grasses, generally with coverage of 20 to 40 per cent;

2) Montane coniferous forest (missing in the extremely arid Kunlun Mountains area): Composed chiefly of *Picea* spp. that is generally distributed on northern slopes, ranging from 2600 to 3600 m asl .

3) Alpine meadow: composed chiefly of Carex spp.

4) Periglacial cushion vegetation.

---

1) "Desert" in Chinese means "wild expanse."

5) Continual snow: elevation of the snow line is 4200 to 4400 m asl. on the northern slope of the Qilian Mountains and the southern slope of the Tianshan Mountains, the snow line lowers to 2700 to 3500 m in the Altay Mountains area and rises to 5000 to 5900 m in the Kunlun Mountains area.

Fauna

Desert fauna like steppe fauna are characterized by the dominance of rodents and hoofed animals., but, owing to much poorer food conditions, much less abundant in population. Their special adjustment to an extremely arid environment (e.g., cave-inhabitating, winter dormancy of rodents, the migrating and speed of the hoofed animals; etc.) is also better developed by necessity. The dominating rodents are *Meriones* spp. and *Allactage* spp. The former live mainly in the gobi, whereas the latter live both in the sandy and the stony deserts. The most commonly seen hoofed animals are the so-called wild goats (*Gazella subgutterosa*), usually running in small herds. Sometimes small groups of wild ass (*Equus hemonius*) may also be seen.

The reptiles, especially *Lacertiformes,* are quite abundant owing to their being well adopted to desert environments. The dominant lacertian species are *Phrynocephalus* spp. and *Eremias* spp.; in a distance of 500 m as many as 40 to 50 of these species might be seen. Commonly seen snakes are *Eryx miliaris* and *Psammophis lineolatus.*

Birds and amphibians are seldom seen. Probably the dominant bird species is *Galerida cristata,* yet usually only one or two of these species appear in a distance of 1000 m; sometimes there are none at all for dozens of kilometers.

Oases stand out conspicuously — not only in vegetation, but also in fauna — amid vast empty expanses of open sand. In these oasis, birds and other vertebrates are much more abundant, and human activities bring in many new faunal elements foreign to the original desert habitat. These include sparrows, swallows, and small domestic rats.

NATURAL REGIONS AND SUBREGIONS

Areal differentiation inside China's desert division is quite conspicuous. This is due chiefly to divergent landforms and ground-surface materials. There are 5 natural regions and 14 subregions that can be identified.

23. The Alashan Plateau — Temperate Desert Region

The Alashan Plateau is the easternmost part of China's desert division, with the Helan Mountains and the Qilian

Mountains as eastern and southern borders. Its northern border is the Sino-Mongolian boundary. Its western border is determined by the 4500°C isotherm of accumulated temperature during the ≥ 10°C period, approximately coincident with the isopleth of 1200 m asl along the southern foot of the Beishan Mountains.

This is exclusively an arid land with an aridity index greater than 4.0 and relative humidity usually less than 10 per cent. Geomorphologically, it is the southwestern part of the immense Nei Mongol Plateau, with denudational middle mountains of 2000 to 2500 m in elevation interspersed with depositional intermontane basins of 1000 to 1500 m. It belongs entirely to the inland drainage basin and contains large tracts without any runoff at all. Soils are mainly composed of yermosols and solonchaks, and vegetation consists chiefly of very sparse shrubs and half-shrubs. Consequently, land types are mostly montane desert, denudational stony gobi, diluvial gravel gobi, and shifting sands, all of them unfavorable to agricultural development and human settlement.

There is conspicuous areal differentiation. From south to north, the elevation decreases gradually, resulting in a drier climate and a diminishing vegetation. From the main peak of the Qilian Mountains (5934 m in elevation) northward to the Black River's terminal lake (Juyan Lake, 820 m) the distance totals only 370 km as linear distance, but the geographical landscape changes tremendously. Figure 13-3 and Landsat image 6 clearly show a SW-NE trending comprehensive physical profile in the Jiayuguan (Photo V-10) area, with land types changing from erosional high mountain and denudational mountain and hill (both mainly distributed in the Qilian Mountains section) to depositional gobi, sandy desert and clay level land (mostly in the Jiuquan and Jinta basins) and finally to denundational hill and stony gobi (in the peneplaned Beishan Mountains). Second, annual precipitation decreases from 150 to 200 mm to less than 50 mm from east to west as the distance from the sea increases. Three natural subregions are, thus, identified.

23(1) Alashan Plateau Proper

Geologically, the Alashan Plateau is an ancient, stable platform that consists of a series of NE-SW or E-W trending ranges interlocked with intermontane basins. The southernmost range is the so-called Heli-Longshou Mountains of the Hexi Corridor. It is peneplaned, with an elevation between 2000 to 2500 m (highest peak, 3440 m). The intermontane basins generally represent the plateau surface and are mostly distributed with denudational stony gobi and diluvial gravel gobi. Three large marginal basins are, however, occupied by extensive sandy deserts: the Badain Jaran, the Tengger, and the Ulan Buh. The Badain Jaran Desert is the second largest shifting sandy desert in China, with an area of about 40000 sq km of which shifting sand occupies about 83 per cent. Its highest sand dune reaches 420m, probably the highest in the world. The Shapotao area in the southeastern margin of the Tengger Desert is famous for its sand-control experimental works (Photo V-9).

Along the lower reaches of large rivers, there also lie ribbonlike fertile alluvial plains that have been utilized since very ancient times. For example, the lower reaches of the Black River and its delta around the Juyan Lake were actively cultivated as early as the second century B.C. and, during the ninth to eleventh centuries A.D., the once powerful Xia Kingdom had its capital at the present Kara Holshun (Black City) ruin. The lower reaches of the Shiyang River is also a prosperous oases, famous for its forest shelterbelts. The northern part of the Ulan Buh Desert, an ancient alluvial plain of the middle Huanghe River is now one of the most important reclamation areas in China.

On the whole, this natural subregion is unfavorable for agricultural development. It should be mainly reserved for pastoral land use. Some exceptions are the local fertile plains where regional planning and scientific farming should be encouraged. Mineral and salt exploitations also have productive prospects.

23(2) The Mazong Mountain

Geologically, the Mazong Mountain subregion (northwestern part of the Beishan Mountains) in a part of the ancient Alashan-Mazong Massif. It is composed of four NE-SW trending mountain ranges en echelon. Most of these ranges have an elevation between 1500 to 2000 m (highest peak, 2580 m), and the process of desiccation dominates. Barren denudational montane desert and stony gobi are widely distributed with darkish desert varnish well developed. It is called the "Black gobi", one of the most wild and barren areas in China. In the intermontane basins, however, natural conditions are somewhat better; depositional gobi are widely distributed, and the basins are covered with sparse vegetation, which can be used as pasture. Rich minerals are also exploited in this natural subregion.

23(3) Eastern and Middle Hexi Corridor

The world-famous Hexi Corridor (Gansu Corridor, Jade Corridor) is sandwiched between the Beishan Mountains (North Mountains), the southern marginal mountain range of the Mongolian Plateau and the Nanshan Mountains (South Mountains) or Qilian Mountains. The corridor plain, stretching nearly 1000 km from southeast to northwest, is blocked and interrupted by Yanzhi Mountain and Black Mountain (Photo V-10) and consequently, is divid-

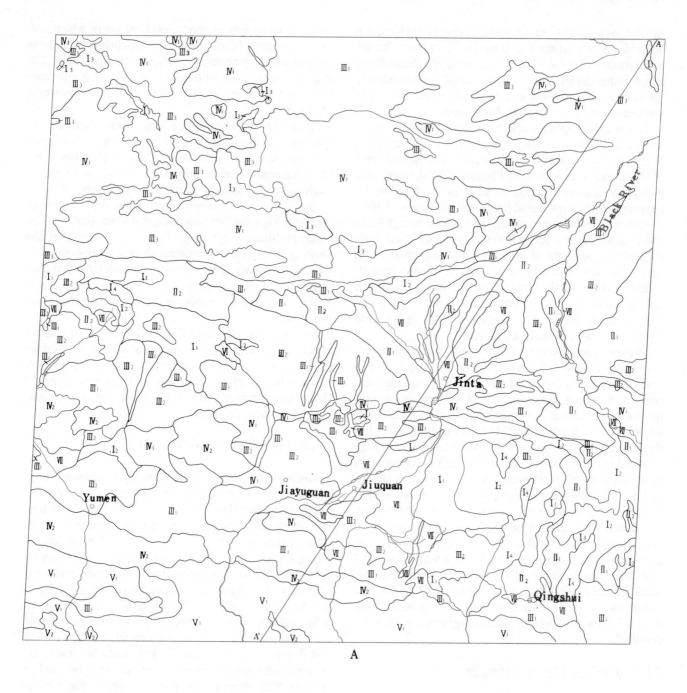

A

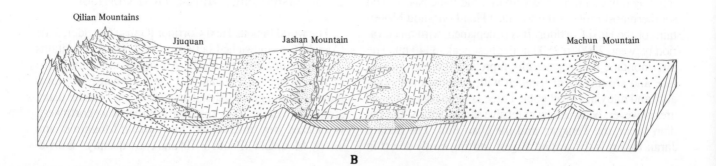

B

Figure 13-3 Land types in the Jiayuguan Area, Hexi Corridor

(A)

I. Clay and silt level terrain
   I (1) Slightly saline clay and silt level terrain; I (2) Saline clay and silt level terrain; I (3) Takyr; I (4) Salt marsh.
II. Sandy level terrain (sandy desert)
   II (1) Shifting sand; II (2) Half-fixed and fixed sand;
III. Stony, gravel level terrain (gobi)
   III (1) Alluvial-diluvial sandy gravel gobi; III (2) Diluvial gravel gobi; III (3) Denudational stony gobi.
IV. Denudational mountain and hill
   IV (1) Montane desert; IV (2) Montane grassland.
V. Erosional high mountain
   V (1) Montane coniferous forest; V (2) Alpine meadow.
VII. Oases

(B)

A-A' A comprehensive physical profile in the Jiayuguan area

ed into eastern, middle, and western sections. The first two sections belong to the Alashan Plateau temperate desert region, whereas the last section is a part of the Tarim Basin warm-temperate desert region.

The corridor plain is mainly composed of diluvial-alluvial deposits with an elevation of around 1500 m, and it slopes gently from the Qilian piedmont northward. A series of landscape types — diluvial gravel gobi, diluvial-alluvial sandy gravel gobi, shifting and fixed sand dunes, clay level land, and oases — appear in succession. Both land and water resources are quite rich. The Black River, Shiyang River, and numerous streams rush down from the Qilian Mountains, most of them percolate underground as soon as they leave the montane area. They reappear on the ground surface along the frontal margin of the diluvial-alluvial piedmont plain, where extensive irrigated oases together with settlements and highways (including the ancient Silk Road) are located. As rain-fall is scanty and decreasing from east to west (Wuwei: elevation 1525 m, annual precipitation 162 mm; Zhangye: elevation 1480 m, annual precipitation 125 mm; Jiuquan: elevation 1470 m, annual precipitation 86 mm), all farming depends on irrigation water, which is furnished by these surface and ground water resources. The rivers are fed by precipitation and snow-melting of the Qilian Mountains.

This natural subregion is probably the most favorable area of the entire desert division for agricultural development. As early as 1134 to 121 B.C., The Huns and other nomadic peoples prospered here. When the Chinese army took over in 121 B.C., the Huns bitterly complained: "To lose our Qilian Mountains means our animals will no more prosper; to lose our Yangtze Mountain means our women will lose their color"[1] Since then, the Chinese farmers have made great advances in transforming the corridor into blooming oases. It is referred to as the "Golden Zhangye,

Silvery Wuwei".

24. The Junggar Basin — Temperate Desert Region

The Junggar Basin is triangular-shaped with the Tianshan Mountains and the Altay Mountains as two limiting sides and the western side remaining relatively open. Its width on the line linking the cities of Ürümqi and Altay reaches 450 km, narrowing eastwardly to practically nothing on the Xinjiang-Gansu border.

Just like the Alashan Plateau, this is a temperate desert, with an aridity index greater than 4.0 and an accumulated temperature during the ≥ 10°C period of between 3200 to 4500°C. Yet, it differs by being rather open to the moist western winds through two gaps, the Ertix Valley and the Junggar Gate, consequently, it is less arid, with annual precipitation of 100 to 300 mm, decreasing from west to east. There is also a much heavier snowfall and snowcover during the winter season, which is quite helpful for the growth of denser vegetation. Geologically and geomorphologically, the Junggar Basin differs from the Alashan Plateau by being an ancient folded system and having a lower elevation (500 to 1000 m), decreasing from northeast to southwest. The lowest point is 189 m near the Ebi Lake. Therefore, in spite of higher latitudes and colder winters, it is favored with warmer summers than the Alashan Plateau and a frost-free season of about 160 days. In the Manas Valley even cotton can be grown. Another difference between this region and the Alashan Plateau is that it is an inland basin type, rather than a plateau type. The distribution of land types consists of depositional gobi around the marginal piedmont plains and the immense Gurbantünggüt Desert in the basin center.

Areal differentiation inside the Junggar Basin temperate desert region is highlighted by the divergence between the Junggar Basin proper with its western border, the Emin Valley, as well as with its eastern border, the Nomin Desert. Thus, three natural subregions can be identified.

24(1) The Junggar Basin Proper

The Junggar Basin proper has an area of about 180000 sq km, of which the stony desert (gobi) and the sandy desert (shamo) occupy 72000 and 60000 sq km respectively. It can be subdivided into two parts. The northern plain, between the Altay Mountains and the Gurbantünggüt Desert, is characterized by rather thin Cenozoic beds and widespread gobi and aeolian landforms. The southern plain, with very thick Cenozoic deposits, includes the extensive

---

1) In Chinese, *Yanzhi* means rouge. It was said that some shrubs grown in the Yanzhi Mountain could be used as rouge.

Gurbantünggüt Desert and the widely distributed gobi on the northern piedmont of the Tianshan Mountains. Between these sandy desert and gobi lies a relatively narrow belt of clayey and silty level land that corresponds roughly to the "spring line" and its neighboring oases of the Hexi Corridor. Most farmlands as well as settlements and highways in the Junggar Basin are concentrated along this fertile and well-watered belt.

Conditions in hydrography, soil and vegetation are also a little bit better here than in the Alashan Plateau. The Ertix River, with an annual runoff discharge of about 10 billion cu m, is the only river in China flowing into the Arctic Sea. All inland rivers together have about another 10 billion cu m annual runoff discharge. In the Gurbantünggüt Desert, the second largest sandy desert in China, shifting sand occupies only 3 percent of the total area, whereas the widely distributed fixed sand has a vegetation coverage of 30 to 50 percent and half-fixed sand of about 20 percent; both areas provide good winter pasture. The extensively distributed yermosols, if well irrigated, can be turned into good farmlands. It is estimated there are about 4 million ha of arable land not yet being used in the Junggar Basin.

On the whole, this natural subregion is mainly suited for pastoral development and has been used for this purpose by many nomadic peoples since very early times. One of the best pastoral systems is the practice of transhumance, with winter pasture located in the basin sandy plains and summer and autumn pasture located in the montane grasslands. Recently, petroleum and other minerals have been greatly developed; Karamay has become a significant oil city. Large-scale agricultural reclamation, which started only 200 years ago, has also been successfuly undertaken since 1950. One of the most outstanding examples is the middle and lower reaches of the Manas Valley which from the northern footslope of the Tianshan Mountains dips northward up to the southwestern margin of the Gurbantünggüt Desert. It was still a pastoral area until 1762, when a small part of the valley was cultivated. In 1955, it had a population of about 50000 and farmland totaling about 30000 ha. By 1977, these figures increased to a population of about 800000 and farmland totaling about 300000 ha (Landsat image 10).

24(2) Nomin Desert

The Nomin Desert subregion includes several small inland basins on either side of the Baytik Mountain (highest peak, 3479 m) along the Sino-Mongolian border. Here, the climate is extremely arid. The lowest point of this natural subregion — Nom Lake (elevation, 500 m) — has an annual precipitation of only 12.5 mm and an aridity index as high as 69.0. The landscape is dominated by denudational stony gobi and diluvial gobi, with glittering desert varnish and sparse vegetation coverage (less than 5 per cent). Under such adverse natural conditions, there are few inhabitants; they are devoted to extensive pastoralism.

24(3) Emin Valley

The Emin Valley is located along the Sino-Russian border, west of the western marginal mountains of the Junggar Basin. The crest lines of the mountain ranges have elevations between 2000 to 3000 m. The Emin Valley has its eastern margin at about 1000 m, then lowers to only 400m in the west near the city of Tacheng. The broad valley plain belongs climatically to the temperate desert-steppe zone, and it has an annual precipitation of about 300 mm and a frost-free season of about 125 days. It is well watered and luxuriantly covered with meadow and swamps, and it can be used for growing wheat and forage crops. On the montane slopes, it is mostly distributed with montane grassland — shrubby-steppe up to 2000 m, meadow steppe up to 2300 m, and alpine meadow above 2300 m. All of these make good pastures.

25. The Altay Mountains — Montane Grassland and Needle-Leaved Forest Region

The Altay Mountains[1], one of the greatest mountain systems in Eurasia, stretch from northwest to southeast more than 2000 km along the Sino-Russian-Mongolian borders. Only the southern slopes of the system's middle section lie inside Chinese territory, with a length of about 500 km and an elevation of more than 3000 m at the crest line. The highest peak, Mount Youyi (Friendship) (4374m) is located on the northwestern margin of the region and its southeastern tip merges into the vast level expanse of black gobi.

This is an ancient folded belt that appeared for the first time after the Caledonian Tectonic Movement. During the Himalayan Tectonic Movement, it was sharply uplifted and faulted, resulting in four prominent topographic steps from the Ertix Valley northeastward to the Sino-Mongolian border. Remnants of multicyclic erosional surfaces are still visible, for which elevations of 2900 to 3000, 2600 to 2700, 1800 to 2000, and 1400 to 1600 m are generally recognized. Therefore, vertical zonation of modern geomorphic processes is quite outstanding:

1) Belt of continual snow and modern glaciation: more than 3200 m in elevation. There are 459 glaciers on this belt, with a total area of 270 sq km.
2) Belt of nivation: 2400 to 3200 m in elevation. This is also a belt of ancient glaciation.
3) Belt of erosion: 1500 to 2400 m in elevation.

---

1) In Mongolian, altay means gold. During the Tang dynasty, the Chinese also called it the Gold Mountain, because of its rich gold mines.

4) Belt of desiccation (denudation); below 1500 m in elevation.

As both latitudinal and altitudinal locations are rather high, the area is climatically characterized by low temperature and comparatively plentiful precipitation. At Zhonghaizi (elevation, 2360 m), heavy frost might occur in early August. On windward (western) slopes of the middle mountains, annual precipitation totals more than 250 to 500 mm. Therefore, the area is the source of many rivers in the Junggar Basin, including the Ertix River and the Ulungur River.

Coinciding with vertical zonation of geomorphic and climatic conditions, vegetation, and soils are arranged in vertical belts that reflect the major areal differentiation in the Altay Mountains natural region. From footslope to mountain top they are as follows:

1) *Artemisia*-graminaceous desert-steppe: distributed from the montane foot up to about 800 to 1450 m in elevation, with its upper limit ascending from northwest to southeast. The dominant species are *Artemisia* spp. and *Stipa effusa*.

2) Shrub-steppe: upper limit ascending from 1200 m (northwest) to 1900 m (southeast). The most commonly seen species are *Spiraea hypericifolia, Artemisia frigida,* and *Festuca ovina.* Because coverage is usually larger than 50 per cent, it is an important spring and autumn pasture.

3) Montane needle-leaved forest: upper limit ascending from 2100 m (northwest) to 2300 m (southeast); vegetation coverage becoming sparser. The predominant tree species are *Larix sibirica* and *Abies sibirica,* with total wooded areas of nearly 1 million ha. This is one of the most important timbering areas in Northwest China.

4) Subalpine grassland and meadow: upper limit ranging from 2300 m (northwest) to 2600 m (southeast). The predominant grass species are *Festuca rubra, F. ovina, F. kirilovii, Poa altaica, P. sibirica,* and *P. alpina,* all yielding high-quality pasture. At the same time, vegetation coverage might be as high as 95 per cent, making this one of the best summer pastures in China, although it is not yet fully used.

5) Alpine meadow: upper limit ranging from 3000 to 3500 m. In northwestern Altay, it is alpine shrub-meadow, dominated by *Betula rotundifolia* and many kinds of lichens and mosses. In southeastern Altay, it is true alpine meadow, composed mainly of *Cobresia* spp, and *Carex spp.,* with vegetation coverage above 90 per cent.

6) Alpine cushion vegetation: distributed immediately below continual snow, composed mostly of lichens and mosses.

Using moisture as the chief diversifying criterion, two natural subregions might be identified in the Altay Mts:

25(1) Northwestern Altay Mountains

The higher elevation and prevalent moisture-bearing winds of the northwestern Altay Mountains result in much heavier precipitation. Hence, lower limits of different vertical vegetation zones are lower and the composition of flora much more complex. Biomass yield is also much higher; for example, timber storage of *Larix sibirica* reaches 500 to 600 cu m here, more than double that of southeastern Altay Mountains.

25(2) Southeastern Altay Mountains

With less precipitation and higher temperatures, the southeastern Altay Mountains are much more desiccated than the northwestern Altay Mountains. For example, the upper limit of Artemesia-graminaceous desert-steppe ascends to 1400 m. Montane needle-leaved forest is also much restricted in area, is sometimes isolated by large patches of grassland, and disappears entirely in the southeastern tip of the Altay Mountains.

26. Tianshan Mountains — Montane Grassland and Needle-Leaved Forest Region

The Tianshan Mountains, one of the greatest mountain systems in the world, stretch from west to east over a total distance of about 2500 km, with a width of from 100 to 400 km. The western section is distributed inside Russian territory; the middle and eastern sections lie in China's largest province (an autonomous region), Xinjiang, dividing it into two distinct natural regions: the Junggar Basin or Beijiang (North Domain) and the Tarim Basin or Nanjiang (South Domain). The part of the Tianshan Mountains inside Chinese territory is composed of more than 20 parallel ranges, interspersed with graben-structured and rhomboid-shaped intermontane basins over a total length of about 1,700 km and a total area of about 244,000 sq km. Their crestlines are usually uplifted above 3000 to 4000 m; the highest peak, the Mount Tomül, towers up to 7435.3 m, according to recent investigation (Photo I-5).

Geologically, this is an ancient folded belt, that was first uplifted during the late Caledonian Tectonic Movement. Through the Indo-China, Yinshan and Himalayan tectonic movements, the area was repeatedly folded, faulted, and uplifted on a large scale, resulting in the eventual formation of the modern, lofty Tianshan Mountains. Just as with the Altay Mountains vertical zonation is quite conspicuous, and there are four belts of modern geomorphic process that appear in succession:

1) Belt of continual snow and modern glaciation: more than 3800 to 4200 m in elevation. This is the largest

modern glacier-distributed area in China; there exist 6896 glaciers, with a total area of 9500 sq km. The terminals of the glaciers usually end at elevations between 3000 to 4000 m, sometimes even moving down to 2500 to 2800 m.

2) Belt of nivation and ancient glaciation; located between 3000 to 3800 m in elevation. Ancient glaciation existed during the Pleistocene Epoch, but has retreated since the Holocene. Nivation is now the dominant geomorphic process, with seasonal snowcover lasting more than 10 months.

3) Belt of fluvial erosion: located at elevations between 1000 to 3000 m this is the vertical zone of heaviest precipitation, with annual precipitation generally more than 500 mm, even more than 1000 mm in the middle Tianshan Mountains. Fluvial erosion ranks first, resulting in very rugged topography. Four or five steps of terraces are well developed, on these, extensive farmlands and pastures as well as important highways are located.

4) Belt of desiccation (denudation): located generally below 1000 m with annual precipitation less than 100 to 200 mm.

Because the Tianshan Mountains are highly uplifted and facing the moisture-bearing westerlies, precipitation is relatively abundant with an average annual precipitation of about 500 mm rising to 1130 mm on the rainiest windward slope. The area appears like a huge elongated "wet island" amid the vast expanse of extremely arid deserts. Precipitation decreases from west (about 800 mm) to east (about 100 to 200 mm). There are also distinct variations between windward and leeward slopes as well as between different elevations.

This gigantic "wet island" together with its relatively dense vegetation cover is a gigantic natural water reservoir from which more than 200 rivers originate. The Ili River (with the largest annual runoff discharge) and the Tarim River (with the largest drainage basin and with its discharge coming chiefly from the Tianshan Mountains originated Aksu River) are the two largest inland rivers in China. There are also innumerable karez (underground irrigation canals), which are fed by surface water in the Tianshan area. According to recent studies, the annual runoff discharge of the Tianshan Mountains' rivers totals 43.9 billion cu m, accounting for 56 percent of the total annual runoff discharge of all of Xinjiang's rivers. Besides, the total water reserve of modern glaciers on the Tianshan Mountains (not including continual snow) is estimated at 360 billion cu m — a huge "solid reservoir".

Distribution of vegetation again shows conspicuous vertical zonation. Six major vertical zones are recognized. The upper limit of each vertical zone varies somewhat according to the direction of the slope as well as longitudinal and latitudinal locations.

1) Temperate montane desert: below 1100 m in elevation; mainly sparse *Artemisia* spp.

2) Montane steppe: located between 1100 to 2000 m, mainly *Artemisia terraealbae*, *Festuca ovina*, and *Stipa* spp.

3) Montane forest-steppe: located around 2000 m in elevation; small patches of *Picea* forest on shady slopes, and grassland of *Poa* spp. and *Artemisia* spp on sunny slopes. This is a good summer pasture area.

4) Subalpine needle-leaved forest: located between 2000 to 2700 m; large patches of *Picea* forest on shady slopes and luxuriant grassland on sunny slopes. The former is good timbering ground, whereas the latter is an excellent summer pasture.

5) Alpine meadow: located between 2700 to 3700 m; dominated by *Carex* spp -*Alchemilla vulgaris* formation, with *Bronmus inermis* and *Poa pratensis* also commonly seen. This is an excellent summer and autumn pasture, frequented by the *Kazakh* nomads who practice transhumance.

6) Alpine cushion vegetation: located between 3700 to 3900 m (snow line, about 3900 m); dominated by lichens and famous for the beautiful snowlotus that blooms in late summer.

Three natural subregions can be identified in the horizontal areal differentiation:

26(1) Middle Tianshan Mountains

The Middle Tianshan Mountains including most parts of the Tiansha Mountains inside Chinese territory. Hence, the above description provides a sketch of this natural subregion.

26(2) Eastern Tianshan Mountains

The Eastern Tianshan Mountains are restricted to a small section of the Tianshan Mountains east of Bogda Mountain, with a width of about 100 km and a crestline elevation below 4000 m. Precipitation decreases gradually from west to east and upper limits of vertical vegetation zones correspondingly ascend higher. For example, the upper limit of *Picea* forest lies at 1700 m on the Bogda Mountain (about 89°E), ascends to 2350 m on the northern slopes of the Barku Mountain (about 91°E), and disappears entirely near Yiwu (about 95°E).

26(3) The Ili Valley

The Ili Valley consists of the upper reaches of the Ili River, which empties into Balkhash Lake inside Russian territory. The broad valley plain consists of several fanlike terraces, composed mainly of quaternary gravels and loessic materials. It is surrounded by mountain ranges, except on the western side, where warm and moist westerlies rush in. Therefore, its climate is less arid than that of the Junggar Basin, warmer in winter, yet cooler in summer (owing to higher elevation). The frost-free season lasts 160 to 210 days and accumulated temperature during the $\geq 10°C$ period is about 3,000°C. Annual precipitation ranges from 280 to 470 mm, decreasing from west to east. Spring and summer are rainy seasons, accounting for about four fifths of the total annual precipitation, which makes dry farming possible and popular. Because the Ili River has an annual runoff discharge of 12.3 billion cu m and is fed by both rainfall and melting snow, there is no lack of irrigation water if it is needed.

On the whole, this natural subregion lies in the desert-steppe zone. The broad, fertile valley plain is very promising for agricultural development. On the surrounding montane slopes, luxuriant montane steppe, montane forest-steppe and alpine meadows are extensively distributed. All of them are excellent pasture and have been famous for their horses ever since the second century B.C.

27. The Tarim Basin — Warm-temperate Desert Region

This is a gigantic, rhomboid inland basin, surrounded by the Tianshan, Beishan, Karakorum, Kunlun, Altun, and other lofty mountains, with a maximum length of nearly 2000 km (including the western Hexi Corridor) and a maximum width of 520 km. This is also the innermost core of China's desert landscape, with annual precipitation of less than 90 mm over all the basin and even less than 10 mm in its central part. Again, owing to its comparatively lower latitudes and altitudes as well as its being closely surrounded by lofty mountains, it is unique as it is the only warm-temperate desert in China. Three crops every two years may be grown if irrigation water is adequate.

This is an ancient massif, encircled and demarcated by many deep faults. According to a recent geomagnetic survey, the ancient massifbase is buried to a depth of 8 to 12 km in its central part; in its marginal downwarping area, terrestrial deposits of Mesozoic and Cenozoic sediments have a maximum thickness of nearly 10 km. Most of them are good ground-water-bearing beds and probably are also excellent petroleum-bearing structures.

Three natural subregions are identified in this natural region:

27(1) The Tarim Basin Proper

The Tarim Basin proper is a nearly all mountain-locked inland basin, with only a narrow corridor plain between the Altun Mountain and the Kuruktag Mountain (about 70 km wide), which merges eastward into the western Hexi Corridor. It has an area of nearly 500000 sq, km, of which sandy deserts occupy about 331,000 sq km and gobi about 105000 sq km. Topography dips generally from west to east as well as somewhat from south to north. The Tarim River is a good indicator of the general topography; its channel has an elevation of about 1300 m in the southwestern Tarim Basin, and it then flows northeastwardly up to its terminal lake, Lop Lake, which is 780 m in elevation, the lowest point in the Tarim Basin.

All rivers originate in the surrounding high mountains, with a total annual discharge of about 37 billion cu m. If there were enough rainfall and runoff, all these rivers might be integrated into one single fluvial system — the Tarim-Konqi-Qargan river system. However, at the present, all rivers originating from the Kunlun Mountains (except two headwaters of the Tarim River — the Yarkant River and the Hotan River) die out amid immense stony and sandy deserts. Even the Tarim River itself is chiefly fed by the Aksu River, which originates in the Tianshan Mountains. The Tarim River flows at last into its terminal lakes of Lop Lake, Karakoshun, and Taitema, known as the "wandering lakes". These lakes are all located in a wide, shallow depression that originated in neotectonic movement during the Quaternary period. They have migrated or "wandered" between 39 to 41° N and 88 to 91° E ever since their formation. In the beginning, the Tarim-Konqi-Qargan river system emptied into the Lop Lake. In the fourth century A.D., however, with the opening of the southern channel of the river system, Karakoshun and Taitema lakes were formed, and the northern channel together with Lop Lake entirely dried out. However, in 1921, the main channel of the Tarim-Konqi-Qargan river system migrated to its northern branch again, resulting in the disappearance of the Karakoshun and Taitema lakes and the reappearance of Lop Lake. In 1952, for the sake of diverting riverine water for irrigation as well as for flood control, the Tarim and Konqi rivers were finally artificially separated, flowing into Taitema and Lop Lake respectively. Thus, these "wandering lakes" have been changing their location and dimensions accordingly, that is, as the Tarim-Konqi-Qargan river system has changed its main channel and discharge. The disappearance or dwindling of one lake will lead to the emergence or enlargment of the other lakes. Yet, under natural conditions, these terminal lakes have never dried out at the same time. It was only after 1972, when the lower Tarim-Konqi-Qargan channel entirely dried out that the Taitma Lake was reduced to a vast expanse of salty marsh. Simultaneously Lop Lake became the empty "Great Ear" as viewed from Landsat[1] (Landsat image 8).

1) National Aeronautic and Space Administration, "Mission to Earth: Landsat Views the World", Plate 309, Washington D. C., 1976.

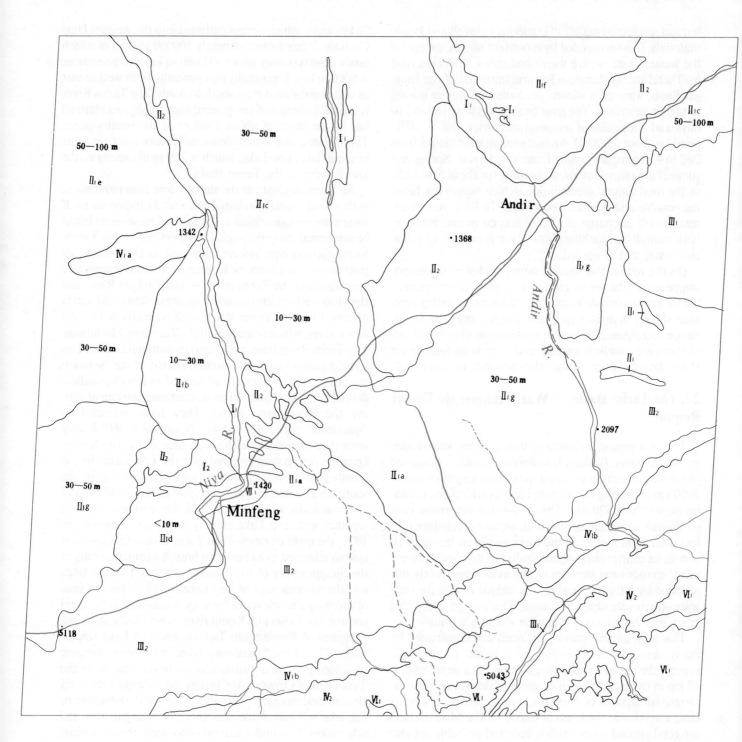

Figure 13-4 Land types of the Minfeng Area, Tarim Basin

I. Clay and silt level land
  I (1) Flood plain (slightly saline or nonsaline); I (2) Old flood plain (saline).

II. Sandy level land (sandy desert, shamo)
  II (1) Shifting sand
    Traverse types: a. barchan dune and barchan Chain; b. composite dune chain; c. Chainlike sandy mount.
    Longitudinal types: d. barchan longitudinal dune; e. composite longitudinal dune; f. composite longitudinal sandy   mount.
    Towering types: g. Pyramid dune
III. Stony, gravel level land (gobi)
  III (1) Alluvial-diluvial sandy gravel gobi; III (2) Diluvial gravel gobi.
IV. Denudational mountain and hill
  IV (1) Montane desert; IV (2) Montane grassland
VI. Nival alpine
  VI (1) Periglacial cushion vegetation
VII. Oases
  VII (1) Irrigated oases

  Figure 13-4 shows the distribution of a series of land types in the Tarim Basin. Immediately below the surrounding mountains diluvial gravel gobi (Photo V-5) and alluvial-diluvial sandy gravel gobi lie on the piedmont plain. Then, there appear the narrow belts of fertile, well-watered, clayey

and silty level lands and oases. Next is the immense Taklimakan Desert, 85 per cent of which consists of shifting sand dunes with narrow ribbons of fixed and half-fixed sand dunes on its margins as well as some corridors of oases that penetrate deeply along large rivers. Finally, there is salty marsh and a eolian yardang (Photo V-6) in the lowest part of the basin.

  From the human point of view, the most important land types are certainly oases and their neighboring arable clayey and silty level lands. According to their reclamation oases in the Tarim Basin might be classified into three categories (Figure 13-5): (a) ancient oasis ruins, such as Loulan and Milan now mostly buried deep inside the desert; (b) old oases, developed before 1949, now still under cultivation and mostly distributed along the Tarim River and its distributaries; (c) new oases, reclaimed since 1949, mostly located in the middle and lower reaches of the Tarim River. In the so called Tarim Valley, more than 140000 ha of farmland have been reclaimed since 1955. For the sake of further agricultural development, great care must be taken in control of salinization and shifting sand as well as in the improved use of irrigation water.

  Another outstanding landscape is the Taklimakan Desert. Taklimakan is an Uigur word whose meaming is "once you get in, you can never get out again". It is the largest sandy desert in China and probably the second

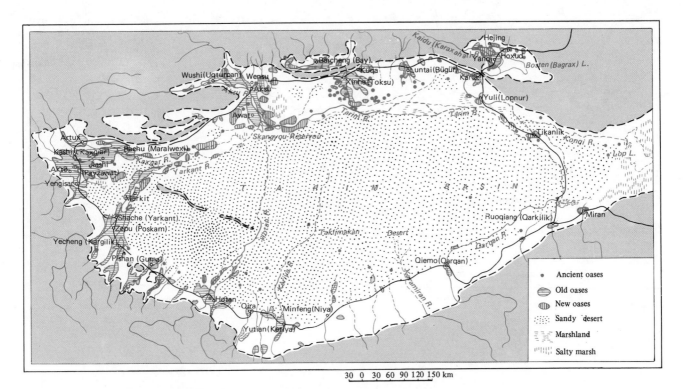

Figure 13-5 Distribution of ancient, old and new oases in the Tarim Basin

Figure 13-6 Land types of the Turpan Basin and its Neighboring Areas

I. Clay and silt level terrain
  I (1) Nonsaline or slightly saline clay and silt level terrain.
I(2) Saline clay and silt level terrain; I (4) Salt marsh.
    a. old; b. new.
  I (5) Yardang
    a. strongly developed; b. Weakly developed.
II. Sandy level terrain (sandy desert) II (1) Shifting sand (longitudinal chain and barchan); II (2) Half-fixed and fixed sand
III. Stony, gravel level terrain (gobi) III (1) Alluvial-diluvial sandy gravel gobi
    a. with overlying loessial material; b. with overlying clay sandy material.
  III (2) Diluvial gravel gobi
    a. old; b. new.
  III (3) Denudational stony gobi
IV. Denundational mountain and hill
  IV (1) Montane desert.
  IV (2) Montane grassland.
V. Erosional high mountain
  V (1) Montane coniferous forest.
  V (2) Alpine meadow.
VI. Nival alpine
  VI (2) Eternal snow
VII. Oases
  VII (2) Irrigated oases

largest shifting sandy desert in the world. It is characterized by a vast expanse of yellowish, shifting sand dunes, mostly 100 to 200 m high (Photo V-4). It enjoys quite variated dune-types (Figure 13-4) and is quite divergent in different parts. Generally speaking, northeasterlies prevail east of the Hotan River and northwesterlies dominate to the west of it. Consequently, shifting sand generally moves southward, on the northern piedmont of the Kunlun Mountains, many ancient oasis ruins and abandoned tracks of the "Silk Road" have been deeply buried in the Taklimakan Desert for the last 1000 to 2000 years.

27(2) The Turpan-Hami Intermontane Basins

The Turpan-Hami intermontane basins area, located between the Alashan Plateau and Tarim Basin and sandwiched between the Tian Mountain and Kuruktag Mountain, is an area of extremes. Toksun in the Turpan Basin has the lowest annual precipitation record in China — only 3.9 mm. Turpan city has the highest maximum temperature record in China — 47.6°C (an unofficial record is 48.9°C), and was famous as the Fire Prefecture more than 600 years ago. It is also famous as a wind reservoir; a maximum wind velocity of 50m/sec was once recorded. Immediately below snow-capped Mount Bogda (5445 m) lies the Ayding Lake in the central part of Turpan Basin. Ayding Lake with an elevation of 155 m below sea level, is the lowest point in China. As denudational stony gobi

and diluvial gravel gobi are extensively distributed both in the Turpan Basin and the Hami Basin and a vast expanse of black gobi dominates the Gaxun Desert, this is probably the most concentrated area of gobi in the world (Figure 13-6, Landsat image 9).

Since very ancient times, this has been a great political and economic center in Northwest China. There are many famous ruins in this district, including several capitals of ancient kingdoms. The Uigur inhabitants are particularly famous for their karez irrigation works. They laboriously build underground irrigation canals, several or even several dozens of kilometers long, which lead and conduct precious surface water from the far away Tianshan Mountains to irrigate and transform the wild stony desert into a green garden of grapes, melons and other crops. Turpan grapes and Hami melons together with Uigur hospitality, are famous throughout China. According to a recent statistic there are 1158 karez canals in the Turpan Basin alone, with a total length of more than 3000 km and a total annual discharge of 777 million cu. m.

27(3) Western Hexi Corridor

The western section of the Hexi Corridor is quite similar to its middle and eastern sections, except that it is somewhat lower in elevation (generally 1000 to 1200 m) and farther away from the sea. Consequently, it is drier (annual precipitation, 40 to 50 mm) and warmer (frostfree season of about 200 days and accumulated temperature during the $\geq 10°C$ period of 3300 to 4000°C). It is a warm-temperate desert in which longfiber cotton can be grown. In spite of surface and ground water resources that are much restricted and wild gobi and yardang that are widely distributed, the western Hexi Corridor has been intensively cultivated for more than 2000 years and has a fertile agricultural soil (although rather restricted in area) of more than 2 m in depth. Here is the location of the westernmost section of the Great Wall and the starting point of the Silk Road which crosses the immense Taklimakan Desert. This subregion is imbued with scenic beauty and ancient cultural splendors. These include the Thousand Buddhas' Cave, Sun Gate, the Jade Gate, the Singing Sand Mount, the New Moon Spring, etc. No doubt this is an area of great interest to the curious tourist.

References

[1] Lanzhou Institute of Desert Research, Chinese Academy of Sciences, 1980, *An Outline of Chinese Deserts,* Science Press, Beijing. (In Chinese).

[2] ———, 1977, *China Tames Her Deserts, A Photographic Record,* Science Press, Beijing. (In Chinese and English)

[3] Zhao Songqiao, 1962, "A Preliminary Discussion on Types of the Gobi in Northwestern Hexi Corridor", *Research in Desert Transformation,* No. 3. (In Chinese. English Translation: JPRS 19,993, OTS. 63—31183)

[4] ————, 1963, "An Integrated Evaluation on Agricultural Natural Resources and Agricultural Natural Conditions in the Ulan-buh Desert", Selected Papers of 1963 Annual Meeting of Chinese Geographical Society. (In Chinese)

[5] ————, 1981, "The Sandy Deserts and the Gobi: A Preliminary Study of Their Origin and Evolution", *Desert Lands of China,* ICASALS Publ.,   No. 81-1, Texas Tech University, Lubbock, Tex.

[6] ————, 1981. "Desertification and De-desertification in China", The Threatened Drylands. Working Group of Desertification in and around Arid   Lands, IGU.

[7] ————, 1983, "Origin and Evolution of the Lop Desert and the Lop Lake" *Geographical Research,* Vol. 2, No. 2. (In Chinese, with English abstract)

[8] Zhao Songqiao & Han Chin, 1981, "Large-scale Agricultural Reclamation in the Tarim Valley and its Impact on Arid Environ-ment", *Desert Lands of China,* ICASALS Publ., No. 81-1, Texas Tech University, Lubbock, Tex. (In English, German translation: Geographische Rundschau, März 3-81)

[9] Integrated Investigation Team of Xinjiang, Chinese Academy of Sciences, 1965, *Soil Geography of Xinjiang,* Science Press, Beijing. (In Chinese)

[10] ————, 1965, *Ground Water of Xinjiang,* Science Press, Beijing. (In Chinese)

[11] ————, 1966, *Surface Water of Xinjiang,* Science Press, Beijing. (In Chinese)

[12] ————, 1978, *Geomorphology of Xinjiang,* Science Press, Beijing. (In Chinese)

[13] M. P. Petrov, 1966, *Deserts of Central Asia,* Moscow. (In Russian)

[14] Sven Hedin, et al., 1928 – 1935, *The Sino-Swedish Expeditions Reports.* Stockholm.

Chapter 14

The Qinghai-Xizang Plateau

The Qinghai-Xizang Plateau, or the Tibetan Plateau, with an average elevation of more than 4000 m asl, is located in southwestern China. It is surrounded by a series of lofty mountains, such the Kunlun, the Altun, and the Qilian Mountains on the north, the Hengduan Mountains to the southeast, the Himalayas to the south, and, the Karakorum Mountains on the west (Figure 14-1). Administratively, it includes all of the Xizang Autonomous Region and Qinghai Province as well as parts of the Xinjiang Uigur Autonomous Region and Gansu, Sichuan, and Yunnan provinces. It has an area of 2.5 million sq km, occupying about one fourth of total land area of China. Nevertheless, it contains only 0.8 per cent of China's total population and 0.8 per cent of China's total cropland.

The recency of the Himalayan Tectonic movement, the violence of recent uplift and the accompanying thermodynamic effect, and the vastness of the land area and the low latitudinal location all combine to make the physical geographical environment complicated and unique. This is the youngest, largest, and highest plateau in the world, with many other extremes, such as the highest cropland and upper forest limit in the world and the largest lake in China. It also exhibits the closest interrelation of horizontal and vertical zonation. Sometimes it is called the "third pole" of the world. It differs distinctly from the other natural regions in China as well as from any other high mountains and plateaus in the world. Yet it exerts very great influence over surrounding regions. Many geographers and other scientists have expressed close interest in this natural realm, and many multidisciplinary scientific investigations have been conducted here since 1949. Nevertheless, scientific information about this area is still far from adequate; we can present only a broad

outline in this textbook. In terms of comprehensive physical regionalization, we identify only one natural division under the natural realm — Qinghai-Xizang Frigid Plateau — although it is subdivided into six natural regions. No subregions are demarcated under these six natural regions.

CHIEF FEATURES OF THE PHYSICAL GEOGRAPHICAL ENVIRONMENT

In chapters 1 and 13, we discussed briefly the origin and evolution of the Qinghai-Xizang Plateau and its huge impact on surrounding regions. We need only repeat here that since the late Miocene Epoch the Qinghai-Xizang Plateau has undergone tremendous changes in terms of the physical geographical environment. The major factor determining such tremendous changes, besides the worldwide Quaternary glaciation and its impact on global climatic fluctuation, is undoubtedly the violent uplift of the area from near sea level to an average of more than 4000 m in elevation. The modern physical geographical environment of the Qinghai-Xizang Plateau is characterized by the features described below.

A Frigid Climate

High elevation results in low temperature, with a decreasing rate of about 0.6°C for every increase of 100 m in elevation. Everywhere in the Qinghai-Xizang Plateau (excluding the southern slopes of the Himalayas, which in reality are not a part of the Qinghai-Xizang Plateau) has an elevation above 2500 to 3000 m and in the northwestern

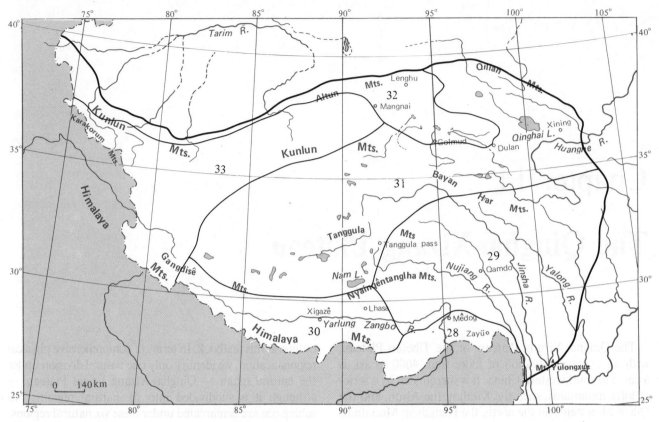

Figure 14-1 A location map of the Qinghai-Xizang Plateau

Qingzang Plateau above 4500 to 5000 m, to say nothing of numerous extremely high mountains towering above 5000 to 6000 m or even above 7000 to 8000 m. Therefore, in spite of rather low latitudes and very rich solar radiation (140 – 190 kcal/cm²/year, the highest in China), the mean January temperature ranges from − 10 to − 15°C and the mean July temperature (the lowest in China) roughly equals the mean January temperature in South China. As a rule, there is no "summer" over the Qinghai-Xizang Plateau, no frost-free season in areas above 4000 m (which is the average elevation of the Tibetan Plateau), and all extremely high mountains are generally wrapped in glaciers and continual snow. When Russian advisers for the Physical Regionalization of China Working Committee of Chinese Academy of Sciences (1953 – 1959) insisted that, owing to its latitudinal or horizontal location, the Qinghai-Xizang Plateau should be classified as subtropical, their suggestion was, not surprisingly, nearly unanimously rejected. The Tibetan Frigid Plateau has been affirmed as one of three natural realms in China ever since.

Another climatic feature is the great diurnal and seasonal variation of temperature. Great thermal differences exist between winter and "summer", with diurnal variation being even greater. During daylight, there are great thermal differences between sunny and shady sites. During summer, Tibetan farmers are often seen working on their farmlands with bare arms under the brilliant sunshine, only to rest at noon in the shade wearing heavy fur coats.

A Nivation Process that Dominates, with Widely Distributed Glaciers and Permafrost

High elevation and a frigid climate together with relics of Quaternary glaciation have led to widely distributed glaciers and permafrost. Existing glaciers on the Qinghai-Xizang Plateau have a total area of about 47,000 sq km, occupying more than four fifths of all the glacial area in China. There are two modern glacier-distribution centers on the Qinghai-Xizang Plateau: the temperate maritime type, located in the southeastern Qinghai-Xizang Plateau, and the cold continental type, distributed in the northwestern Qinghai-Xizang Plateau. There are also numerous isolated alpine valley glaciers on the Himalayas, Qilian Mountains, and other mountainous areas.

Permafrost is well developed in the central and northwestern Qinghai-Xizang Plateau. On the highly uplifted Qingzang Plateau, permafrost with a total thickness of 80 to 90 m and a seasonal active layer of 1 to 4 m. occurs in a vast continuous expanse, forming a huge "permafrost island" within the low latitudinal permafrost-free belt.

Thus, the nivation process dominates on the Qinghai-Xizang Plateau. Great diurnal and seasonal thermal variations make freezing and thawing action occur frequently and extensively. Mechanical weathering plays a heavy role in denudation. Different types of periglacial landforms are also commonly seen.

A Young Stage of Geomorphologic and Soil Development

Recency of the tectonic movement and the accompanying violent uplift also bear importantly on modern physical geographical processes, particularly modern geomorphic processes and soil formation.

The ancient, low, and undulating late Tertiary erosional surface, which was once widely distributed in Eurasia and other continents, has now been uplifted to 4500 to 5000 m asl and has become the high plateau surface of the modern Qinghai-Xizang Plateau. Consequently, the geomorphic process and landforms have been repeatedly rejuvenated; in fact, this is the youngest and largest rejuvenated geomorphic realm in the world. In its southeastern parts, different kinds of rejuvenated landforms occur. On nearly all riparian longitudinal profiles there appear three conspicuous knickpoints, which represent three cyclic stages of rejuvenation. In the middle and lower reaches of many large rivers, the valley-within-a-valley landform is commonly seen. Owing to prominent areal differentiation in moisture condition from southeast to northwest and, consequently, great divergence in the intensity of erosion, the deeply dissected southeastern parts of the Qinghai-Xizang Plateau where the fluvial erosional process dominating contrast strongly with the central and northwestern parts. In these latter areas, an undulating and undissected high plateau surface prevails under frigid and arid conditions, and the drainage system turns from oceanic to interior.

The soil-forming process is also characterized by its youngness with soil profiles slightly developed and soil minerals weakly weathered. Very coarse soil texture and very strong freezing-thawing action are other outstanding features. All these are unfavorable conditions for agricultural production.

A Special Floral and Faunal Distribution

Both flora and fauna in the Qinghai-Xizang Plateau are represented by two distinct systems. In the frigid and arid central and northwestern parts, fauna pertain to the paleoarctic realm, flora to the Tibetan Plateau flora subregion of the panarctic region. No more forests exist; shrubs are dwarfed. In the warmer and more humid southeastern parts, fauna pertain to the oriental realm and flora to the Chinese Himalayan forest subregion of the panarctic region. Vegetation becomes much more luxuriant, and the needle-leaved forests are distributed in large stands.

The surrounding lofty mountains, according to their orientation, geographical location, and other features, play quite an important role. The extremely high Himalayas with their westeastward orientation and southern marginal location, serve as an effective barrier between the Qinghai-Xizang Plateau and the Indian subcontinent. They bar practically all tropical and subtropical faunal and floral elements. The latitudinal, high Kunlun-Altun-Qilian mountain system is also a barrier between the Qinghai-Xizang Plateau and Northwest China, yet owing to somewhat lower relative relief and numerous mountain passes this barrier is not highly effective. There are many common species of fauna and flora appearing in both the Qinghai-Xizang Plateau and Northwest China; the Qaidam Basin in the northern marginal Qinghai-Xizang Frigid Plateau realm might also be considered a part of the northwestern Arid China Realm. The Hengduan Mountains in the Southeastern marginal Qinghai-Xizang Plateau, consisting of a series of parallel longitudinal high mountain ranges and deep river gorges, serve as a highway along larger river channels and facilitate the migrating and intermingling among different species of fauna and flora. Furthermore, owing to the Qinghai-Xizang Plateau's great relative relief and consequently very conspicuous vertical zonation as well as its having been an excellent asylum for fauna and flora during Quaternary glaciation, many pre-Tertiary relic species survive in the area. This is also the distributing center for many modern species, such as *Rhododendron* spp. in flora and *Carruian* spp. in fauna. No wonder this area is sometimes called as the richest natural museum in the world.

AREAL DIFFERENTIATION AND NATURAL REGIONS

As stated above, both horizontal and vertical zonation are conspicuous in the Qinghai-Xizang Plateau, and these two factors are closely interrelated.

The horizontal differentiation is chiefly characterized by moisture divergence that, in turn, is chiefly caused by the combined effect of mountain-barrier action and atmospheric circulation. As mountain ranges on the Tibetan Plateau follow mostly a west-eastward direction and serve as a series of barriers against moisture-bearing maritime monsoons, annual precipitation decreases successively from more than 1000 mm in the southeast to less than 100 mm in the northwest. correspondingly, aridity increases from

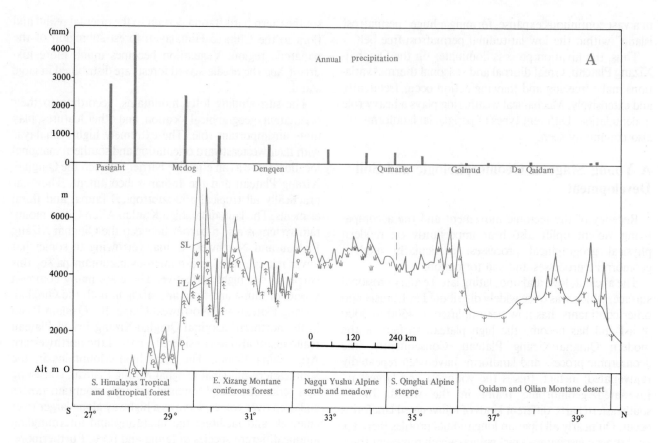

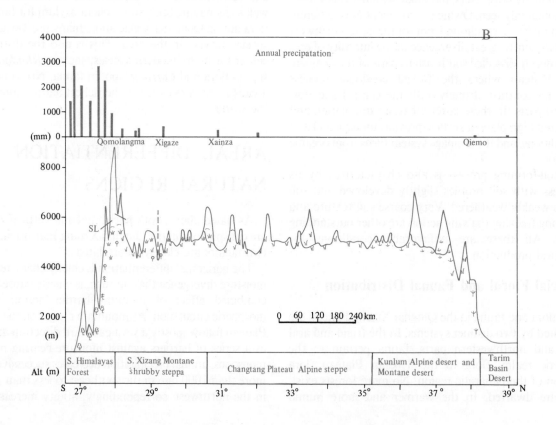

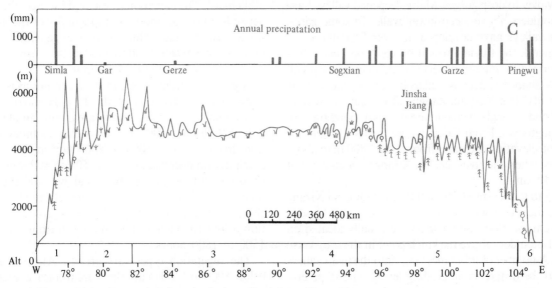

Figure 14-2 Comprehensive physical profiles of the Qinghai-Xizang Plateau

(A) Along 95°E; (B) Along 87°E; (C) Along 32°N.

humid to subhumid and from semiarid to extremely arid. Figure 14-2 shows elevation and annual precipitation along three physicogeographical profiles (95°E, 87°E, and 32°N).

The Qinghai-Xizang Plateau stretches more than 12° of latitude from south to north; hence, latitudinal divergence of temperature exists as well. For example, Bamda (30°14'N; elevation, 4115 m) has a mean coldest month temperature of −8.4°C and a mean warmest month temperature of 11.1°C; Mori (38°15' N; 4091 m), at about the same elevation, records −16.9°C and 5.6°C respectively. The upper limit of the forest belt and the elevation of the modern snow line on northern slopes of the middle Qilian Mountains are both lower by 600 to 1000 m than those on the southern slopes of the middle Himalaya Mountains. However, such latitudinal horizontal zonation on the Qinghai-Xizang Plateau is usually overshadowed by much stronger vertical zonation.

The vertical zonation is certainly the most significant areal differentiation on the Qinghai-Xizang Plateau, although it is also stamped with features of horizontal zonation. Two systems of altitudinal zones can be identified: the maritime (humid and subhumid) and the continental (semiarid and arid). The former occurs mainly in the southeastern Qinghai-Xizang Plateau and features intense fluvial erosion, strong biochemical weathering, acid soils, and mesophytic types of vegetation. The spectrum of altitudinal zones is composed chiefly of a montane needle-leaved forest belt, with alpine scrub and meadow belts distributed above it. By contrast, the continental system appears extensively in the central and northwestern Qinghai-Xizang Plateau, and dominated by desiccation and

nivation processes, intense mechanical weathering, alkaline soils, and xerophytic types of vegetation. The spectrum of altitudinal belts is composed mainly of montane steppe and montane desert, with alpine meadow on higher slopes.

By integration of all these vertical and horizontal areal differentiations, six natural regions can be identified in the Qinghai-Xizang Plateau. They are briefly outlined as follows.

28. Southern Himalayan Slopes — Tropic and Subtropic Montane Forest Region

The southern Himalayan slopes region includes the southern slopes of the East Himalayas and the Kangrigarbo Mountain with deep gorges and high ridges. It comprises the southern flankes of the Qinghai-Xizang Plateau, and extends from east to west.

The Himalayan ranges have crest lines of more than 6000 to 7000 m asl., with numerous summits over 7000 to 8000 m. The main range of the East Himalayas is remarkably lower — about 7756 m in Namjagbarwa. To the east, there are a number of glaciers of the oceanic type. On the Kangrigarbo Mountain, there is the famous Abgyag glacier. It runs through the montane needle-leaved forest belt and ends with a glacial tongue in the montane evergreen broad-leaved forest belt at about 2500 m asl.

Erosion and deposition landforms associated with glaciation are found mainly in the valleys below the snow line. In the upper reaches of the rivers and their tributaries, U-shaped glaciated valleys are well developed. As a result of recent river sculpture, the valley form changes from wide trough valley to deep gorges. Many deepened valleys have level shoulders above steep lower walls. In some cases, hanging valleys have developed in several tributaries.

In general, the fluvial erosion process dominates. It is characterized by intense dissection and deep gorges with typical V-shaped valleys as well as by the steepness of the valley walls. The Yarlung Zangbo River cuts its way in a great bend through the mountain ranges a round the foot of Namjagbarwa. From Paiqu, where the river enters the gorge, the elevation decreases from about 2800 m to about 600 m asl when the river reaches Xirang — a distance of about 240 km.

Owing to the recent violent uplift of the Qinghai-Xizang Plateau, the valley-within-a-valley landform may be seen at different locations. Settlements are usually located on the level shoulders above the knick point in the transverse profiles. A number of tributaries "fall" into the main valleys owing to the difference in eroding levels. The slope erosion is intense in the rainy season; hillcreeps and mud flows occur frequently.

The flat-topped crests, usually developed at lower altitudes, form the foothills of the southern slopes of the East Himalayas. The alluvial-diluvial platforms, terraces, and fans are also developed in the broad valley section of the rivers, especially those of the Zayü River.

The lofty Himalayas, extending along the southern rim of the Tibetan Plateau, are an effective climatic barrier. They stop the northern cold air masses from invading southward. The strong, moisture-laiden southern monsoons bring abundant precipitation to the southern slopes of the Himalayas. The climate is warm and humid which differs entirely from the dryness and cold on the Qinghai-Xizang Plateau proper.

In most of the valleys and hills with an elevation below 2500 m asl the mean temperature of the warmest month varies from 18 to 25°C and that of the coldest month from 2 to 16°C. There is a permanent frost-free season below 1000 to 1200 m asl. The temperature decreases with increases in elevation, resulting in conspicuous vertical differences in climate. Above the upper forest limit, there are the alpine scrub and meadow belts followed by subnival and nival belts.

The southern monsoons from the Indian Ocean find their way up the gorges of the Yarlung Zangbo and other rivers. The region. especially at the foot of hills, is distinguished by high humidity throughout the year and by a lack of seasonal contrast in precipitation and temperature. The monsoon lasts from May to October. During March and April, the rains are not so regular, with copious orographic precipitation. In wintertime, dense fog and heavy dew occur often, and there is no dry period. On the whole, the region is the most humid section of the Himalayas. Mean annual precipitation varies from 1000 to 4000 mm in the districts with altitudes below 2500 to

3000 m asl. The elevation of the highest precipitation belt depends on the position and topography of the mountain ranges. In the East Himalayas it is generally found at altitudes of between 1500 to 2500 m asl. where the annual rainfall is about 2000 to 3000 mm.

Areal differentiation in moisture conditions is quite obvious. The eastern end of the East Himalayas and its foothill belt is the most humid area, with annual rainfall as high as 2000 to 4000 mm; conversely, annual precipitation drops to about 1000 to 2000 mm in the west. The precipitation decreases to about 600 to 800 mm, with increasing altitude above the cloud belt (or the highest precipitation belt). Because of cloudiness and fog, the sunshine conditions are unfavorable, with gross annual radiation below 100 kal/cm^2 and annual sunshine of less than 1500 to 1800 hours.

The vegetation types vary greatly. At lower altitudes below 1000 to 1200 m asl, there are tropical evergreen and semievergreen rainforests in which trees of the genus of *Dipterocarpus, Dysoxylum, Terminalia,* and *Shorea* are outstanding. Once the primeval rainforest is cleared on these slopes, dense bamboo jungle may extend and remain as secondary growth. In the valley of the Yarlung Zangbo, the tropical evergreen rainforest goes at least as far up as Singing (450 m asl.) and the semievergreen rainforest to Mêdog (1000 m, 29°N).

The lower montane belt of evergreen broad-leaved forest occupies the largest share in the vertical zonation. It consists chiefly of evergreen Fagacae, among which the genera of *Castanopsis* and *Cyclobalanopsis,* heavily clogged with mosses, are dominant. This belt is usually characterized by mist forest or mossy forest, corresponding with the wettest belt of plentiful precipitation. Tree ferns are common; the number of lianas and epiphytes is very great. Zayü, bordered by the Yunnan Plateau on the southeast, has a wide expanse of pine forest (consisting mainly of *Pinus yunnanensis*) because of lighter precipitation and an obvious dry season.

Various types of vegetation, such as the sclerophyllous forest of *Quercus semecarpifolia,* the hygrophilous forest of *Chuka dumosa,* and the Himalayan endemic forest of *Pinus griffithii,* occur in the montane needle and broadleaved mixed forest belt. A good number of species and genera of *Abies* (*A. spectabilis, A. delarayi, A. georgei*) dominate the montane coniferous forest belt. The forest of *Betula utilis* and the stunted scrub of *Rhododendron* are frequently found near the upper forest limit, which varies in altitude from 3700 to 4100 m. Above the forest belt, there are alpine scrubs and meadows, the former consisting chiefly of *Rhododendron, Salix,* and *Cassiope;* the latter of *Carex, Juncus, Kobresia,* and *Polygonum.* A variety of alpine flowers bloom, and during their short growing season, they create a strangely colorful landscape. Under this forest vegetation the lateritic red earth, montane yellow earth, brown earth, and podzols are developed.

Forest resources, including various kinds of timber and rare plant resources, abound in the region. Tropical and subtropical fruits, such as oranges and bananas, grow at lower altitudes. Tea as well as temperate fruits, such as apples, pears, and peaches, can be planted in the region.

The farmlands are limited by rugged topography, stretching mainly along the valleys, especially in the drainage of the Zayü River. Rice, African millet, maize and wheat are the leading crops. The area is still characterized by relatively primitive farming methods, relying on slash and burn cultivation. The yaks and sheep are pastured in the alpine belt, whereas cattle graze in the middle-and lower altitudes.

In the future, we are looking forward to better use of the tropical and subtropical forests and other biological resources as well as extensions in the planting of rice, tea and other crops.

29. Southeastern Qinghai-Xizang Plateau — Montane Needle-Leaved Forest and Alpine Meadow Region

The southeastern part of the Qinghai-Xizang Plateau comprises mainly the middle-northern parts of the Hengduan Mountains. The region consists of a series of high mountain ridges sandwiched between deep river gorges. The mountain ranges trend nearly north to south. From west to east they are. The Nyainqêntanglha-Boxoila, the Taniantaweng, Ningjing, Chola, Shaluli, Daxue, Zheduo, and Qionglai mountains. The Nujiang River (upper reaches of the Salween River), the Lanchang River (upper reaches of the Mekong River), the Jinsha River (upper reaches of the Changjiang River), and their numerous tributaries occur in between. All of them cut deeply into the southeastern Qinghai-Xizang Plateau in parallel gorges with the elevation of valley floors varying between 2000 to 4000 m asl. On the ridges, peaks tower as high as 5000 to 6000 m, where glacial and periglacial landforms are commonly seen. The undulating residual surfaces of the southeastern Qinghai-Xizang plateau are fragmentary, with altitudes about 3500 to 4500 m asl and generally tipping from north to south.

In the eastern Nyainqêntanglha Mountains, peaks rise to over 6000 to 6500 m asl; the snow line is located at about 5100 to 5300 m asl. Maritime glaciers, with a total glacial area of more than 5000 sq km are concentrated here. The Karchin glacier, with a length of 32 km (ending at 2530 m asl in Yi'ong), is the largest oceanic-type glacier in China.

On the whole, the region inclines from north to south. The northern or upper reaches of rivers are generally located above the knickpoint of fluvial rejuvenation and are characterized by a slightly dissected plateau with gentle slopes. In the middle and southern part, the river systems feature deep gorges, steep slopes, and narrow valleys as well as alternatively parallel gorges and ridges. In the upper reaches of the Yalong River on the east and the Huanghe River on the northeast, however, the elevation is remarkably lower and glaciers do not exist.

This southeastern section of the Qinghai-Xizang Plateau reflects geological control and is characterized by sharp fluvial landforms. In the northern part (northwards of 30°N), the rivers run in a nearly northwest-southeast direction with slight gradient. Terraces and floodplains occur in a number of broad valleys. Southwards from 30°N, the rivers turn to a north-south orientation characterized by deep cut gorges, steep valley walls, swift torrents, and steep river gradients. The terraces and floodplains almost disappear in the bottom, while mudflows, landslides and slopeslips occur frequently. The annual depth of runoff varies from between 350 to 600 mm.

Further up the rivers the valleys open out. The headwaters and upper reaches of the rivers are located in the middle-e astern part of the Qinghai-Xizang Plateau, with an altitude of about 3500 to 4500 m asl. The broad valley and basins are studded amid hills with a relative relief of about 200 to 500 m. In the Nagqu-Yushu region northeast of Lhasa, with an altitude of about 4000 to 4500 m asl. The periglacial geomorphological features occur extensively and permafrost exists in isolated spots. Meanders occur frequently in the broad valleys and basins that are scattered with a number of oxbow lakes. Owing to the existence of permafrost, long-term thawing-freezing processes, and gentler relief, the lowlands are unfavorable for drainage, and vast marshy meadows and swamps extend throughout them.

During the summer, usually from June until September, the region is under the influence of monsoons, both from the southeast and southwest. Annual precipitation totals 400 to 1000 mm, decreasing northwestward from the periphery to the interior. For example, the southern periphery receives an annual precipitation as high as 800 to 1000 mm in Kangdin and Bome. This decreases northwestward to only 400 mm or less in Nagqu and Yushu.

Vertical changes in temperature are obvious. The mean temperature of the warmest month is 12(10) to 18°C in the valley, with an altitude about 2500 to 4000 m asl and only 6 to 10°C in high ranges or plateau surfaces of 4000 to 4500 m asl. The gross annual radiation is 120 to 160 kcal/cm^2, less than that of the Qinghai-Xizang Plateau proper. The aridity varies between 0.8 to 1.5. In summer, Owing to the control of east-west shearlines located through the northern part of the region. there is more precipitation than in other parts on the high plateau surface; hail storms occur as often as 20 to 35 days per year, resulting in the heaviest hail hazards in China.

A number of warm, dry valleys are located in the gorge area, resulting from the foehn and rainshadow effect. The mean temperature of the warmest month is as high as 18 to 20°C or more with annual precipitation of between 250 to 400 mm.

In addition, the altitudinal changes of soils are quite prominent. The montane drab soils are developed on narrow terraces, diluvial fans, and lower slopes of the dry valleys with an elevation of under 3000 to 3600 m. These are characterized by a thin horizon with coarse texture and alkaline to neutral reaction. The montane brown soils are found in the montane needle-and broad-leaved mixed forest belt and montane needle-leaved forest belt. They are characterized by thick horizons with a high content of organic matter and an acid reaction. Above the upper forest limit occur alpine shrubby-meadow soils and alpine meadow soils, which have compact sod and a high content of organic matter. Owing to the important influence of alternate thawing and freezing, solifluction and soil creep occur very often.

The region is one of the richest areas of alpine flora in the world. The vegetation types vary greatly through distinct vertical changes. At the floor of dry valleys in eastern Xizang, the thorny shrubs consist chiefly of *Sophora viciflora* and *Ceratostigma griffithii*. In western Sichuan the thorny and succulent shrubs, consisting mainly of *Opuntia monacantha, Acacia farnesiana, and Pistacia weinmannifolia,* are found in the bottom of dry valleys at an altitude of below 1800 m.

With the exception of valleys below 2400 m asl, a number of montane evergreen broadleaf forests exist. Most areas are covered by montane mixed needle and broad-leaved forests and needle-leaved forests, the former consisting of *Pinus densata, Tsuga dumosa,* and *Quercus aquifolioides,* the latter consisting of numerous trees of the genera *Picea, Abies, Sabina,* and *Larix.* Another feature is that only one or a few species predominate in these forests. From the periphery to the interior, the forests are covered continuously or in patches. The upper forest limit of *Picea balfouriana* is 4400 m asl. on the shady side and that of *Sabina tibetica* 4600 m on the sunny side, this constitutes the highest forest limit in the world. Above the upper timber line, alpine scrubs, and meadows are found usually occupying the divides and ridges between the gorges. The alpine scrub consists of *Rhododendron microphyll, Salix, Dasiphora arbuscula,* and *Caragana jubata.* The alpine meadow consists of *Kobresia, Polygonum gentiana,* and *Saussurea.*

In the northern part of the region the predominant vegetation is alpine scrub and meadow. The lowlands of the broad valleys and basins are covered by marshy meadows and swamps, consisting mainly of *Kobresia littleḋalei, K. tibetica, Carex lanceolata, C. muliensis,* and *Blysmus sinocompressus.*

The southern part of the region is rich in forest resources. The volume of timber is about 500 to 800 m³/ha. The native products include such medicinal commodities as musk, the tuber of elevated gastrodia (*Gastrodia elata*), the bulb of fritillaria, Chinese caterpillar fungus (*Cordyceps sinensis*) as well as mushrooms. Fruit trees such as apple, pear, peach, and walnut may be grown at lower altitudes. There are rich plant resources such as aromatic azalea, colophony, and rosin, which could be more fully and economically utilized.

Farmlands are relatively few and are chiefly concentrated on the terraces and alluvial fans in the valley areas. *Qingke* (highland barley), wheat, peas, and potato are the main crops. In addition, corn may be grown at lower altitudes.

Animal husbandry still plays an important role in the region. Yak, sheep, and goat make up the bulk of the herds. Transhumance is practiced. The herds are driven into the alpine belt after the melting of the snow and are returned to the valleys and foothills in the cold season. Water resources abound in the gorge areas; the further development of small, scattered irrigation projects and hydroelectric stations is promising.

The natural pastures, stretching extensively on the plateau surface in the northern part of the district, are good for grazing yak and sheep. Because of the insufficiency of the winter pasture, rotation grazing must be employed, and the mowing of grassland for winter has been established. In addition, the highland caterpillar damages the pastures seriously and should be exterminated as soon as possible.

30. Southern Qinghai-Xizang Plateau — Shrubby Grassland Region

The southern Qinghai-Xizang Plateau lies between the Gangdisê-Nyainqêntang1ha ranges in the north and the Himalayas to the south. Its drainage is by means of the Yarlung Zangbo and the Pum river systems (Landsat image 11).

The Himalayan range, bordering the Qinghai-Xizang Plateau in the south and uplifted since the Pliocene Epoch, is highest in its middle section where there are numerous lofty peaks. Besides Mount Qomolangma (8848 m), 5 peaks rise above 8000 m and more than 40 peaks to above 7000 m. All these high peaks provide a favorable basis for glacial action. Owing to the intensive uplifting, the Himalayan range has one steep-sloped side on the south and much gentler slopes on the north. The southern slopes of the mountain system tower abruptly to 6000 to 7000 m from the Ganges Plain, whereas the northern slopes descend gradually to the Qinghai-Xizang Plateau and its basins.

On the northern slopes of the Himalayas, there appear a series of fault basins and valleys. The major lakes, from east to west, include Zhegu, Yamzhoyum, Puma, Doqên, and Paikü. They have no outlet and as a result contain slightly saline water. The eastward-trending Pum Valley

— the upper reaches of the Arun River — traverses through the basins before turning south to cross the Himalayas. The depressions and basins are about 4300 to 4600 m in elevation.

Skirting the southern rim of the basin are enormous glaciofluvial fans, morainic platforms, and other glaciated landforms that border the northern foot of the Himalayas.

The Yarlung Zangbo River, traversing from west to east along the great graben at the southern foot of the Gangdisê-Nysinqêntanglha ranges, is a large river with the highest altitude in the world. In the broad valley of its upper stream — the Maquan River (Danqog Zangbo) — with an altitude of more than 4500 m, there stretch broad alluvial plains scattered with barchan sand dunes.

In the middle reaches of the Yarlung Zangbo, broad valleys and depressions alternate with narrow gorges. In the broad valleys, there are innumerable braided channels. The piedmont plains, diluvial fans, and several steps of terraces occur in the valleys; deposits of aeolian sands are also widely spread.

The Gangdisê-Nyainqêntanglha ranges trend from west to east. Like the great Himalayas, they have a steep sloped side on the south and gentle slopes on the northern side that border the Qinghai-Xizang Plateau.

The climate is conditioned by its topography and the atmospheric circulation. Owing to the southerly latitude and the lower altitude of about 3500 to 4500 m asl, the region has a higher temperature than other parts of the Qinghai-Xizang Plateau. The mean temperature of the warmest month ranges from 10 to 16°C, the coldest from 0 to 10°C. Average duration with a daily temperature of above 5°C varies from 100 to 220 days.

As a result of the great climatic barrier of the main Himalayan range, annual precipitation decrease from 500 mm in the east to 200 mm in the west and aridity index increases from 1.5 to 3.0. At the northern foot of the Himalayas, the rainshadow area, precipitation drops to 200 to 300mm. In the valley of the middle reaches of the Yarlung Zangbo River in 70 to 80 percent of the precipitation occurs at night, resulting in abundant sunshine, which is favorable for crop growing. The gross annual radiation is 160 to 190 kcal/cm², this is one of the highest records in China. Sunshine abounds with 3000 to 3400 sunshine hours. Lhasa is known as the City of Sunlight.

The predominant vegetation is montane shrubby-steppe and alpine steppe. The broad valleys, basins and lower slopes — at altitudes below 4400 to 4600 m asl — are mainly covered by montane steppe, consisting chiefly of *Stipa bungeana, Aristida triseta, Pennisetem flaccidem,* and *Orinus thoroldii.* In the middle reaches of the Yarlung Zangbo River, the montane shrubby-steppe occurs. It consists of *Sophora moorcroftiana* with other shrubs, such as *Leptorchermis microphylla* and *Ceratostigna minus.*

Above altitudes of 4400 to 4600 m asl, the alpine steppe

prevails. It consists chiefly of *Stipa purpurea, Artemisia wellbyi, A. younghusbandii,* and *A. stracheyi.* At the basin and valley of the northern foot of the Himalayas dwarf shrubs of *Caragana tibetica* and *C. versicolor,* are scattered in the alpine steppe belt. The upper limit of the alpine steppe belt reaches an altitude of about 5000 to 5200 m asl. Above this occurs the alpine meadow and cushion vegetation, the former consisting chiefly of *Kobresia pygmaea* and *Carex montiis-everestii,* the latter of *Androsace, Arenaria,* and *Astragalus.*

In the valley of the Yarlung Zangbo, the alpine meadow which consists chiefly of *Kobresia pygmaea,* covers areas above the montane steppe belt. In the eastern part of the district, the deciduous shrubs are composed of mesophilous *Berberis, Leptodermis, Rosa,* and *Sabina. Rhododendron* shrubs appear in the alpine belt. In addition, a number of woodlands, consisting chiefly of *Betula utilis* or *Sabina tibetica,* can be seen around the Yamzhoyum Lake and the Mamxong Pass to altitudes of 4300 to 4500 m asl.

The main types of soils are montane shrubby steppe soil and alpine steppe soil, which are characterized by an accumulation of calcium carbonate and alkaline reaction. Land types are mainly alpine meadow-high mountain, shrubby steppe-high plain, and meadow-level land.

The middle reaches of Yarlung Zangbo River together with its larger tributaries, such as the Nyang River and the Lhasa River constitute one of the main farming and settlement areas in the Qinghai-Xizang Plateau. The farmlands make up about 60 per cent of the total farmlands of the Xizang Autonomous Region. The farms stretch on the terraces along the river and on the lower part of the alluvial-diluvial fans skirting the rims of the basins. The altitude varies between 3300 to 4200 m asl, and there are good soil textures and some irrigation facilities.

The main crops are qingke, winter and spring wheat, peas, and rape, with one harvest per year. The upper limit for crops is very high on the northern slopes of the Himalayas — 4750 m for qingke and 4200 m for winter wheat.

An analysis of the ratio of potential evapotranspiration to precipitation shows that 4500 m³/ha of water is needed to irrigate winter wheat in a normal year. Deficiency of soil fertility is another limiting factor. Both amelioration of the soil condition with fertilizers and extension of irrigated fields are necessary. It is also advisable to plant forest belts for the conservation of soil and water and to protect farmlands from shifting sand. In the western and higher pastures, the land is chiefly used for grazing sheep and yaks.

31. Central Qinghai-Xizang Plateau — Montane and Alpine Grassland Region

The region stretches from southwest to northeast in the

central part of the plateau and includes the Qingzang Plateau, southern Qinghai and eastern Qinghai-Qilian mountain areas.

The lofty Qingzang Plateau (literally, "northern upland") is the major part of the plateau proper. Situated between the Kunlun Mountains in the north and the Gangdisê-Nysinqêntanglha Mountains to the south, this desolate upland, with an altitude of 4500 to 4800 m asl, is characterized by the immense intermontane basin (plateau surface) sandwiched between the higher northern and southern mountain regions. It belongs to the interior drainage system and is studded with numerous lakes, such as Nam Lake, and Siling Lake. Both the lacustrine plains and the piedmont depositions are quite extensive. The glaciers appear around high peaks above 6000 m asl, although a continental ice sheet did not occur. Most of the lake basins are of fault origin. Owing to the diminishing precipitation of the climate, the lakes have declined gradually. In the southern part, they are saline lakes of the sulphate and carbonate type, whereas in the northern part, they are the highly mineralized chloride type.

The upper reaches of the Changjiang River and the Huanghe River, with an altitude of 4200 to 4700 m, are a slightly dissected plateau that stretches between Tanggula Range in the south and the eastern range of the Kunlun Mountains in the north. The mountain ranges trend generally east-west and have a relative relief below 500 m. The area is a rolling plateau, with rivers flowing eastward, as do the headwaters of the Huanghe River.

The eastern Qilian Mountains, which rise to nearly 4000 to 5000 m above the Hexi Corridor, consist of a series of parallel ranges that extend from WNW to ESE. There are many intermontane basins and valleys, with an altitude of about 2500 to 3500 m asl in the eastern Qinghai, including the Qinghai Basin, the broad valleys of the Huanghe River and its main tributaries, such as the Huangshui River.

The Qinghai Lake or Koko Nor lies at an elevation of 3200 m, with an area of 4400 sq km. It is the largest lake in China.

The eastern region of Qinghai, with an altitude of 2000 to 3000 m, is covered by a mantle of loess. Because of intensive fluvial activity, erosional terraces occur extensively.

As a whole, the region has a semiarid climate. Owing to the difference of location and topographic configuration, areal differentiation is obvious. The northeastern part is better watered than the southwestern and is thus less arid.

In the Qingzang Plateau and southern Qinghai Plateau, the mean temperature of the warmest month varies between 6 to 10°C. The lofty, nearly parallel ranges in the southern Qinghai-Xizang Plateau operate as an effective rain barrier. The annual precipitation varies between 100 to 400 mm, decreasing from east to west. Aridity increases from 1.5 to 6.0. Strong winds blow frequently in winter and spring.

In the eastern Qinghai and Qilian Mountains areas, Owing to the southerly warm and moist currents encountering cold air masses from the north and, thus, producing precipitation on a larger scale, the annual precipitation varies from 250 to 600 mm, with an aridity index of 1.0 to 3.0. Because of the lower elevation, the mean temperature of the warmest month is about 12 to 18°C.

In the Qingzang Plateau and the upper reaches of the Changjiang River the main type of vegetation is alpine steppe, consisting chiefly of *Stipa purpurea, S. subsessiliflora* var. *basiplumosa,* and *Carex moorcroftiana.* Above the alpine steppe belt occurs the alpine meadow of *Kobresia pygmaea.* Valleys or basins are covered by marshy meadows consisting of *Kobresia littledalei* and *Kitibetica.* The halophilous meadow, consisting of *Aneurolepidium dasistachyum* and *Polygonum sibiricum,* occurs in the northern and western part of the Qangtang plateau. In addition the prostrate shrubs of *Hippophae* and *Myricaria* are scattered on the gravel gobi.

The montane steppe prevails in the eastern Qilian Mountains. It consists of *Stipa krylovii, S. breviflora,* and *Artemisia frigida.* The montane forest steppe belt contains needle-leaved forest, consisting of *Picea crassifolia* and *Sabina przewalski.* Above the montane forest steppe belt there are alpine scrubs and meadows, the former composed of *Rhododendren, Caragana jubata, Potentilla fruticosa, and Salix,* the latter composed of *Kobresia* and *Polygonum.* Around the Qinghai Basin the alpine steppe consists of *Stipa purpurea.*

The zonal soil is known as alpine steppe soil and is characterized by a thin layer, coarse texture, and alkaline reaction. Common features are the retarded circulation of the bio-matter, slight humification with on organic content of 1–2 per cent, weak seasonal eluviation, strong cryo-weathering, and significant alternation of freezing and thawing. In addition, chestnut soils and brown soils are found in the eastern Qilian Mountains.

In the Qingzang Plateau, the land is chiefly used for grazing sheep and yaks. Owing to low productivity, strong seasonal differntials in tempersture and moisture, and insufficient winter pastures, better land and herd management is necessary to combat the damage of gales or drought.

Because of the deficiency of moisture, the Qingzang and southern Qinghai plateaus are not good for dry farming, except in small local patches where highland barley can be planted.

At lower altitudes in the northeastern part of the Central Qinghai-Xizang Plateau, such as in the Huangshui Valley, physical conditions are more favorable for agricultural development and the land has been cultivated since the second century B.C. Spring wheat, qingko, and rape are the leading crops. The surrounding area has ex-

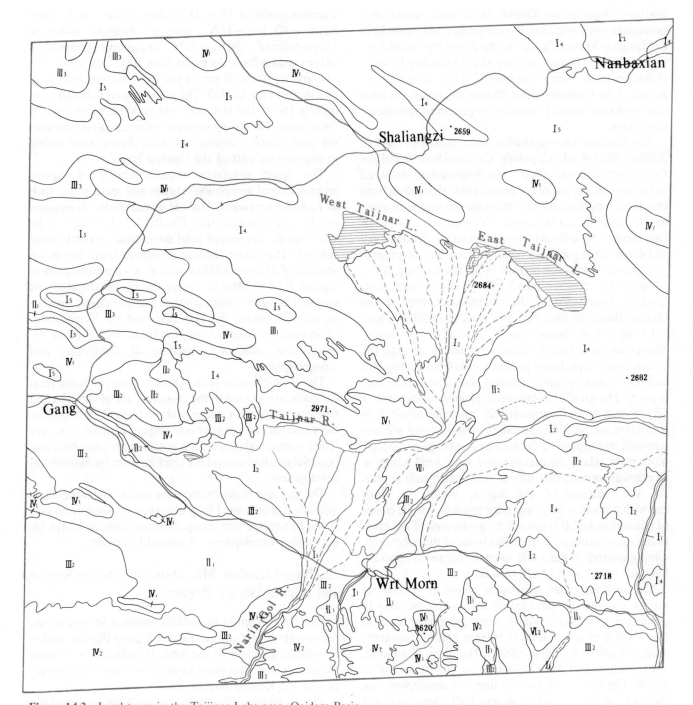

Figure 14-3 Land types in the Taijinar Lake area, Qaidam Basin

I. Clay and silt flatland;   II. Sandy desert;   III. Gravel desert;
IV. Hilly and montane desert.

tensive pastures, and sheep, yak, and horses make up the bulk of the herds. Future development should include better grassland management, such as rotation grazing, preparation for winter-forage protection of pastures from overgrazing, and a program to combat rat damage.

32. The Qaidam Basin and Northern Kunlun Mountains Slopes-Desert Region

The region, which includes the Qaidam Basin, the western Qilian Mountains, the Altun Mountains, and the northern slopes of the Kunlun Mountains, extends in a nearly east -west direction. It is the transitional region from the Qinghai-Xizang Plateau to the desert regions of Northwest China. The region, especially the Qaidam Basin, is characterized by a temperate desert, but it is still considered as part of the Qinghai-Xizang Plateau, mainly on the basis of its geological formation and its temperature and moisture conditions.

The Qaidam, an oval-shaped basin at an elevation of 2600 to 3000 m asl, is a graben. On the whole, it inclines from northwest to southeast. The Basin can be visualized in the form of a series of concentric belts, proceeding from the outer mountain barriers through the piedmont plain to the playa lakes at the center. Consequently, a series of land types — denudational mountain and hill, denudational and depositional gobi, sand dune, yardang, and salty marsh — appear in succession (Figure 14-3). The yardang, formed by aeolian erosion on the Tertiary loose stratum, is extensively distributed in the northwestern part of the Qaidam Basin, the most extensive yardang distributed area in China. The southeastern part is covered by the diluvial-alluvial facies of the Quaternary period. In general, the diluvial-alluvial piedmont plain stretches from the foothills and is composed mainly of gravel gobi. The central part is covered by an alluvial-lacustrine plain of sandy loam or clay; it is an area of comparatively low relief, about 2600 to 2700 m asl. Shifting and half-fixed sand dunes are also scattered on the gravel gobi. At the outer fringe of the piedmont plain, where the underground water level is high or emerges as springs; salty soils or salty marshlands appear. At the bottom of the basin, which is studded with numerous playa and salty marshes, lies the center of mineralization. It is covered by a vast area of salt crust of the sulphate chloride type. This is one of the richest salt-mining grounds in the world; even the highways of 60 km long are paved with blocks of solid salt.

The western Qilian Mountains consist of a series of ranges and intermontane valleys and basins, trending in a nearly northwestern-southeastern direction, with a crest line altitude of about 4000 to 5000 m and intermontane valleys of between 3000 to 3500 m. The Altun Mountains, with an altitude of 3600 to 4000 m asl, run from ENE to WSW. The Kunlun Mountains turn to a nearly west-east direction, and in their middle section, high ridges reach over 5000 to 6000 m asl.

The region is controlled by westerlies year round and influenced by the Siberian High. The lofty, nearly parallel ranges in the southern part of the Tibetan Plateau cut off the maritime air masses entirely; almost none of the moisture crosses the Kunlun Mountains into the Qaidam Basin. This is the driest area in the Qinghai-Xizang Plateau.

Climatically, this region is also characterized by a cool summer and a severe cold winter, with frequent strong winds in spring and winter. The mean temperature of the warmest month is 10 to 18°C; the coldest month varies from −10 to −15°C, with an absolute minimum temperature of −30 to −35°C. The annual precipitation of the Qaidam Basin decreases from 100 to 200 mm in the east and to 10 to 20 mm in the west. Annual aridity increases from 6.1 to 50.0. The gross annual radiation is as high as 160 to 180 kcal/cm^2, and sunshine totals 3000 to 3600 hours. The strong westerlies blow almost throughout the year, hence, yardang and sandy dunes occur widely in the western part of the Qaidam Basin.

The desert vegetation consists chiefly of superxerophilous and xerophilous shrubs and semishrubs, such as *Ephedra przewalski, Salsola abrotanoides, Sympegma regelii,* and *Artemisia* spp. The bare lands of hills, gobi, salty marsh, and barren sand dunes also appear in large patches. The upper limit of montane desert lies at an altitude of 2500 to 3000 m asl and is sometimes even as high as 3600 to 3800 m. Above this occurs the montane steppe. The alpine desert, consisting of *Ceratoides compacta,* prevails around the Har Lake and the Yema South Mountain. A number of lowlands are covered by halophilous meadow, consisting of *Phragmites and Aneurolepidium.*

The principal oases of the Qaidam Basin, as in the Hexi Corridor, are located in the margin of the piedmont plain where groundwater emerges and can be used for irrigation. Only one harvest per year is possible. The main crops are spring wheat, highland barley, potato, rape, sugar beet, and peas. Salinization control is a chief measure for agricultural development.

The vast area of desert, montane steppe, and halophilous meadow is chiefly used for grazing camels, sheep, and yak. It will be necessary to set up a base of artificial forage for the further development of animal husbandry.

33. Ngari-Kunlun Mountains — Desert-Steppe and Alpine Desert Region

The Ngari-Kunlun mountains region is located in the northwestern part of the Qinghai-Xizang Plateau, and it borders on Kashmir in the west. It includes the western Ngari area and the southern sides of the middlewesten section of the Kunlun Mountains.

The Ngari area is composed of the upper reaches of the Indus River and the broad valley of Bangong Lake, with altitudes varying between 3800 to 4500 m asl. The area is encircled by the West Himalayas and the Gangdisê and the Karakorum Range. All of these mountains have crest lines of more than 5500 to 6000 m asl. Because of the extremely arid climate, only small glaciers appear on a few peaks.

The Kunlun Mountains, with an altitude of more than 6000 m asl, follows a nearly west-east direction trending to the northern rim of the Qinghai-Xizang Plateau. The west Kunlun area is a center of glaciation, with glacial area of more than 4000 sq km.

Between the Kunlun and the Hoh Xil Mts, there stretches a series of east-west trending lake basins with altitudes of 4800 to 5100 m asl. The intermontane basins, are distributed in the west and are distinct from the eastern table plateau with its gentle relief. The denudational piedmonts are connected with the lacustrine plains or terraces surrounding the lake basins. The nival process prevails and permafrost occurs everywhere. At the bottom of ancient lakes or lacustrine basins, the alpine desert soil contains residual carbonate or chloride salt.

The Ngari area is rather warm in summer. The mean temperature of the warmest month varies from 10 to 14°C and that of the coldest month from −10 to −14°C. Owing to the climatic barrier of parallel ranges in the southwest, the annual precipitation is less than 50 to 150 mm, the aridity index is 6.1 to 15.0, and the total hours of sunshine exceed 3000 to 3400. In spring and winter, strongwinds occur frequently.

Because of its higher elevation and inland location, the climate of the Kunlun Mountains is arid and cold. The mean temperature of the Warmest month varies from 4 to 6°C, with the daily minimum temperature below 0°C throughout the warm season. According to a short-term observation, a minimum temperature of −18°C occurs in August at the side of the Yunpo Lake (4880 m asl) north of the Hoh Xil Mountain. The absolute minimum temperature, on the other hand, is as low as −40°C. The annual precipitation varies from 20 to 100 mm, with an aridity index of about 6.1 to 20.0.

The montane desert steppe and the montane desert prevail in the Ngari. The area consists chiefly of *Stipa glareosa, Ceratoides compacta,* and *Ajania fruticulosa.* In the lower part of the Xiangquan (Langqên Zangbo) Valley, there are *Artemisia salsoloides, A. sacrorum,* and *Capparis spinosa,* which may be considered as an infiltration of the *Artemisia* steppe from the Kashmir Valley in the west. The upper limit of the montane desert, consisting of *Ceratordes latens,* reaches an elevation as high as 4600 to 5200 m asl. In the alpine belt of the southern part of the region, there is alpine shrubby-steppe, consisting of *Stipa purpurea* and Caragana versicolor.

On the vast lacustrine plains of the Kunlun Mountains, the zonal type of vegetation is alpine desert, consisting of *Ceratoides compacta* and *Ajania tibetica,* which are species endemic to the Qinghai-Xizang Plateau. The diluvial fans, with a more gentle relief and a sand-gravel soil texture, are mainly covered by *Carex moorcroftii,* and are also scattered with *Ceratoides compacta* and *Stipa subsessiliflora* var. *basiplumosa,* which pertain to alpine desert steppe.

Alpine desert, consisting of *Ajania tibetica,* occurs in the intermontane basin (4700 to 5200 m) in the northern part of the Kunlun Mountains. Alpine desert is characterized by sparse cover and dwarf plants. The Akesayqin (literally the white desert) is a vast bare area without any flowering plants.

The main soil types are montane desert-steppe soil and montane desert soil. They are characterized by indistinct differentiation in profile, marked gravel-pavement at the surface horizon, low content of organic matter ($< 1.0\%$) and slight eluviation of $CaCO_3$. Owing to the intensive processes of recent deposition, many profiles remain undeveloped.

On the whole, the region is used for grazing sheep and goats, with the exception of the lower valleys in the southern part, where small areas of farmland have been developed with irrigation. The great obstacles for the development of animal husbandry are lack of water, insufficient winter and spring pasture and a long severe, winter.

References

[1] Symposium on Qinghai-Xizang (Tibetan) Plateau, Beijing, 1980, *Environment and Ecology of Qinghai-Xizang Plateau,* 3 Vols. Science Press, Gordon and Breach, Science Publishers Inc. (In Chinese and English)

[2] Zhang Rongzu et al., 1981, *Physical Geography of Tibet,* Science Press, Beijing. (In Chinese)

[3] Zheng Du et al., 1979, "On the Natural Zonation in the Qinghai-Xizang Plateau", *Acta Geographica Sinica,* Vol. 34, No. 1. (In Chinese, with English abstract)

[4] Lee Xingchung et al., 1979, "A Discussion on the Age, Amplitude and Type of Qinghai-Xizang Plateau Uplift", *Scienta Sinicia,* Vol., No. 6. (In Chinese and English)

[5] Xu Shuying et al., 1962, "Monsoon on the Qinghai-Xizang Plateau", *Acta Geographica Sinica,* Vol. 28, No. 2. (In Chinese, with English abstract)

[6] Zhang Jinwei et al., 1966, *Vegetation in Central Tibet,* Science Press, Beijing. (In Chinese)

[7] Integrated Investigation Team of The Qinghai-Xizang Plateau, Chinese Academy of Sciences, 1966 – 1968, *Scientific Reports on Qomolangma Mount Area,* 9 vols, Science Press, Beijing. (In Chinese)

[8] Schweinfurth, U., 1957, "Die Horizontale und Vertikale Verbreitiung der Vegetation im Himalaya", *Bonn. Geogr. Abl.,* H. 20. (In German)

[9] Jin Zhou (ed.) 1981. *Tibet: No Longer Medicval.* Foreign Language Press, Beijing.

[10] Zhang Mingtao et al., 1982, *The Roof of the World: Exploring the Mysteries of the Qinghai-Xizang Plateau,* Foreign Language Press, Beijing.

Appendix I
A Glossary of Geographic Place names

A

Aba	阿坝
Abag banner	阿巴嘎旗
Abgyag glacier	阿札贡拉冰川
Aihui	爱辉
Ailao Mountain	哀牢山
Aksayqin (Akesaichin)	阿克赛钦
Aksu	阿克苏
Aksu River	阿克苏河
Alashan (Alxa) Plateau	阿拉善高原
Ali (Ngari)	阿里
Ali Mountain	阿里山
Allung	阿龙
Altay Mountains	阿尔泰山
Altun Mountains	阿尔金山
Anda	安达
Andir River	安迪尔河
Anhui	安徽
Ankang	安康
Anqing	安庆
Anshan	鞍山
Anshun	安顺
Anxi	安西
Aomen (Macao)	澳门
Awa Mountain	阿瓦山
Ayding Lake	艾丁湖

B

Back Elbow Plain	后套平原
Badain Jaran Desert	巴丹吉林沙漠
Baicheng	白城
Baihe River	白河
Baiku Lake	佩枯湖

Baillingmiao	百灵庙
Bailong River	白龙江
Baingoin	班戈
Baiyang Lake	洋淀
Bangong Lake	班公湖
Banpo Village	半坡村
Banyan Gol	巴彦高勒
Baoding	保定
Baoji	宝鸡
Baoshan	宝山
Baotou	包头
Barku Mountain	巴尔库山
Barkan	巴尔康
Batang	巴塘
Bayan Har Mountains	巴颜喀喇山
Bayan Obo	白云鄂博
Baita Mountain	白塔山
Bei'an	北安
Beihai	北海
Beijiang River	北江
Beijing (Peking)	北京
Beilun River	北仑河
Beishan (Mazong) Mountains	北山（马鬃山）
Bengbu	蚌埠
Bijie	毕节
Black River (Rueshui)	黑河（弱水）
Bogda Mountain	博格多山
Bohai Sea	渤海
Bome	波密
Bose	百色
Boshula Mountain	伯舒拉山
Bosten Lake	博斯腾湖
Buyun Mountain	步云山

C

Central China	华中
Changbai Mountains	长白山
Changchun	长春
Changde	常德
Chanjiang	昌江
Changjiang (Yangtze) River	长江（扬子江）
Changsha	长沙
Changshu	常熟
Changtang (Qingzang) Plateau	羌塘高原
Changzhi	长治
Chaohu Lake	巢湖
Chaobai River	潮白河
Chaor River	绰尔河
Chaoyang	朝阳
Chaozhou	潮州
Chengde	承德
Chengtu	成都
Chenxi	辰溪
Chenzhou	彬州
Chifeng	赤峰
Chin xian	晋县
Chola Mountain	雀儿山
Chongming Island	崇明岛
Chonqing	重庆
Chongzuo	崇左
Chung-hai-tze	中海子

D

Daba Mountains	大巴山
Dabie Mountains	大别山
Dadang	大塘
Dadu River	大渡河
Da Hinggan (Great Hinggan) Mountains	大兴安岭
Daiyun Mountain	戴云山
Daliang Mountains	大凉山
Dalian (Dairen)	大连
Dalou Mountain	大娄山
Dandong	丹东
Dan xian	儋县
Danxong	当雄
Da Qaidam	大柴旦
Daqing	大庆
Daqing Mountains	大青山
Datong (Shanxi province)	大同（山西省）
Datong (Anhui province)	大通（安徽省）
Daxa River	大夏河
Daxiang Mountain	大相岭
Daxue Mountains	大雪山
Dayu Mountain	大庾岭

Dedu	德都
Delun	吉伦
Dengkou	澄口
Dengqen	丁青
Dezhou	德州
Dianchi Lake	滇池
Dinghai	定海
Diangye	定日
Dongfang	东方
Donghai (East China) Sea	东海
Dongsha Archipelago	东沙群岛
Dongsheng	东胜
Dongting Lake	洞庭湖
Dondjiang River	东江
Doqen Lake	多庆湖
Dojiang Dam	都江堰
Duan	都安
Dukou	渡口
Dulan	都兰
Dunhuang	敦煌
Duolun	多伦
Dupang Mountain	都庞岭
Duyun	都匀

E

East China Sea (Dong Hai)	东海
Ebi Lake	艾比湖
Ejin banner	额济纳旗
Ejin (Juyun) Lake	居延海
Elbow (Hetau) Plain	河套平原
Emei Mountain	峨嵋山
Emin River	额敏河
Erenhot	二连
Ergan River	额尔古纳河
Erlang Mountain	二郎山
Ertix River	额尔齐斯河

F

Fangshan	房山
Fankou	樊口
Feihe River	淝河
Fencheng	汾城
Fenghuo Mountain	风火山
Fengman	丰满
Fengning	丰宁
Feng xian	凤县
Fenhe River	汾河
Five-Linked Lake	五大连池
Foping	佛坪
Foziling Reservoir	佛子岭水库
Friendship (Youyi) Peak	友谊峰
Front Elbow Plain	前套平原

Fuchun River	富春江
Fuding	福鼎
Fuhe River	抚河
Fujian	福建
Fuling	涪陵
Fujiang River	涪江
Funiu Mountain	伏牛山
Fuqing	福清
Fushun	抚顺
Fuxian Lake	抚仙湖
Fuyun	富蕴
Fuzhou	福州

G

Gandise Mountains	冈底斯山
Ganhe River	甘河
Ganjiang River	赣江
Ganguan Island	甘泉岛
Gansu	甘肃
Ganzhou	赣州
Gaoligong Mountains	高黎贡山
Gaoxiang	高雄
Gaoyou Lake	高邮湖
Gar	噶尔
Garze	甘孜
Gaxun Desert	嘎顺戈壁
Gaxun Lake	嘎顺湖
Gejiu	个旧
Geladangdong Mount ains	格拉丹冬山
Gerze	改则
Gold-Silver Island	金银岛
Golmud	格尔木
Gongga Mountain	贡嘎山
Great Hinggan (Da Hinggan) Mts.	大兴安岭
Guangdong	广东
Guangxi	广西
Guangyuan	广元
Guangzhou (Canton)	广州
Guanting Reservoir	官厅水库
Guan xian	灌县
Guiji Mountain	会稽山
Guco	古措
Guilin	桂林
Guiyang	贵阳
Guizhou	贵州
Guohe River	涡河
Gurbantüngüt Desert	古尔班通古特沙漠
Gutian River	古田溪
Guyuan	沽源
Gyangze	江孜
Gyaring Lake	札陵湖

H

Haifeng	海丰
Haihe River	海河
Haikang	海康
Haikou	海口
Hailar	海拉尔
Hainan Island	海南岛
Haiyan Mountain	海洋山
Hami	哈密
Handan	邯郸
Hangzhou	杭州
Hanjiang River	韩江
Hanshui River	汉水
Hanzhong	汉中
Haoshaoliao	火烧寮
Hara Lake	哈拉湖
Harbin	哈尔滨
Heaven Pond	天池
Hefei	合肥
Hegang	鹤岗
Heihe (Black) River	黑河（弱水）
Heilong (Amur) River	黑龙江
Hekou	河口
Hengduan (Traverse) Mountains	横断山
Hengshan Mountain	恒山
Hengyang	衡阳
Hebei	河北
Hepu	合浦
Himalayan Mountains	喜马拉雅山
Hobq Desert	库布齐沙漠
Hohhot	呼和浩特
Hoh Xil Mountains	可可西里山
Helan Mountains	贺兰山
Hemudu (Culture)	河姆渡（文化）
Henan	河南
Hong Kong	香港
Hongze Lake	洪泽湖
Hexi Corridor	河西走廊
Hotan	和阗
Hotan River	和阗河
Huade	化德
Huaihe River	淮河
Hualian	花莲
Huanghai (Yellow) Sea	黄海
Huangshan Mountains	黄山
Huangshui River	湟水
Huangyan Reef	黄岩岛
Huanxian	环县
Huashan Mountain	华山
Huaying Mountain	华蓥山
Huayuankou	花园口
Hubei	湖北
Hukou	湖口

Hulan River	呼兰河
Hulunbuir	呼伦贝尔
Hulun Lake (Dalai Nor)	呼伦池（达赉湖）
Huma	呼玛
Hunan	湖南
Hunshandake Sandy Land	浑善达克沙地
Hutao River	滹沱河

I, J

Ili River	伊犁河
Inner Mongolia (Nei Mongol)	内蒙古
Jashan Mountain	夹山
Jarga Plateau	若尔盖高原
Jartai	吉兰泰
Jarud Banner	札鲁特旗
Jialiang River	嘉陵江
Jiamusi	佳木斯
Jian	吉安
Jiangsu	江苏
Jiangxi	江西
Jiangyin	江阴
Jiao-Lai Plain	胶莱平原
Jiaozuo	焦作
Jiayuguan	嘉峪关
Jilin	吉林
Jilong	基隆
Jinan	济南
Jinggan Mountain	井冈山
Jinghong	景洪
Jingle	静乐
Jingpo Lake	镜泊湖
Jinhua	金华
Jining	集宁
Jinsha River	金沙江
Jinta	金塔
Jin xian	金县
Jinzhou	锦州
Jishou	吉首
Jiufeng Mountain	鹫峰山
Jiujiang	九江
Jiuling Mountain	九岭山
Jiulong River	九龙江
Jinquan	酒泉
Jixi	鸡西
Junggar Basin	准噶尔盆地
Junggar Gate	准格尔门
Juyan Lake (Ejin Nor)	居延海

K

Kaidu River	开都河
Kaifeng	开封
Kanas Lake	喀纳斯湖

Kangdin	康定
Kangrigarbo Mountain	岗日嘎布山
Kangrinboque Mount	康仁波钦峰
Kara Holshun (Black City)	黑城
Karamay	克拉玛依
Karchin glacier	卡钦冰川
Kashi (Kashgar)	喀什
Kaxgar River	喀什噶尔河
Keriya River	克里雅河
Kolshin Grassland	科尔沁草原
Kolshin Sandy Land	科尔沁沙地
Konqi R.	孔雀河
Korla	库尔勒
Kunlun Mountains	昆仑山
Kunming	昆明
Kuecang Mountain	括苍山
Kuqa	库车
Kurug Mt. (Kurugtag)	库鲁塔格

L

Laizhou Gulf	莱州湾
Lancang	澜沧
Lancang (Mekong) River	澜沧江（湄公河）
Lanzhou	兰州
Laobie Mountain	老别山
Laoshan Mountain	崂山
Ledong	乐东
Leizhou Peninsula	雷州半岛
Lenghu	冷湖
Lhasa	拉萨
Lianyungang	连云港
Liaodong Peninsula	辽东半岛
Liaohe River	辽河
Liaoning	辽宁
Lijiang	丽江
Linfen	临汾
Lingjiang River	灵江
Lingling	零陵
Lianyun Mountain	连云山
Lingwu	灵武
Lishi (loess)	离石(黄土)
Litang	里塘
Liupan Mountains	六盘山
Liuyang	浏阳
Liuzhou	柳州
Loess Plateau	黄土高原
Longchuan	陇川
Longling	龙陵
Longmen Mountain	龙门山
Longyan	龙岩
Lop Lake (Lop Nor)	罗布泊
Luanhe R.	滦河
Lufeng	陆丰

Luliang Mountains	吕梁山	Nansha Archipelago	南沙群岛
Lunan Mountain	路南山	Nantong	南通
Luntai	轮台	Nanwei Island	南威岛
Luohe River	洛河	Nanxiong	南雄
Luoyang	洛阳	Nanyang	南阳
		Naqu	纳曲
M		Nei Mongol (Inner Mongolia)	内蒙古
		Neijiang	内江
Magong	马公	Nenjiang River	嫩江
Malan loess	马兰黄土	Nganglaling Lake	昂拉仁湖
Manas River	玛纳斯河	Ngali (Ali)	阿里
Mangnai	茫崖	Ngoring Lake	鄂陵湖
Mangshi	芒市	Ningbo	宁波
Mangui	满贵	Ningxia	宁夏
Manzhouli	满洲里	Ninjing Mountain	宁静山
Maoming	茂名	Niuzhuang	牛庄
Maquan River	马泉河	Nomin Desert	诺明戈壁
Markit	麦盖提	Nom Lake	淖毛湖
Mazong (Beishan) Mountains	马宗山	Northeast China (Dongbei)	东北
Mapam Yum Lake	玛旁雍湖	Northeast China Plain	东北平原
Medog	墨脱	North China (Huabei)	华北
Meitan	湄潭	Northwest China	西北
Mei xian	梅县	Nujiang (Salween) River	怒江 （萨尔温江）
Mengding	孟定	Nyainqentanglha Mountains	念青唐古拉山
Mengla	勐腊		
Mengzi	蒙自	**O**	
Miaodao Archipelago	庙岛群岛		
Miaoling Mountains	苗岭	Ondor-miao	温都尔庙
Micang Mountain	米仓山	Ordos Plateau	鄂尔多斯高原
Minjiang River	岷江	Orqen	鄂伦春
Mingjiang River	闽江	Otog Banner	鄂托克旗
Minfeng (Niya)	民丰（尼雅）	Oujiang River	瓯江
Minquin	民勤		
Minshan Mountain	岷山	**P**	
Mohe	漠河		
Mori	木里	Pacific Ocean	太平洋
Mudanjiang	牡丹江	Paiku Lake	佩枯湖
Mufu Mountain	幕府山	Pamir Plateau	帕米尔高原
Mu-us Sandy Land	毛乌素沙地	Pearl (Zhujiang) River	珠江
Muztataga Mount	慕士塔格峰	Penghu	澎湖
		Pindinshan	平顶山
N		Pingdong	屏东
		Pingjiang	平江
Nagqu	那曲	Pinliang	平凉
Namcha Barwa Mountain	南迦巴瓦山	Pingshan	平山
Nam Lake (Nam Co)	纳木湖	Pingluo	平罗
Nanchang	南昌	Pishan	皮山
Nanchong	南充	Poyang Lake	鄱阳湖
Nandu River	南渡江	Pum Yum Lake	普莫雍湖
Nan-hai Sea (South China Sea)	南海	Pum-qu Valley	朋曲谷地
Nanjing (Nanking)	南京		
Nanling Mountains	南岭		
Nannin	南宁		
Nanping	南平		

Q

Qaidam Basin	柴达木盆地
Qamdo	昌都
Qarhan Salt Lake	察尔汗盐池
Qarqan River	车尔成河
Qianhe River	泾河
Qiemo	且末
Qingdao (Tsingtao)	青岛
Qianligang Mountain	千里岗山
Qianshan Mountain	千山
Qilian Mountains	祁连山
Qinhe River	沁河
Qinghai Lake (Kuku Ňor)	青海湖
Qinghai-Xizang (Tibetan) Plateau	青藏高原
Qingzang (Changtang) Plateau	羌塘高原
Qinling Mountains	秦岭
Qinzhou	钦州
Qionghai	琼海
Qiaonglai Mountain	邛崃山
Qiongzhong	琼中
Qiongzhou Strait	琼州海峡
Qiqihar	齐齐哈尔
Qomolangma Mount	珠穆朗玛峰
Qumarled	曲麻莱
Qujiang River	渠江
Quzhou	衢州
Quwu Mountain	屈吴山

R

Raoho	饶河
Riyue (Sun Moon Mountain)	日月山
Riyue Lake	日月潭
Rueshui River (Black R.)	弱水（黑河）
Ruijin	瑞金
Ruoqiang	若羌

S

Sanggan River	桑干河
Sanjiang (Three Rivers) Plain	三江平原
Sankeshu (Three Trees)	三棵树
Sanmen Gorge	三门峡
Sanshu	三水
Second Songhua River	第二松花江
Serling Lake	奇林湖
Saluli Mountain	砂鲁里山
Shandong	山东
Shanghai	上海
Shanhaiguan	山海关
Shantou	汕头
Shaanxi	陕西

Shanxi	山西
Shaoquan	韶关
Shaoyang	邵阳
Shapotao	沙坡头
Shenhang Island	深航岛
Shennongjia Mount	神农架
Shenyang	沈阳
Shidao Island	石岛
Shihezi	石河子
Shijiazhuang	石家庄
Shiquan River	狮泉河
Shiyang River	石羊河
Shizuishan	石咀山
Shuangyashan	双鸭山
Shule River	疏勒河
Shunde	顺德
Sichuan Basin	四川盆地
Simao	思茅
Singing	西金
Siping	四平
Small Hinggan (Xiao Hinggan) Mts.	小兴安岭
Small Wutai (Xiao Wutai) Mountain	小五台山
Sog xian	索县
Songhua River	松花江
Songshan Mountain	嵩山
Sonid banner	苏尼特旗
South China (Hua Nan)	华南
South China Sea (Nan Hai)	南海
Southwest China	西南
Suide	绥德
Suifenhe	绥芬河
Suzhou	苏州

T

Taba Mountains	大巴山
Table Mountains	桌子山
Tacheng	塔城
Taching Oil Field	大庆油田
Taibai Mount	太白山
Taibei	台北
Taihang Mountains	太行山
Taiho	泰和
Taihu Lake	太湖
Taijinar Lake	台吉纳尔湖
Tainan	台南
Taiping Island	太平岛
Taishan Mountain	泰山
Taitima Lake	台特马湖
Taiwan Island	台湾岛
Taiwan Strait	台湾海峡
Taiyuan	太原
Taizhong	台中
Taklimakan Desert	塔克拉玛干沙漠

Xin xian	忻县	Yingkou	营口
Xinjiang	新疆	Yiwu	伊吾
Xinyang	信阳	Yongchun	永春
Xiong'er Mountain	熊耳山	Yongding River	永定河
Xiqing Mountain	西倾山	Yongjiang River	甬江
Xirang	希浪	Yonle Archipelago	永乐群岛
Xisha Archipelago	西沙群岛	Yongning	永宁
Xishuangbanna	西双版纳	Yongxing Island	永兴岛
Xiushui River	修水	Youjiang River	右江
Xixabangma Mount	希夏邦马峰	Youyi (Friendship) Mount	友谊峰
Xizang (Tibet)	西藏	Yuanjiang River (Red R.)	元江（红河）
Xuande Archipelago	宣德群岛	Yuanling	沅陵
Xuefeng Mountain	雪峰山	Yuecheng Mountain	域城岭
Xuwen	徐闻	Yueyang	岳阳
Xuzhou	徐州	Yulin	榆林（陕西）。玉林（广西）
		Yulung Mount	玉龙山
Y		Yument	玉门
		Yunling Mountain	云岭
Ya xian	崖县	Yunmong Swamp Area	云梦泽
Yalong River	雅砻江	Yunnan	云南
Yalu River	鸭绿江	Yushan Mountain	玉山
Yamzhog Yum Lake	羊卓雍错	Yushu	玉树
Yan'an	延安	Yutian	于阗
Yanding	永定		
Yangzhuo	阳朔	**Z**	
Yancheng	盐城		
Yanchi	盐池	Zayü	察隅
Yangtze (Changjiang) River	长江	Zhangbei	张北
Yangqing	延庆	Zhangguancai Mountain	张广才岭
Yanji	延吉	Zhangjiakou (Kalgan)	张家口
Yanshan Mountain	燕山	Zhangye	张掖
Yanzhi Mountain	焉支山（胭脂山）	Zhanjiang	湛江
Yantai	烟台	Zhedo Mountain	折多山
Yarkant River	叶尔羌河	Zhegu Lake	哲古湖
Yarlung Zambo River	雅鲁藏布江	Zhejiang	浙江
Yecheng	叶城	Zhenghe Reefs	郑和群岛
Yellow (Huanghe) River	黄河	Zhengzhou	郑州
Yellow Sea (Huang Hai)	黄海	Zhenjiang	镇江
Yellow Dragon Mountain	黄龙山	Zhongba	仲巴
Yenchang	运城	Zhongsha Archipelago	中沙群岛
Yibin	宜宾	Zhouhe River	沾河
Yihe River	伊河	Zhongtiao Mountain	中条山
Yichun	伊春	Zhoushan	舟山
Yilan	宜兰	Zhujiang River (Pearl R.)	珠江
Yilehuli Mountain	伊勒呼里山	Zhumadian	驻马店
Yinchuan	银川	Zigong	自贡
Yingde	英德	Zishui River	资水
Yingjiang	盈江	Ziwu Mountain	子午岭
Yingtan	鹰潭	Zomo Lake	淖毛湖
Yining	伊宁	Zuli River	祖厉河
Yinshan Mountains	阴山	Zunyi	遵义
Yinghe River	颍河		

Appendix II
Climatic Statistics for 20 Selected Stations

	Jan.	Feb.	Mar.	Apr.	May	Jun.	Jul.	Aug.	Sep.	Oct.	Nov.	Dec.	Annual
Harbin (45°41′N, 126°37′E; Elevation: 172m; 1951 – 1970)													
Temperature (°C)	−19.7	−15.4	−5.1	6.1	14.3	20.0	22.7	21.4	14.3	5.9	−5.8	−15.5	3.6
Humidity (%)	73	69	59	51	51	65	77	79	72	65	66	72	67
Precipitation (mm)	4.3	3.9	12.5	25.3	33.8	77.7	176.5	107.0	72.7	26.6	7.5	5.7	553.5
Sunshine (%)	61	67	65	57	57	56	51	53	59	60	61	58	59
Shenyang (41°46′N, 123°26′E; Elevation: 47m; 1951 – 1970)													
Temperature (°C)	−12.7	−8.6	−0.3	9.1	17.0	21.4	24.6	23.7	17.2	9.6	−0.3	−8.7	7.7
Humidity (%)	65	59	54	56	55	68	78	79	72	68	65	65	65
Precipitation (mm)	8.3	7.7	13.1	36.4	57.7	87.9	217.7	180.3	86.0	36.0	19.0	9.4	755.4
Sunshine (%)	59	63	64	61	60	55	48	53	65	63	59	56	58
Beijing (39°48′N, 116°28′E; Elevation:51m; 1951 — 1970)													
Temperature (°C)	−4.7	−2.3	4.4	13.2	20.2	24.2	26.0	24.6	19.5	12.5	4.0	−2.8	11.6
Humidity (%)	44	49	52	48	51	60	77	80	70	66	59	50	59
Precipitation (mm)	2.6	7.7	9.1	22.4	36.1	70.4	196.6	243.5	63.9	21.1	7.9	1.6	682.9
Sunshine (%)	69	67	64	63	64	62	53	54	66	66	64	67	63
Qingdao (36°09′N, 120°25′E; Elevation:17m; 1951 — 1970)													
Temperature (°C)	−2.6	−0.5	4.6	10.9	16.7	20.9	24.7	25.4	20.5	14.3	7.4	0.5	11.9
Humidity (%)	67	68	66	67	70	79	88	86	78	74	73	70	74
Precipitation (mm)	7.6	11.4	21.5	33.3	48.7	92.2	209.7	155.2	108.2	45.5	34.3	9.7	777.4
Sunshine (%)	63	59	60	59	59	57	43	54	60	67	59	60	58
Xi an (34°18′N, 108°56′E; Elevation:397m; 1951 — 1970)													
Temperature (°C)	−1.3	2.1	8.0	14.0	19.2	25.3	26.7	25.4	19.4	13.6	6.5	0.6	13.3
Humidity (%)	66	68	67	70	70	57	71	74	80	79	78	72	71
Precipitation (mm)	7.6	10.3	24.7	53.0	62.3	57.6	105.9	80.1	100.2	61.5	34.0	7.1	604.2
Sunshine (%)	48	42	41	43	47	54	51	56	42	43	46	46	46

Lanzhou (36°03′N, 103°53′E; Elevation:1517m; 1951 — 1970)

	Jan	Feb	Mar	Apr	May	Jun	Jul	Aug	Sep	Oct	Nov	Dec	Year
Temperature (°C)	−7.3	−2.5	5.3	11.7	16.7	20.5	22.4	21.0	15.9	9.4	1.6	−5.7	9.1
Humidity (%)	59	53	49	49	51	52	60	65	68	69	65	66	59
Precipitation (mm)	1.4	1.8	7.4	19.0	40.0	33.0	59.3	85.6	51.0	26.9	4.9	1.5	331.9
Sunshine (%)	63	64	60	58	58	60	58	59	54	59	62	63	60

Shanghai (31°10′N, 121°26′E; Elevation:4.5m; 1951 — 1970)

	Jan	Feb	Mar	Apr	May	Jun	Jul	Aug	Sep	Oct	Nov	Dec	Year
Temperature (°C)	3.3	4.6	8.3	13.8	18.8	23.2	27.9	27.8	23.8	17.9	12.5	6.2	15.7
Humidity (%)	74	78	78	80	82	84	83	82	81	77	78	77	80
Precipitation (mm)	44.3	63.0	80.5	111.1	129.3	156.6	142.4	116.0	145.9	46.8	53.4	39.2	1128.5
Sunshine (%)	45	39	38	30	38	40	57	63	48	50	47	45	46

Wuhan (30°38′N, 114°04′E; Elevation:23m; 1951 — 1970)

	Jan	Feb	Mar	Apr	May	Jun	Jul	Aug	Sep	Oct	Nov	Dec	Year
Temperature (°C)	2.8	5.0	10.0	16.0	21.3	25.8	29.0	28.5	23.6	17.5	11.2	5.3	16.3
Humidity (%)	76	78	81	81	80	78	79	78	77	77	79	77	79
Precipitation (mm)	35.5	60.5	104.0	144.4	161.2	218.0	179.0	133.4	80.6	53.2	56.6	33.5	1260.1
Sunshine (%)	40	36	33	38	42	51	61	67	55	49	45	43	47

Chengdu (30°40′N, 104°04′E; Elevation:506m; 1951 — 1970)

	Jan	Feb	Mar	Apr	May	Jun	Jul	Aug	Sep	Oct	Nov	Dec	Year
Temperature (°C)	5.6	7.6	12.1	17.0	21.1	23.7	25.8	25.1	24.1	16.7	12.0	7.3	16.3
Humidity (%)	79	81	78	78	77	81	85	85	85	86	83	83	82
Precipitation (mm)	5.0	11.4	21.8	51.1	88.3	119.4	228.9	265.8	113.5	47.9	16.5	6.4	976.0
Sunshine (%)	24	21	25	28	32	32	39	42	25	19	21	21	28

Kunming (25°01′N, 102°41′E; Elevation:1891m; 1951 — 1970)

	Jan	Feb	Mar	Apr	May	Jun	Jul	Aug	Sep	Oct	Nov	Dec	Year
Temperature (°C)	7.8	9.8	13.2	16.7	19.3	19.5	19.7	19.2	17.6	15.0	11.5	8.3	14.8
Humidity (%)	68	62	58	56	64	78	83	84	82	82	76	73	72
Precipitation (mm)	10.0	9.8	13.6	19.6	78.0	181.7	216.4	195.2	122.9	94.9	33.7	15.9	991.7
Sunshine (%)	73	74	76	73	59	37	36	45	47	44	65	68	57

Guangzhou (23°08′N, 113°19′E; Elevation:63m; 1951 — 1970)

	Jan	Feb	Mar	Apr	May	Jun	Jul	Aug	Sep	Oct	Nov	Dec	Year
Temperature (°C)	13.4	14.2	17.7	21.8	25.7	27.2	28.3	28.2	27.0	23.8	19.7	15.2	21.8
Humidity (%)	69	78	83	84	85	86	84	83	80	72	69	68	78
Precipitation (mm)	39.1	62.5	91.5	158.5	267.2	299.0	219.6	225.3	204.4	52.0	41.9	19.6	1680.5
Sunshine (%)	42	27	21	25	38	40	56	55	54	61	59	50	44

(continuance)

	Jan.	Feb.	Mar.	Apr.	May	Jun.	Jul.	Aug.	Sep.	Oct.	Nov.	Dec.	Annual
Taibei (25°02'N, 121°31'E; Elevation:9.0m; 1956 — 1967)													
Temperature (°C)	14.6	15.5	18.9	21.7	24.9	26.9	28.6	28.6	26.8	23.4	20.8	17.1	22.3
Humidity (%)	84	84	83	84	81	82	78	78	79	79	83	82	82
Precipitation (mm)	100.0	116.3	136.1	135.9	152.0	263.3	255.4	336.0	284.5	86.3	72.3	109.4	2047.5
Sunshine (%)													
Haikou (20°02'N, 110°21'E; Elevation:14m; 1951 — 1970)													
Temperature (°C)	17.1	18.2	21.5	24.8	27.4	28.0	28.4	27.7	26.8	24.8	22.0	18.7	23.8
Humidity (%)	85	88	87	85	83	84	82	86	86	83	83	84	85
Precipitation (mm)	26.4	37.8	51.4	95.4	188.7	243.4	188.6	233.0	337.3	143.8	105.4	38.5	1689.6
Sunshine (%)													
Xisha Archipelago (16°50'N, 112°20'E; Elevation: 4.9m; 1958 — 1970)													
Temperature (°C)	22.8	23.2	25.1	27.0	28.8	28.9	28.6	28.5	27.8	26.8	25.6	24.1	26.4
Humidity (%)	78	80	81	81	81	84	84	84	85	81	81	78	82
Precipitation (mm)	19.0	14.2	24.7	25.6	62.9	167.2	231.3	205.1	215.7	221.0	153.6	51.9	1392.2
Sunshine (%)													
Hohhot (40°49'N, 111°41'E; Elevation: 1063m; 1951 — 1970)													
Temperature (°C)	-13.5	-9.3	-0.4	7.7	15.2	20.0	21.8	19.9	13.8	6.5	-3.0	-11.4	5.6
Humidity (%)	57	54	43	44	42	50	64	70	63	61	58	58	56
Precipitation (mm)	2.4	6.1	10.1	19.9	28.4	46.2	104.4	136.9	40.4	24.1	5.9	1.4	426.1
Sunshine (%)	70	72	69	66	66	64	60	62	69	72	71	68	67
Yinchuan (38°29'N, 106°13'E; Elevation:1112m; 1954 — 1970)													
Temperature (°C)	-9.2	-4.9	2.9	10.4	17.0	21.4	23.5	21.1	16.1	8.0	0.6	-7.0	8.5
Humidity (%)	60	53	51	48	49	54	64	71	67	65	66	66	59
Precipitation (mm)	1.0	2.2	6.6	16.1	17.3	22.4	38.2	55.8	27.1	12.9	5.0	0.8	205.5
Sunshine (%)	76	74	67	63	66	68	64	64	66	71	73	76	68
Hami (42°49'N, 93°31'E; Elevation:738m; 1951 — 1970)													
Temperature (°C)	-12.3	-5.6	4.8	13.1	20.3	25.5	27.7	26.2	19.4	9.8	-1.0	-9.1	9.9
Humidity (%)	66	50	35	26	26	30	32	32	34	39	50	62	40
Precipitation (mm)	1.6	0.8	1.2	2.9	2.0	5.2	6.4	5.2	3.0	2.0	1.3	1.7	33.4
Sunshine (%)	72	77	74	72	74	75	72	77	84	83	77	72	76

Hotan (37°8′N, 79°56′E; Elevation: 1375m; 1954 — 1970)

Temperature (°C)	−5.7	0.1	9.2	16.1	20.0	23.7	25.5	24.1	19.6	12.2	3.4	−3.2	12.1
Humidity (%)	52	47	34	30	36	37	41	44	42	40	44	53	42
Precipitation (mm)	1.6	2.0	1.1	4.1	6.5	7.8	3.4	3.7	3.8	0.1	0.6	0.4	35.0
Sunshine (%)													

Golmud (36°12′N, 94°38′E; Elevation:2808m; 1956 — 1970)

Temperature (°C)	−11.8	−7.6	−0.6	6.1	11.0	15.0	17.6	16.3	11.1	3.0	−5.5	−10.8	3.7
Humidity (%)	42	32	27	26	29	31	36	36	34	33	38	41	34
Precipitation (mm)	0.8	0.4	1.1	1.5	5.4	5.5	6.9	9.0	5.8	1.0	0.8	0.3	38.3
Sunshine (%)													

Lhasa (29°42′N, 91°08′E; Elevation: 3658m; 1954 — 1970)

Temperature (°C)	−2.3	0.8	4.3	8.3	12.6	15.5	14.9	14.1	12.8	8.1	1.9	−1.9	7.5
Humidity (%)	28	27	30	35	40	53	67	70	65	48	37	35	45
Precipitation (mm)	0.2	0.1	1.5	4.4	20.6	73.1	141.7	149.1	57.3	4.8	0.3	0.3	453.9
Sunshine (%)	78	72	63	62	69	62	51	53	66	83	85	82	68

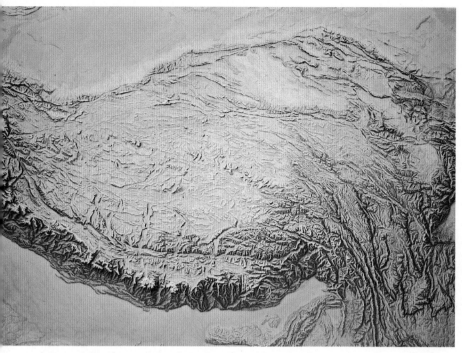

I-1   A bird's eye view of the Qinghai–Xizang Plateau.

I-2   Mount Qomolangma, the highest peak in the world, 8848m.

I-3   Mount Xixabangma, 8012m.

I-4   Mount Kangrinboque, main peak of the Gangdise Mountains, 6714m.

I-5   Mount Tomul, 7435m; the highest peak of the Tianshan Mountains.

I-6   The Qilian Mountains, northeastern border of the Qinghai-Xizang Plateau.

I-7   The Liupan Mountains, one of the most important N-S trending mountains in China.

I-8   The Great Wall and The Yanshan Mountains, northern suburbs of Beijing.

I-9   Mount Taibai, main peak of the Qinling Mountains, 3105m.

I-10   Mount Shennonjia, 3165m, in western Hubei Province, where "wild man" is supposed to exist.

I-11   The scenic Wuyi Mountain, stretching between Fujian and Jiangxi provinces by several hundred kilometers.

II-1   Oil drilling in the Bohai Sea.

II-3   The icy Do-Do River, main source of the Changjiang (Yangtze) River.

II-2   The Qinghai Lake (Koko Nor), the largest Lake in China.

II-4   The famous suspension bridge over the Dadu River.

II-5   The Wu Gorge, 40 km in length, a part of the Yangtze Gorges (Three Gorges).

II-6   The Ngoring Lake, source of the Huanghe (Yellow) River.

II-7   The Liu-jia Gorge of upper reaches of the Huanghe River, the site of the largest hydro-electric power station in China.

II-8   The middle reaches of the Huanghe River, on the Loess Plateau.

II-9   By the Sunghua River near Harbin.

II-10   A large-scale hydro-electric station in upper reaches of the Qiantang River.

II-11   The Taihu Lake, located in the center of the very densely populated Changjiang Delta.

II-12   Riyue (Sun-Moon) Pond in the central hilly area of the Taiwan Island.

II-13   The largest waterfall in China — the Huanggaoshu waterfall.

II-14   The majestic Yarlung Zangbo River, flowing slowly west-eastward on southern Qinghai-Xizang Plateau with an elevation about 4000m asl.

III-1　On the western slope of the Da Hinggan Mountains.

III-2　On the ridge of the Xiao Hinggan Mountains.

III-3　A primeval swamp in the Sanjiang (Three Rivers) Plain.

III-4　A deer farm near the city of Kirin.

III-6　The Shanhaiguan, or the Gate between Mountain and Sea．It is the first fortress of the Great Wall.

III-5　Autumn in the Changbai Mts.

III-7　The North China Plain along the Taihang Mountains piedmont.

III-8　A large-scale irrigation project in the middle Huaihe River.

III-9   A loessic "yuan" (plain) in northern Shaanxi Province.

III-10   Loessic "Liang" (ridge) and "mao" (gentle slope) in northern Shanxi Province.

III-11   The Lava flow around the Fire-Burned Mountain, which located in the northern Northeast China Plain, is the "youngest" volcano in China (latest eruption: 1720 A.D.)

III-12   The best grassland and pasture in China-Hulun Buir Steppe.

IV-1   Shanghai suburb, a part of the fertile and densely populated "watery country".

IV-2   The famous Dujiang Dam, an irrigation project started more than 2,000 years ago.

IV-3   Bamboo grove in Zhejiang Province.

IV-4   Lichis are ripe in Guangdong Province.

IV-5   Terraced farmlands in western Yunnan Province where mountainous lands occupy more than 95% of the total land area.

IV-6   The so-called "Stone Forest" in eastern Yunnan Province, an example of typical karst topography.

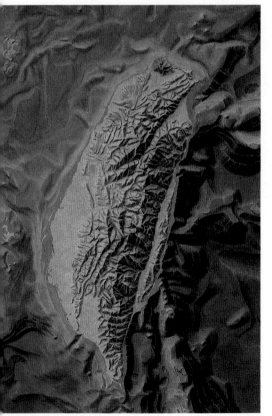

IV-8   A landscape of the humid, tropical Xishuangbanna in southern Yunnan Province.

IV-7   Topography of Taiwan Is. and its surrounding sea.

IV-9   Terrace landscape in northeastern Hainan Island.

IV-10   Coral-reefs build up all South China Sea Islands.

IV-11   In the northern Hainan Island, paddy rice fields are usually interwoven and protected by palm groves.

IV-12   The Peacock Lake with its beautiful lotus flowers is located amid dense tropical monsoon forest in the Xishuangbanna area, southern Yunnan.

V-1   Meadows along a river on the Nei Mongol Plateau.

V-2   A typical steppe landscape of eastern Nei Mongol Plateau.

V-3   The fertile Ningxia Plain (West Eibow Plain); cultivated more than 2,000 years.

V-4   Active sanddunes in the Tarim Basin.

V-5   Depositional gobi (gravel desert) on southern flank of the Tianshan Mountains.

V-6   Yardang in the dried Lop Lake basin.

V-7   Dam and trunk canal of the Manas River which enlarge the irrigated croplands by 14 times.

V-8   A new oasis in the southwestern margin of the Gurbantunggut Desert.

V-9   Grass checkerboard in the Shapotou area (Ningxia) effectively fix the shifting sands.

V-10   Jiayukuan, with the Black Mt. in background, is the last fortress of the Great Wall and the starting point of a vast expanse of sandy and gravel deserts.

V-11   The shifting sands in the steppe zone might be quickly turned into productive oases if good resources management is taken, such as in southern Mu-Us Sandy Land.

VI-1　Natural pasture immediately below continual snow on the northern slopes of the Himalayan Mts.

VI-2　The lofty and wild Qingzang Plateau, with an elevation above 4,500m and a vegetation of sparse desert-steppe.

VI-3　A group of hot springs in the southern Qinghai-Xizang Plateau.

VI-4　The source area of the Huanghe River, with colorful cushion vegetation.

VI-5　Virgin forests on the Hengduan Mountains.

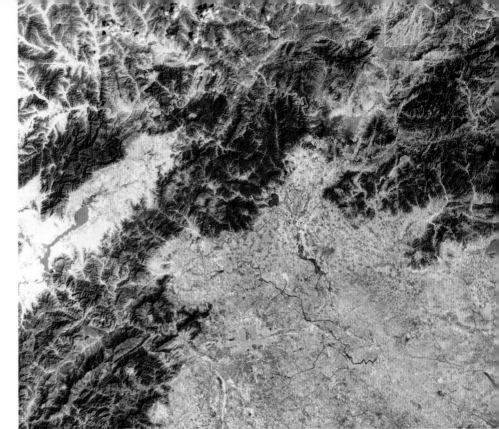

Around Beijing. (Landsat TM image; 3, October, 1984)

The Huanghe River has changed its course and flows into the sea
via the Gingshui River. (Landsat TM image; 14, March, 1985)

Landsat image 1 The North China plain.

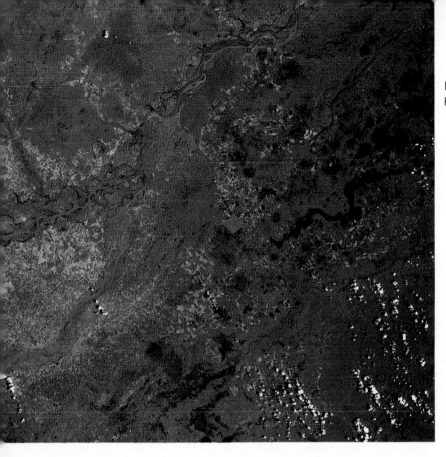

Landsat image 2 The Sanjiang Plain,
Heilongjiang Province.

Landsat image 3 Around the Dongting Lake.

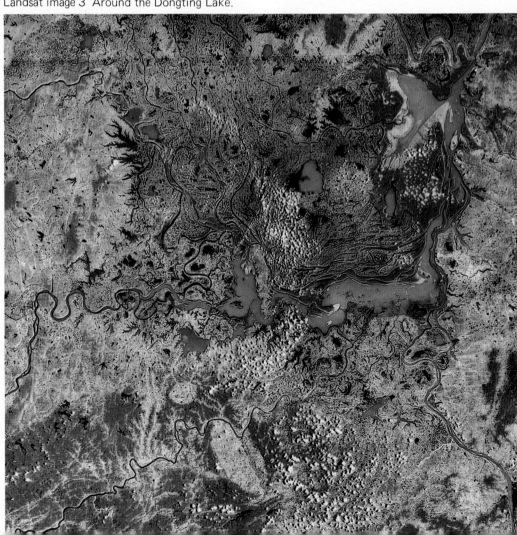

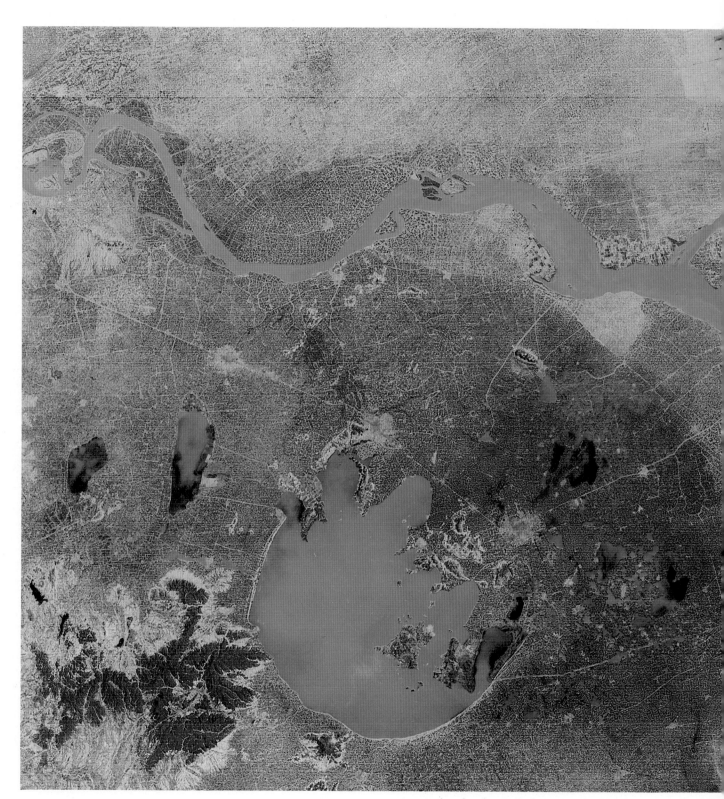

Landsat image 4 Around the Taihu Lake.

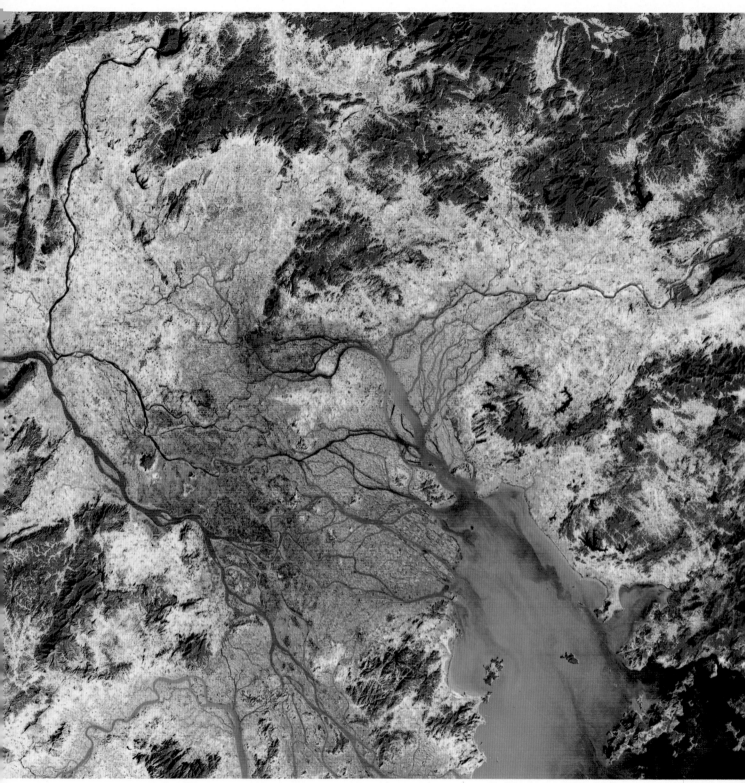

Landsat image 5 The Zhujiang (Pearl R.) Delta.

Landsat image 6 The Hainan Island.

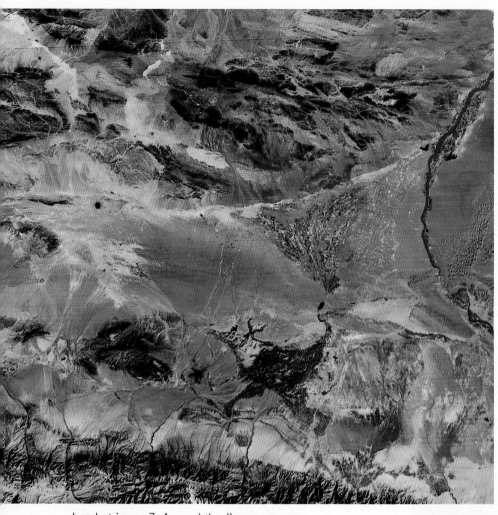

Landsat image 7 Around the Jiayuguan area.

Landsat image 8 The dried Lop Lake basin, the famous "Great Ear" in the Landsat image.

Landsat image 9 Around the Turpan Basin.

Landsat image 10 Development of agricultural land in northern piedmont of the Tianshan Mountains.

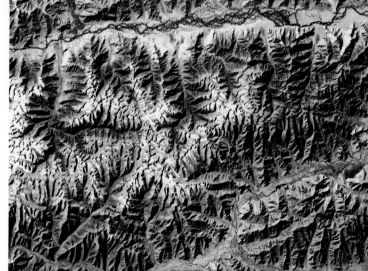

Landsat image 11 Around the Lhasa area.

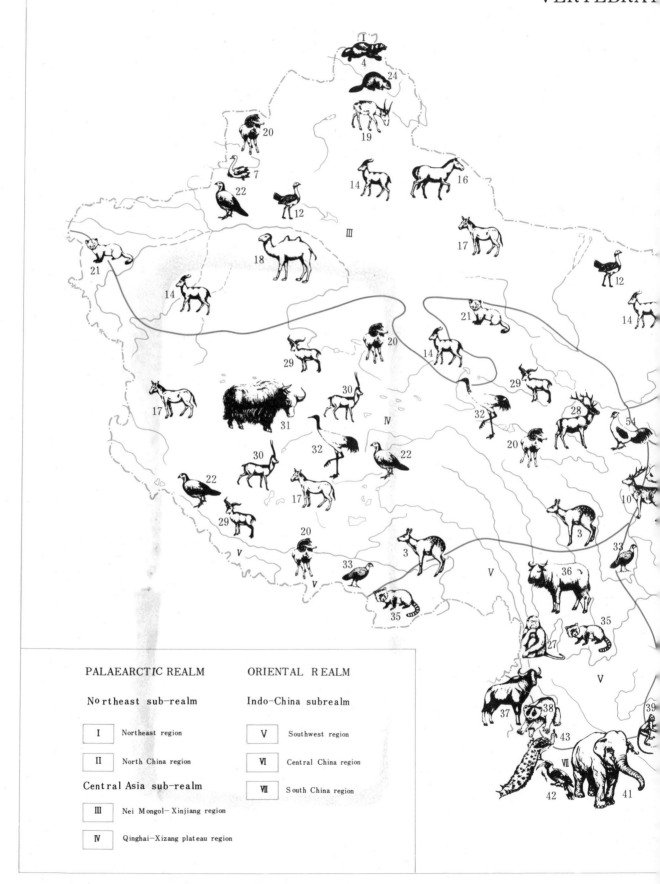

DISTRIBUTION OF F

VERTEBRAT